Y0-CCT-878

This workbook is part of a professional grade program designed to cover the fundamentals as well as the more advanced applications of geometric tolerancing. It covers what you need to know in order to function and work as a professional in an organization utilizing geometric tolerancing. It is programmed to accompany the Geometric Dimensioning and Tolerancing Video Series to the ASME Y14.5-2009 & ASME Y14.5.1M-1994 (R2004)).

A typical unit consists of the participant viewing the video and following along in the workbook. During and after units, a workshop exercises are completed. The solutions to the workshop exercises are presented, reviewed and discussed by a program leader. A leader's guide as well as a wood and plastic model set are available to assist the program leader. The model set is comprised of parts, assemblies and inspection equipment shown in this workbook.

The workbook is written in 17 units that follow along with the video units. It is suggested the participant review the unit in the workbook prior to the delivery session. The participant should follow along and turn the pages in the workbook as topics are presented in the video program. The workbook will provide a good reference to the material and concepts covered. The intent of the video program is to lay the basic ground work so the participant may begin building knowledge on a solid foundation.

The workbook can also be used in a classroom lecture situation without the video series. It is written in a simple and easy to understand language with extensive 3D graphics. The workbook is designed as a simple practical hands-on approach to the subject. The workbook can be used for basic as well as advanced programs. The program leader will decide which material is relevant for a particular group. The remaining workbook material can be used as reference. As an option, there is also a color animated MS PowerPoint presentation available for program leaders.

The workbook takes a global, practical approach to explaining how the geometric system works (see unit 3). Parts are shown in functional assemblies and are fully toleranced illustrating actual on-the-job applications. Inspection, gaging and verification procedures are also shown.

Make sure you read and understand the first six units very thoroughly. After you understand the basic principles, the remaining units can be read in any order that strikes your interest. Geo Tol is a new language, and like any new language, it will take time and practice. The more you use it, the more you will understand it. Suggestions for improvement are welcome and encouraged. Send us an e-mail with your comments and visit our website at www.geotol.com. Good Luck.

Note: The information in this text is based on ASME and ISO standards and though deemed to be correct must be considered as advisory and to be used at the discretion of the user. The figures in this book are intended only as illustrations to aid the user in understanding the principles and methods of geometric dimensioning and tolerancing.

The absence of a figure illustrating a desired application is neither reason to assume inapplicability. In some instances, figures show added detail for emphasis. In other cases, figures are incomplete by intent. Numerical values of dimensions and tolerances are illustrative only. Many examples in the text are in metric but the inch system can be used as well without prejudice. Be sure to consult national and international standards for more information on geometric tolerancing.

About the Authors

Scott Neumann

Scott Neumann is a senior partner in Technical Consultants Inc. He graduated from University of Florida in Mechanical Engineering. He specializes in geometric tolerancing and tolerance stack-up and analysis.

Scott is a member of the American Society of Mechanical Engineers (ASME) and is a senior level ASME certified, geometric dimensioning and tolerancing professional. He also represents the USA in the International Standards Organization, ISO TC10 committee. He regularly attends the USA, ASME Y14.5 subcommittee on dimensioning and tolerancing. He is a member on the ASME Y14.45, Measurement Data Reporting, committee.

Scott presents geometric tolerancing training programs to engineering, manufacturing and quality personnel at major corporations in the USA and around the world. He was a major contributor to the Geometric Tolerancing Applications with Stacks workbook and the Geometric Tolerancing Stacks and Analysis workbook. He is also an expert in measurement and inspection techniques and has produced a video program on geometric tolerancing.

Al Neumann

Al Neumann is President/Director of Technical Consultants Inc (TCI), a training and consulting firm specializing in geometric tolerancing. Al Neumann, is a graduate of Central Michigan University. He has worked in a variety of engineering and manufacturing firms related to automotive, aircraft, medical as well as the defense industry.

Al Neumann established Technical Consultants Inc (TCI) in 1985. TCI is devoted to education and training in geometric tolerancing and product definition. Al Neumann is a recognized international expert in geometric dimensioning and tolerancing. He has written many books and has produced four video training programs on the subject. Mr. Neumann has presented thousands of training programs to thousands of people throughout the US and worldwide over the past 25 years.

He has been a member on the ASME Y14.5, Dimensioning and Tolerancing subcommittee for the past 25 years. He served as section sponsor on the Datums, section for the ASME Y14.5-2009 standard. He currently serves as vice chairman on the ASME Y14.5 committee. Mr. Neumann is also a member on the ASME Y14.5.1 subcommittee on Mathematical Definitions of Dimensioning and Tolerancing. In the past, he has served as chairman on the US Technical Activities Group to ISO on Dimensioning and Tolerancing and Mathematical Definitions. He has served on national and international standards committees for the past 25 years. Mr. Neumann is a senior member of the Society of Manufacturing Engineers and the American Society of Mechanical Engineers.

Al and Scott Neumann travel worldwide presenting on-site training and consulting on geometric tolerancing and related subjects. They work with large and small companies customizing training and education to meet particular needs.

Acknowledgments

The authors would like to express their appreciation and gratitude to the many professional associates and friends on the standards committees, schools, colleges, universities, companies and professional organizations, both nationally and internationally, who provided technical input and draft review of this workbook. We would like to acknowledge the support, patience and understanding from wife and mother, Kathy Neumann.

Acknowledgement and credit is gratefully given for certain references and definitions derived from the ASME Y14.5-2009 and ASME Y14.5.1M-1994 (R2004) standards published by the American Society of Mechanical Engineers (ASME), New York, NY.

Credit is also given to The Society of Manufacturing Engineers (SME) who provided support and information over the past 25 years to make this book a success.

Special credit must be given to my long term friend Jim Jones. Jim has provided me with tremendous support and guidance over the years to help refine and improve our presentations and training materials.

Acknowledgment is also given to the many professionals in the field who helped review this book and improve its content:

Brian Okerlund - Jet Propulsion Lab - Pasadena, CA
John Balmer - Jet Propulsion Lab - Pasadena, CA
Tony Cimabue - Tech-Knowledgy for Design - Sante Fe, NM
John Eavey - Documentation Quality Assurance - Fort Wayne, IN
Tim Anderson - Freightliner Corp. - Portland, OR

Geo Tol Pro
A Practical Guide to Geometric Tolerancing

Table of Contents
and
Training Program Outline

Unit 1

Introduction - Symbols and Terms

Introduction to Geometric Tolerancing
ASME Standards related to Product Definition and Metrology
ISO Standards related to Product Definition and Metrology
Tolerances are like Bacteria
Common Symbols
Common Symbol Application
Features of Size and Features without Size
Geometric Tolerancing Characteristics
Geometric Tolerance Zone Shapes
Geometric Tolerancing Categories
Feature Control Frame
Symbols Associated with a Feature Control Frame
Symbols Related to Datum Identification
Index Plate - Practical Example
Index Plate - Matching a Perfect DRF to an Imperfect Part
Index Plate - Practical Example in 3D
Definitions
Fundamental Rules
Material Conditions - MMC and LMC
Workshop Exercise 1.1

Introduction to Geometric Tolerancing

Geometric Dimensioning and Tolerancing (GeoTol) is an international engineering language that is used on engineering drawings. It establishes a Cartesian coordinate system, also know as the datum reference frame, on the part. Geo Tol uses a series of internationally recognized symbols and feature control frames, rather than words, to describe the product. These symbols and frames are applied to the features of a part and provide a very clear definition of design intent.

Geo Tol is a very precise mathematical language that describes the form, orientation and location of part features in zones of tolerance. The Geo Tol system has a strong mathematical base which is essential in today's modern computerized world. The language was designed by engineers, and is used by engineers to describe a product.

This language is similar to any other language such as French, German, Japanese, Chinese or Spanish. There are certain grammar and punctuation rules that must be learned and practiced. This is important because we want to avoid "slang" being introduced which could lead to misunderstandings or incorrect interpretations of the design intent.

Like any language, it takes a while to learn the geometric tolerancing system. People from various companies and backgrounds possess different knowledge levels. There are some personnel who might be considered conversational, others can read it but not write it, and others might be experts. It takes time, practice and effort to thoroughly understand the system.

In the early days, manufacturing a product was a father to son tradition handed down through the ages. There were no tolerances. People made parts the best they could to the dimensions given. Parts were generally not interchangeable.

In the late 1700's, Eli Whitney, who is well known as the inventor of the cotton gin, was also a firearms manufacturer who supplied muskets to the United States government. He solicited and won a contract promoting the idea of interchangeable parts for the manufacture of firearms. Whitney won the contract, by building ten guns, claiming the same exact parts and mechanisms, and assembled and disassembled the guns before the United States Congress. The idea was simple, all dimensions had a tolerance. Workers had gages to verify a feature was the correct size or dimension. This is the plus/minus or the directly toleranced dimensioning system.

This type of 2D tolerancing was used solely until the late 1930's. About that time, Stanley Parker from the Royal Torpedo Factory in Scotland realized there were some problems using the directly toleranced system. Parker was one of the first people to initiate the use of geometric tolerancing, using position tolerancing to create round, rather than square, tolerance zones to locate holes.

The use of geometric tolerancing system gradually increased worldwide and came to the United States back in the 40's, mostly for military parts, or where interchangeability is essential. The geometric tolerancing system has evolved in the U.S. over the years by some early military standards, beginning with Mil Std 8 in the 1950's to the current ASME Y14.5-2009 standard.

The concept of a datum reference frame was first introduced in the USASI Y14.5- 1966 standard. It has been expanded on in subsequent standards and is now mandatory to achieve interchangeable parts.

Evolution of Dimensioning and Tolerancing Standards in the USA:

Mil Std 8   1950's
Mil Std 8A
Mil Std 8B
Mil Std 8C-1963
ASA-Y14.5-1957
USASI Y14.5-1966
ANSI Y14.5-1973
ANSI Y14.5M-1982
ASME Y14.5M-1994
ASME Y14.5-2009

The latest American National Standard on dimensioning and tolerancing is the ASME Y14.5- 2009 standard. There have been many updates, clarifications and new concepts over past standards. In the appendix of this book you will find a list of the major changes from the previous ASME Y14.5M-1994 (R2004) standard.

To provide a solid mathematical base, there is also a standard on Mathematical Definitions of Dimensioning and Tolerancing. ASME Y14.5.1-1994 (R2004). This document is primarily used in the development of software and a clear mathematical definition. There is also a certification program and standard on geometric tolerancing provided by ASME, it is the ASME Y14.5.2-2000 Certification of Geometric Dimensioning and Tolerancing Professionals. Information on these standards and others are available from The American Society of Mechanical Engineers. www.asme.org.

There is also a series of international standards on geometric dimensioning and tolerancing, the introductory document being ISO 1101-2004. ISO (the International Organization for Standardization) is a world-wide federation of national standards institutes (ISO member bodies). The American National Standards Institute (ANSI) is the primary source and official sale agent for ISO standards in the United States. ANSI is also a source for American National Standards and national standards from other countries. See also www.asme.org

Drawings in this book contain a combination of both metric and inch values, recognizing the fact that both units are common within industry. Inch or Metric can be used without predjuce to the geometric tolerancing concepts illustrated in this text.

Note: The information in this text is the authors interpretation of geometric dimensioning and tolerancing based on ASME and ISO standards and though deemed to be correct must be considered as advisory and to be used at the discretion of the user.

In some instances, figures show added detail for emphasis. In other cases, figures are incomplete by intent. Numerical values of dimensions and tolerances are illustrative only. Most examples in the text are in metric but the inch system can be used as well without prejudice. Be sure to consult national and international standards for more information on the subject.

The following is a reference list of American National Standards and ISO Standards that are related to geometric tolerancing, product definition, metrology and verification. These are some of the most common standards. Standards are in constant flux and change periodically. For more detailed information on current available standards contact:

The American Society of Mechanical Engineers (ASME) www.ASME.Org
ISO Standards   www.ds.dk/isotc213/

ASME Reference Standards

ASME Y14.5.1M-1994 (R2004), Mathematical Definition of Dimensioning and Tolerancing Principles
ASME Y14.5.2-2000 Certification of Geometric Dimensioning and Tolerancing Professionals
ASME Y14.41 - 2003 (R2008) Digital Product Definition Data Practices
ASME Y14.43 - 2003, (R2008) Dimensioning and Tolerancing Principles for Gages and Fixtures
ASME Y14.8-2009, Castings and Forgings
ANSI B4.2-1978 (R2004), Preferred Metric Limits and Fit
ANSI B4.1-1967 Preferred Limits and Fits for Cylindrical Parts
ASME Y13.38M-2007, Abbreviations
ASME Y14.100-2004, Engineering Drawing Practices
ASME Y14.3M-1994 (R2008), Multiview and Sectional View Drawings
ASME Y14.1M-2005, Drawing Sheet Size and Format
ASME Y14.2M-2008, Line Conventions and Lettering
ANSI/IEEE 268-1992,2 Metric Practice
IEEE/ASTM SI 10-2002 ERRATA 2005, Standard for Use of the International System of Units (SI) — The Modern Metric System
ASME B5.10-1994, Machine Tapers — Self Holding and Steep Taper Series
ASME B46.1-200, Surface Texture, Surface Roughness, Waviness, and Lay
ASME Y14.36M-1996, Surface Texture Symbols
ANSI B89.3.1-1972 (R2003), Measurement of Out-of-Roundness
ANSI B92.1-1996,1 Involute Splines and Inspection, Inch Version
ANSI B92.2M-1980,1 Metric Module, Involute Splines
ASME B94.11M-1993, Twist Drills
ANSI Y14.6-2001 (R2007), Screw Thread Representation
ANSI Y14.6aM-1981 (R1998), Screw Thread Representation (Metric Supplement)
ANSI Y14.7.1-1971 (R1998), Gear Drawing Standards — Part 1: For Spur, Helical, Double Helical, and Rack
ANSI Y14.7.2-1978 (R1999), Gear and Spline Dwg Standards - Pt 2: Bevel and Hypoid Gears
ANSI/ASME B1.2-1983, Gages and Gaging for Unified Inch Screw Threads
ANSI B4.4M-1981 (R1987), Inspection of Workpieces
ASME B89.7.2-1999 Dimensional Measurement Planning
ASME B89.7.3.1-2001 Guidelines for Decision Rules: Considering Measurements Uncertainty in Determining Conformance to Specifications
ANSI/ASME B89.6.2-1973 (R2003), Temperature and Humidity Environment for Dimensional Measurement
ANSI/ASME B94.6-1984 (R2003), Knurling

ISO Standards Related to Product Definition and Metrology

ISO Reference Standards

ISO 1101-2004 Geometrical Product Specifications (GPS) — Geometrical tolerancing — Tolerances of form, orientation, location and run-out

ISO 1-1975, Standard reference temperature for industrial length measurements

ISO 286 Part 1, 1998, IS0 system of limits and fits -Bases of tolerances, deviations and fits

ISO 286 -Part 2, 2006, ISO system of limits and fits — Tables of standard tolerance grades and limit deviations for holes and shafts

ISO 1660:1987, Technical drawings — Dimensioning and tolerancing of profiles

ISO R 1938-1971, IS0 system of limits and fits, Part II : Inspection of plain workpieces

ISO 2692:—1), Geometrical Product Specification (GPS) — Geometrical tolerancing — Maximum Material Requirement (MMR) and Least Material Requirement (LMR)

ISO 2768-Part1 & 2, 1989, General tolerances -Tolerances for linear and angular dimensions without individual tolerance indications

ISO 5458:1998, Geometrical Product Specifications (GPS) — Geometrical tolerancing — Positional tolerancing

ISO 5459:1981, Technical drawings — Geometrical tolerancing — Datums and datum-systems for geometrical tolerances specifications, operators and uncertainties

ISO ISO/TR 5460-1985Technical drawings - Geometrical tolerancing -Tolerancing of form, orientation, location and run-out Verification principles and methods – Guidelines

ISO 8015:1985, Technical drawings — Fundamental tolerancing principle

ISO 10578:1992, Technical drawings — Tolerancing of orientation and location — Projected tolerance zone

ISO 10579:1993, Technical drawings — Dimensioning and tolerancing — Non-rigid parts

ISO/TS 12180-1:2003, Geometrical Product Specifications (GPS) — Cylindricity — Part 1: Vocabulary and parameters of cylindrical form

ISO/TS 12180-2:2003, Geometrical Product Specifications (GPS) — Cylindricity — Part 2: Specification operators

ISO/TS 12181-1:2003, Geometrical Product Specifications (GPS) — Roundness — Part 1: Vocabulary and parameters of roundness

ISO/TS 12181-2:2003, Geometrical Product Specifications (GPS) — Roundness — Part 2: Specification operators

ISO/TS 12780-1:2003, Geometrical Product Specifications (GPS) — Straightness — Part 1: Vocabulary and parameters of straightness

ISO/TS 12780-2:2003, Geometrical Product Specifications (GPS) — Straightness — Part 2: Specification operators

ISO/TS 12781-1:2003, Geometrical Product Specifications (GPS) — Flatness — Part 1: Vocabulary and parameters of flatness

ISO/TS 12781-2:2003, Geometrical Product Specifications (GPS) — Flatness — Part 2: Specification operators

ISO 14660-1:1999, Geometrical Product Specifications (GPS) — Geometrical features — Part 1: General terms and definitions

ISO 14660-2:1999, Geometrical Product Specifications (GPS) — Geometrical features — Part 2: Extracted median line of a cylinder, cone, extracted median surface, extracted local size feature

ISO/TS 17450-2:2002, Geometrical product specifications (GPS) — General concepts — Part 2: Basic tenets, specifications, operators and uncertainties

Tolerances placed on engineering drawings can be very small and difficult to keep in perspective. Small tolerances can be compared to bacteria. Our mothers always told us to wash our hands to stay free of bacteria so we don't get sick. But when we look at our hands, we do not see any bacteria. Bacteria is so small you need a microscope to view the little germs. Tolerances are also often very small and not visible with the naked eye.

Bacteria

When you look at manufactured parts, they look flat, square and straight, but if you view the parts with precise inspection equipment such as micrometers, calipers, height gage, indicators or CMM's, you would find that there are imperfections on the parts. These variations or imperfections are allowed within the tolerance limits or constraints placed on the produced part. Depending on requirements of the part, these tolerance constraints can be very small. In order to understand geometric tolerancing, you must think of the parts being imperfect. The engineer designs the parts with perfect geometry in CAD, but the produced parts are never perfect.

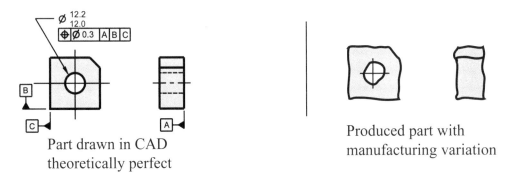

Part drawn in CAD
theoretically perfect

Produced part with
manufacturing variation

Parts always vary. To verify the produced parts, quality control matches the perfect CAD geometry with the imperfect geometry of the produced part. The produced parts shown in this book are always drawn with error, to help the user understand how to match perfect geometry with the imperfect geometry of the part.

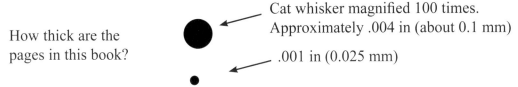

How thick are the
pages in this book?

Cat whisker magnified 100 times.
Approximately .004 in (about 0.1 mm)

.001 in (0.025 mm)

The illustration above attempts to provide a perspective of tolerances. An average cat whisker from a common house cat in cross-section is about .004 inch (or about 0.1 mm in diameter). The graphic represents a cat whisker magnified 100 times. The engineer or designer should strive to keep tolerances as large as possible while still preserving the function of the part. Small tolerances can increase cost in the manufacture, inspection and tooling of parts. Tooling tolerances are generally 30% of part tolerance and inspection equipment is 10% of part tolerances. Tolerances applied to the design travel in increments all through the manufacturing and verification process. Tight tolerances are sometimes necessary, but it is just important to keep them in perspective. Generally to a point, larger tolerances make the part more economical to produce. Economical production is also an important functional requirement for engineering.

Common Symbols

This is a list of the most common symbols that are used on the face of a drawing. Note the comparison with the ISO standards. The symbols are used to replace text on the drawing and provide clear communication of design intent. The symbols marked with an "X" indicate new or revised from the previous ASME Y14.5M-1994 (R2004) standard.

	Term	Symbol ASME Y14.5	Symbol ISO
	Basic Dimension Theoretically Exact Dimension (ISO)	12	12
	Diameter	∅	∅
	Spherical Diameter	S∅	S∅
	Radius	R	R
	Controlled Radius	CR	None
	Spherical Radius	SR	SR
	Square	□	□
	Statistical Tolerance	⟨ST⟩	⟨ST⟩
	Reference Dimension	(12)	(12)
	Number of Places	3X	3X
	Counterbore	⊔	None
X	Spotface	⌴SF⌴	None
	Countersink	⌄	None
	Deep/Depth	⤓	None
	Envelope Principle	None	Ⓔ
X	Independency Principle	Ⓘ	None
	Dimension not to Scale	<u>23</u>	<u>23</u>
	Arc Length	23̂	23̂
	Slope	◹	◹
	Conical Taper	▷	▷
X	Continuous Feature	⟨CF⟩	None
	Dimension Origin	⊶→	⊶→
	First Angle Projection	⊟⊕	⊟⊕
	Third Angle Projection	⊕⊟	⊕⊟

Common Symbol Application

The drawing below is shown with many of the common symbols applied. The symbols are a universal method of specifying requirements without the use of notes or words. The symbols are designed to be very intuitive and look like the requirements that they are identifying.

Repetitive features such as holes, slots and tabs, which are repeated often, can be specified by stating the number of features or places and an "X" and then followed by the requirement. A space is used between the "X" and the requirement as shown below. Where used with a basic dimension, the number of places and the X may be placed either inside or outside the basic dimension frame.

The symbol for diameter is a circle with a slash. The symbol for radius is the letter R. The symbol for square features is a square box. The symbol for counterbore and spot face are similiar, except the spotface symbol will also have the letters SF inserted in the symbol. The countersink symbol is shown as a 90° Vee. The deep or depth symbol will identify the depth of the indicated feature. There is no space between the value or other specified symbols.

Drawing with practical application of common symbols

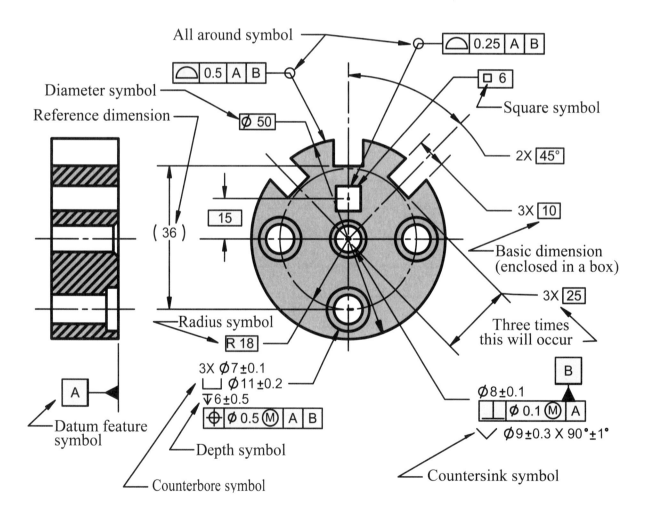

1.8

Features of size and Features without Size

Geometric tolerancing is a feature based system. A feature is a general term applied to a physical portion of a part such as a surface, pin, tab, hole or slot. Parts are composed of many features.

There are two types of features: Features of size and features without size. This is an important distinction in geometric tolerancing. Features of size can have feature modifiers such as MMC, LMC and RFS applied. Features without size can not have feature modifiers applied.

Examples of features without size: (Surfaces)

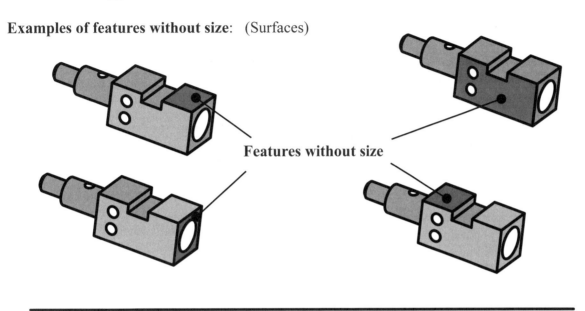

Features without size

Examples of features of size

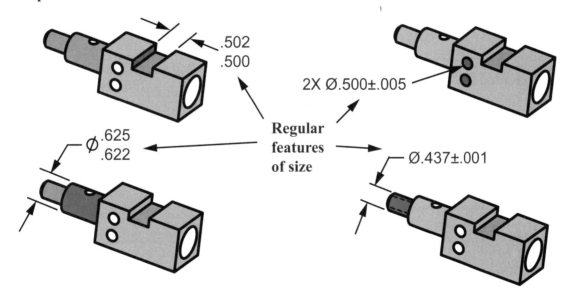

.502
.500

2X Ø.500±.005

ϕ .625
.622

Regular features of size

Ø.437±.001

Feature of Size. There are two types of features of size.
Regular Feature of Size. One cylindrical or spherical surface, a circular element, a set of two opposed parallel elements or opposed parallel surfaces, each of which is associated with a directly toleranced dimension.
Irregular Feature of Size There is additional information on irregular features of size later in unit 2.

Geometric Tolerancing Characteristics

There are 14 geometric tolerancing characteristics. The symbols that represent these characteristics are shown below. These characteristic symbols are placed in the first compartment of a feature control frame. They define the type of tolerance that is to be applied to the feature.

The characteristics are grouped together into types of tolerance: form, orientation, location, runout, and location of derived median points. The general primary use and description of each characteristic is also shown. A more complete definition and description of the concepts can be found later in the text.

SYMBOL	GEOMETRIC CHARACTERISTIC	TYPE OF TOLERANCE	PRIMARY CONTROL
◻	FLATNESS	Form No relation between features	Controls form (shape) of size and non-size features.
—	STRAIGHTNESS		Datum reference is not allowed
⌭	CYLINDRICITY		Controls form (shape) of size features
○	CIRCULARITY (ROUNDNESS)		Datum reference is not allowed
⊥	PERPENDICULARITY	Orientation No relation between features	Controls orientation (tilt) of surfaces, axes, or median planes for size and non-size features Datum reference required *Optional: Angularity symbol may be used for all orientation controls*
//	PARALLELISM		
∠	ANGULARITY		
⊕	POSITION	Location	Locates center points, axes and median planes for size features. Can also control orientation.
⌒	PROFILE OF A SURFACE		Locates surfaces Can also be used to control size, form, and orientation of surfaces based on datum reference
⌒	PROFILE OF A LINE		
⟗	TOTAL RUNOUT	Runout	Controls surface coaxiality Can also control form and orientation of surfaces.
↗	CIRCULAR RUNOUT		
◎	CONCENTRICITY	Location of derived median points.	Locates derived median points of a feature
═	SYMMETRY		*Not common, consider position, runout, or profile.*

1.10

Geometric Tolerance Zone Shapes

Geometric tolerancing is a three dimensional language. A datum reference frame (Cartesian coordinate system) is established on a part. Geometric tolerancing is applied to the part features. The applied geometric specifications create zones of tolerance within which the feature must lie.

The tolerance zones are portrayed in a variety of shapes, usually defined by the geometric characteristic applied or information shown in the individual feature tolerance compartment of the feature control frame. There are three dimensional tolerances, as well as two dimensional tolerances. The three dimensional tolerances are certainly the most common shaped zones applied to features.

Two dimensional tolerance zones are usually used for two dimensional extruded type features, where geometric control is often more important in one direction than another. Two dimensional tolerances are often used to further refine the cross sectional aspects of a three dimensional control.

Common geometric tolerance zone shapes

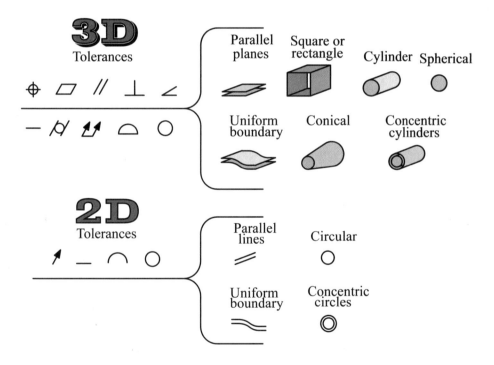

The shapes shown above are the most common shaped tolerance zones. The zones are parallel planes, squares, cylinders, uniform boundaries etc. The engineer can also create jelly bean zones, amoeba shaped zones or just about any shape desired. The use of geometric tolerancing can be simple or complex, depending on the necessary requirements. There are many examples shown later in the book.

Creating a simple drawing with common requirements can be very easy. However, the geometric tolerancing system is also a very powerful language and can be used to define the most complex design criteria for the most discriminating engineer. To use geometric tolerancing to its fullest takes time, study and experience.

Geometric Tolerance Categories

Geometric tolerancing is a feature-based system. Parts are composed of features. Geometric tolerances are applied to features by feature control frames. Geometric tolerances are divided into three main categories: form, orientation, and location.

Form tolerances control the "shape" of features and are often used as a refinement of size.

Orientation tolerances control the "tilt" of features and are always associated with basic angle dimensions, often used as a refinement to location. If applied to surfaces, orientation tolerances also control form.

Location tolerances control location are always associated with basic linear dimensions. Position locates and orients the median plane or axis of features of size. Profile locates feature surfaces. Profile is the most powerful characteristic of all, and also controls orientation and form.

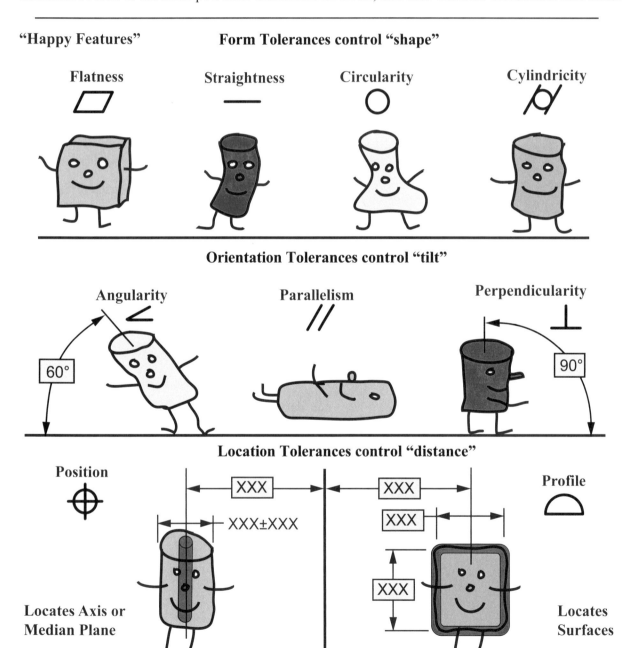

1.12

Feature Control Frame

The feature control frame states the requirements or instructions for the feature to which it is attached. As its name implies, the feature control frame controls features. Each feature control frame will state only one requirement or one message. There is only one set up or one requirement for each feature control frame. If it is necessary to have two requirements for a feature, it requires two feature control frames.

The feature controlled can be a surface or a feature of size, like a hole shown in the example below. If the feature has size, the size requirements for the feature will be defined.

To help interpret a feature control frame, it is important to note that there are always two words that are implied to precede a feature control frame. The two words are "This feature", as in "This feature" flat or as read below, "This feature" positioned.

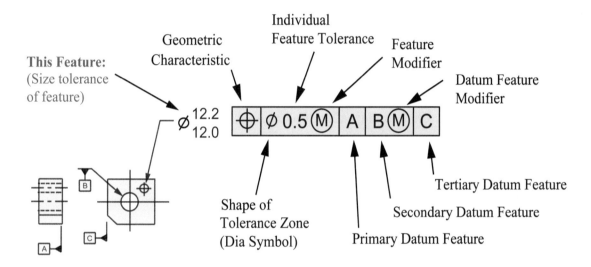

The first compartment of a feature control frame contains one of the 14 geometric characteristic symbols. There can never be two geometric characteristic symbols placed in a feature control frame. If there are two requirements for a feature, there must be two feature control frames or a composite tolerance. The geometric characteristic symbol will specify the requirement for the feature such as: This feature must be flat, or this feature must be positioned, etc.

The second compartment of a feature control frame contains the total tolerance for the feature. The feature tolerance is always a total tolerance. It is never a plus/minus value.

If the tolerance is preceded by a diameter symbol (Ø), the tolerance is a diameter or cylindrical shaped zone, as in the position of a hole. If the tolerance is preceded by a spherical diameter symbol (SØ), the tolerance will be a spherical shaped zone, as in the location of a ball or sphere. If the tolerance zone is preceded by the square symbol, then the tolerance zone is square shaped. If there is no symbol preceding the tolerance, the default tolerance zone shape is parallel planes or a total wide zone, as in the position of a slot or profile of a surface.

Feature Control Frame

Following the feature tolerance in the feature control frame, a material condition modifier such as MMC or LMC may be specified if the feature has size, such as for a hole, slot, tab, pin etc. If the feature has size and no modifier is specified, the default modifier is RFS. (See material condition modifier rules in unit 3 for more information on this subject.) If the feature has no size, such as a plane surface, then a feature modifier is not applicable.

If necessary, additional feature modifiers may be specified after the material condition modifiers, These modifiers can be projected tolerance zone, free state, tangent plane and statistical tolerance.

The third and following compartments of the feature control frame contain the specified datum feature reference(s) if datum references are required. For example, if a form tolerance such as flatness or straightness is specified, then no datum feature reference is allowed. However, if a location tolerance like position is specified, then datum feature references are usually specified. The alphabetical order of the datum references has no significance. The significance is their order of precedence reading from left to right as primary, secondary and tertiary. (See datum section unit 6 for more information.)

In some cases, the datum feature modifiers Maximum Material Boundary (MMB) or Least Material Boundary (LMB) may be applied to the datum feature. The default modifier is Regardless of Material Boundary (RMB). (See material condition modifier rules in unit 3 for more information.)

Basic Dimensions

Basic dimensions are theoretically exact numerical values or mathematical expressions used to define the form, size, orientation or location of a part or feature. Basic dimensions are enclosed in a box. They may also be invoked by a note on the drawing or by referencing a standard that contains these requirements. The mathematical data set in a CAD model may also be defined as basic. Permissable variations from basic dimensions are usually defined in a feature control frame or by other notes on a drawing. **The default tolerances stated in the title block of a drawing do not apply to basic dimensions.**

Symbols associated with a Feature Control Frame

This table lists symbols that are commonly found in or associated with a feature control frame. It also lists a brief description of the term or specifications associated with each symbol. A more complete discussion of each specification can be found later in the text.

The symbols marked with an "X" are new or revised from the previous ASME Y14.5M-1994 standard. The terms Maximum Material Boundary (MMB), Least Material Boundary (LMB) and Regardless of Material Boundary (RMB) are new in the ASME Y14.5-2009 standard. ISO does not use these terms, instead MMC, LMC and RFS are used for both features and datum features.

	TERM	SYMBOL ASME Y14.5	SYMBOL ISO
X	At Maximum Material Condition (MMC) (When applied to a feature) At Maximum Material Boundary (MMB) (When applied to a datum feature)	Ⓜ	Ⓜ
X	At Least Material Condition (LMC) (When applied to a feature) At Least Material Boundary (LMB) (When applied to a datum feature)	Ⓛ	Ⓛ
X	Regardless of Feature Size (RFS) (When applied to a feature) Regardless of Material Boundary (RMB) (When applied to a datum feature)	None Implied for features and datum features	None Implied for features and datum features
	Diameter	$\emptyset$	$\emptyset$
	Spherical Diameter	S$\emptyset$	S$\emptyset$
	Square	□	□
	Projected Tolerance Zone	Ⓟ	Ⓟ
	Free State	Ⓕ	Ⓕ
	Tangent Plane	Ⓣ	None
X	Translation	▷	None
X	Unequally Disposed Tolerance	Ⓤ	UZ (Proposed)
	Statistical Tolerance	⟨ST⟩	⟨ST⟩
	All Around	⌒	⌒
X	All Over	⌾	Proposed
	Between	↔	Proposed

Symbols Related To Datum Identification

This table identifies terms and symbols that are commonly associated with establishing datums from datum features. The datum feature symbol represents contacting the high points of the datum feature. The datum target symbols represent datum simulator contact points, lines or areas on the datum feature. A more complete definition and description of each these specifications can be found later in units 6 and 8. The movable datum target symbol marked with an "X" is new for the ASME Y14.5-2009 standard. It is a carry over from the Y14.8 Casting and Forging standard.

TERM	SYMBOL ASME Y14.5	SYMBOL ISO
Datum Feature	A ▲	A ▲
Datum Feature ANSI Y14.5M-1982 (former standard)	– A –	None
Datum Target	Ø10 / A1	Ø10 / A1
Datum Target Point	A1 ✕	A1 ✕
Datum Target Line	A1	A1
Datum Target Area	Ø20 / A1	Ø20 / A1
X Movable Datum Target	A1	Proposed
Dimension Origin	⊕→	⊕→

Index Plate - Practical Example

The establishment of a Datum Reference Frame (DRF) and the application of geometric tolerances are applied to parts based on functional design requirements. An index assembly is shown below. The top index plate clamps on its face and is centered by the pilot pin. The plate is rotationally aligned in the assembly with the side planar surface. The index plate is then bolted down in the assembly with three screws.

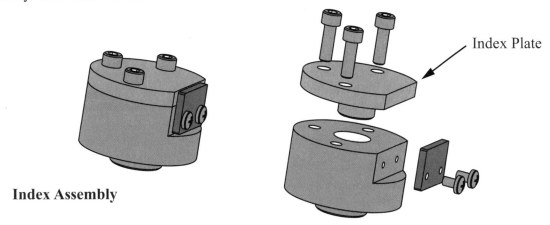

Index Assembly

Index Plate

The engineering drawing below reflects geometric tolerance applied to the index plate based on functional assembly conditions. The A, B, C, datum reference frame is established and qualified from features on the part based on functional mating conditions. The primary datum feature A, is the large planar face; its form is qualified with a flatness specification. The secondary datum feature B is the OD of the pilot, and it is oriented with a perpendicularity specification with respect to datum A.

The tertiary datum feature planar surface, is located with respect to the A,B - DRF with a profile specification. The remaining features on the part are located with respect to the A,B,C - DRF with position and profile tolerances that are appropriate to their importance in the assembly. The outside surface of the part is located with a relatively large profile tolerance. The 3 holes have more restrictive position tolerances applied.

Engineering drawing with applied geometric tolerancing based on functional requirements

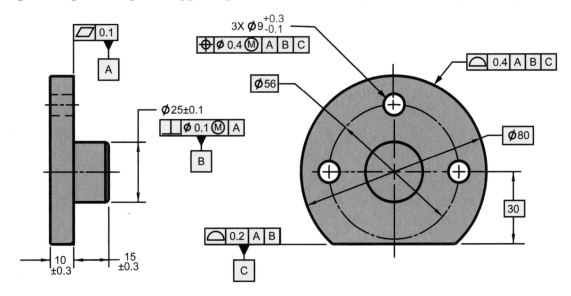

Index Plate - Matching a Perfect DRF to an Imperfect Part

The produced index plate has imperfect geometry because of variation in manufacturing. This illustration shows how the theoretical perfect geometry of the design A, B, C - DRF is mated to the imperfect geometry of the produced part. The geometry, sequence of events and corresponding terminology in the mating process are the basics to understand geometric tolerancing and are used throughout this text. Detailed definitions of these terms can be found on page 2.15 and in the ASME Y14.5-2009 standard. All dimensions originate from the Datum Reference Frame (DRF) and not the part.

Establishment of A, B, C - DRF on the produced index plate.

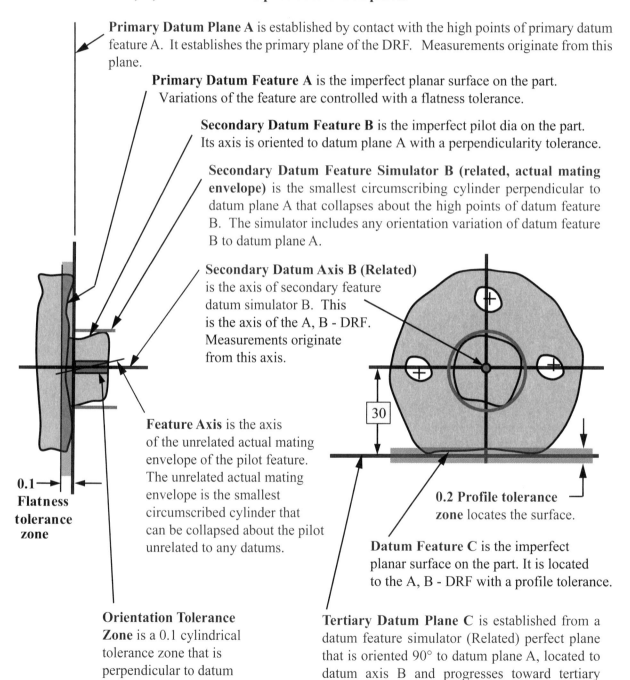

Primary Datum Plane A is established by contact with the high points of primary datum feature A. It establishes the primary plane of the DRF. Measurements originate from this plane.

Primary Datum Feature A is the imperfect planar surface on the part. Variations of the feature are controlled with a flatness tolerance.

Secondary Datum Feature B is the imperfect pilot dia on the part. Its axis is oriented to datum plane A with a perpendicularity tolerance.

Secondary Datum Feature Simulator B (related, actual mating envelope) is the smallest circumscribing cylinder perpendicular to datum plane A that collapses about the high points of datum feature B. The simulator includes any orientation variation of datum feature B to datum plane A.

Secondary Datum Axis B (Related) is the axis of secondary feature datum simulator B. This is the axis of the A, B - DRF. Measurements originate from this axis.

Feature Axis is the axis of the unrelated actual mating envelope of the pilot feature. The unrelated actual mating envelope is the smallest circumscribed cylinder that can be collapsed about the pilot unrelated to any datums.

0.1
Flatness tolerance zone

30

0.2 Profile tolerance zone locates the surface.

Datum Feature C is the imperfect planar surface on the part. It is located to the A, B - DRF with a profile tolerance.

Orientation Tolerance Zone is a 0.1 cylindrical tolerance zone that is perpendicular to datum plane A. The unrelated feature axis of the pilot must fall within this zone.

Tertiary Datum Plane C is established from a datum feature simulator (Related) perfect plane that is oriented 90° to datum plane A, located to datum axis B and progresses toward tertiary datum feature C to make maximum possible contact with the high points on the datum feature. This plane sets the orientation planes of the DRF.

1.18

3D Model Tolerancing

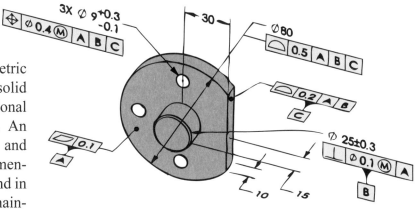

This is an example of geometric tolerancing applied to the solid model based on the functional requirements of the assembly. An A,B,C - DRF is established and qualified. All untoleranced dimensions are basic and can be found in the CAD math data. The remaining features on the part are located to the DRF. Features on the model can also be queried for specifications.

See digital data file 2324 for hole and surface locations.

The drawing above is an example of geometric tolerancing applied to a solid model. This drawing is per the ASME Y14.41-2003 (R2008) standard. This standard shows how to symbolically apply geometric tolerancing to the digital model, but does not replace the ASME Y14.5 standard, it just allows and extends the geometric tolerancing principles to digital models. For definitions and technical definitions of geometric tolerancing principles, the ASME Y14.5-2009 standard is the primary document.

Notice on the drawing above that many dimensions are not shown but can be found in the digital file. The digital file dimensions are basic. Geometric tolerances are applied to the features by the use of feature control frames. As you can see, it would be impossible to use directly toleranced or plus/minus type dimensions on this type of product definition. For this new technology, and the eventual goal of a paperless system to advance, it is critical to use geometric tolerancing.

Definitions

The definitions below are terms used in this text and the ASME Y14.5-2009 standard.

Feature - A physical portion of a part such as a surface, pin, hole, or slot or its representation on drawings, models, or digital data files.

Feature Axis - The axis of the unrelated actual mating envelope of a feature.

Feature Center Plane - The center plane of the unrelated actual mating envelope of a feature.

Note: When the term "feature axis" or "feature center plane" is used, it refers to the axis or center plane of the unrelated actual mating envelope unless specified otherwise.

Actual Mating Envelope - This envelope is outside the material. A similar perfect feature(s) counterpart of smallest size that can be contracted about an external feature(s) or largest size that can be expanded within an internal feature(s) so that it just contacts with the surface(s) at the highest points. There are two types of actual mating envelopes:

> 1. **Unrelated Actual Mating Envelope** - A similar perfect feature(s) counterpart expanded within an in internal feature(s) or contracted about an external feature(s) and **not constrained** to any datum reference frame.

> 2. **Related Actual Mating Envelope** - A similar perfect feature(s) counterpart expanded within an in internal feature(s) or contracted about an external feature(s) **while constrained** either in orientation or location or both to the applicable datum reference frame.

Actual Minimum Material Envelope
This envelope is within the material. A similar perfect feature(s) counterpart of largest size that can be expanded within an external feature(s) or smallest size that can be contracted about an internal feature(s) so that it coincides with the surface(s) at the lowest points. There are two types of actual minimum material envelopes.

> 1. **Unrelated Actual Minimum Material Envelope** - A similar perfect feature(s) counterpart contracted about an internal feature(s) or expanded within an external feature(s), and not constrained to any datum reference frame.

> 2. **Related Actual Minimum Material Envelope** - A similar perfect feature(s) counterpart contracted about an internal feature(s) or expanded within an external feature(s) while constrained in either orientation or location or both to the applicable datum(s).

Derived Median Plane - An imperfect (abstract) plane formed by the center points of all line segments bounded by the feature. These line segments are normal (perpendicular) to the center plane of the unrelated actual mating envelope.

Derived Median Line - An imperfect (abstract) line formed by the center points of all cross sections of the feature. These cross sections are normal (perpendicular) to the axis of the unrelated actual mating envelope.

Fundamental Rules

Dimensioning and tolerancing shall clearly define engineering intent and conform to the following fundamental rules. See ASME Y14.5-2009 for additional information.

1. Each dimension shall have a tolerance, unless specifically identified as reference, maximum, minimum, or stock. (commercial stock size). The tolerance may be applied directly to the dimension (or indirectly in the case of basic dimensions) or indicated by a general note in the title block.

2. Dimensioning and tolerancing shall be complete so there is full understanding of the characteristics of each feature. Values may be expressed in an engineering drawing or in a CAD product definition data set.

3. Dimensions and tolerances shall be selected and arranged to suit the function and mating relationship of a part and shall not be subject to more than one interpretation. Each necessary dimension of an end product shall be shown in true profile views and refer to visible outlines. No more dimensions than those necessary for complete definition shall be given. The use of reference dimensions on a drawing should be minimized.

4. The drawing should define a part without specifying manufacturing methods. Thus, only the diameter of a hole is given without indicating whether it is to be drilled, reamed, punched, or made by any other operation. However, in those instances where manufacturing, processing, quality assurance, or environmental information is essential to the definition of engineering requirements, it shall be specified on the drawing or in a document referenced on the drawing.

5. A 90° angle with tolerance applies where center lines and lines depicting features are shown on a 2D orthographic drawing when defined by directly toleranced dimensions at right angles and no angle is specified.

6. A 90° basic angle applies where center lines of features or surfaces shown at right angles on a 2D orthographic drawing when defined by basic dimensions and no angle is specified.

7. A zero basic dimension applies where axes, center planes, or surfaces are shown coincident on a drawing, and geometric tolerances establish the relationship among the features.

8. Unless otherwise specified, all dimensions and tolerances are applicable at 20°C (68°F) in accordance with ANSI/ASME B89.6.2.

9. Unless otherwise specified, all dimensions and tolerances apply in a free-state condition.

10. Unless otherwise specified, all tolerances apply for full depth, length, and width of the feature.

11. Dimensions and tolerances apply only at the drawing level where they are specified. A dimension specified for a given feature on one level of drawing (e.g., a detail drawing) is not mandatory for that feature at any other level (e.g., an assembly drawing).

12. Where a coordinate system is shown on the drawing, it shall be right-handed unless otherwise specified. Each axis shall be labeled and the positive direction shall be shown.

Material Conditions - MMC and LMC

A concept in geometric tolerancing is the use of feature material condition modifiers. In specifying geometric controls, there is often need to state that a tolerance applies to a feature at a particular feature size. The terms Maximum Material Condition (MMC) and Least Material Condition (LMC) allow an engineer to clearly specify this design intent.

These material condition modifiers are used in a feature control frame in the feature tolerance compartment and follow the feature tolerance. They are applicable when referring to features of size such as holes, slots, tabs, pins, etc. These modifiers are not applicable to non-features of size, such as plane surfaces. The application of the MMC and LMC modifiers provide additional geometric tolerance beyond the specified tolerance as the features depart from the specified condition. See unit 3 later in the text for the details.

Maximum Material Condition- Acronym - MMC
The condition where the feature contains the maximum material within the stated limits of size- for example, the largest pin or the smallest hole.

Least Material Condition - Acronym - LMC
The condition where the feature contains the least material within the stated limits of size- for example, the smallest pin or largest hole.

MMC and LMC examples;

The drawing below identifies the MMC and LMC for each feature of size.

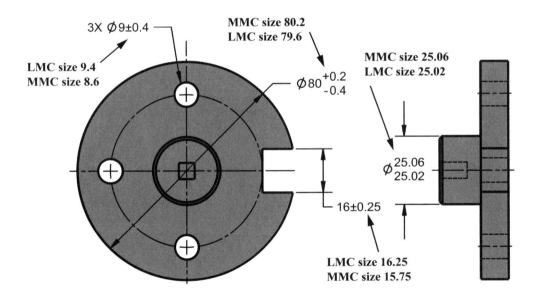

Workshop Exercise 1.1

1. What is the name and date of the current American National Standard on dimensioning and tolerancing?

2. What is the name and date of the current American National Standard on mathematical definitions of dimensioning and tolerancing?

3. What is the name and date of the current American National Standard on certification of dimensioning and tolerancing professionals?

4. In the International Standards Organization (ISO) there are many documents that cover dimensioning and tolerancing. What is the number and date of the ISO standard that covers the generalities of geometric tolerancing? Tolerancing of form, orientation, location and runout?

On the drawing below, find and label the following symbols.

5. Feature control frame
6. Radius symbol
7. Basic dimension
8. Two places designation
9. Square symbol
10. All around symbol

11. Countersink symbol
12. Counterbore symbol
13. Datum feature symbol
14. Reference symbol
15. Depth symbol
16. Diameter symbol

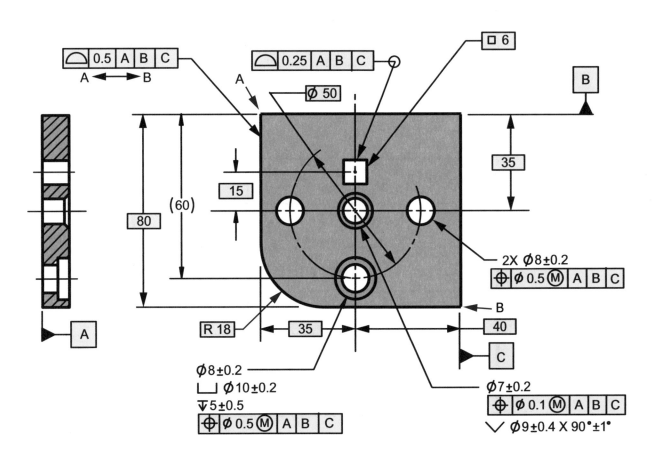

17. In the table below, sketch the proper symbol next to the geometric characteristic term. Select from the symbols shown below. Also, identify the type of tolerance such as: form, orientation, location, runout and location of derived median points.

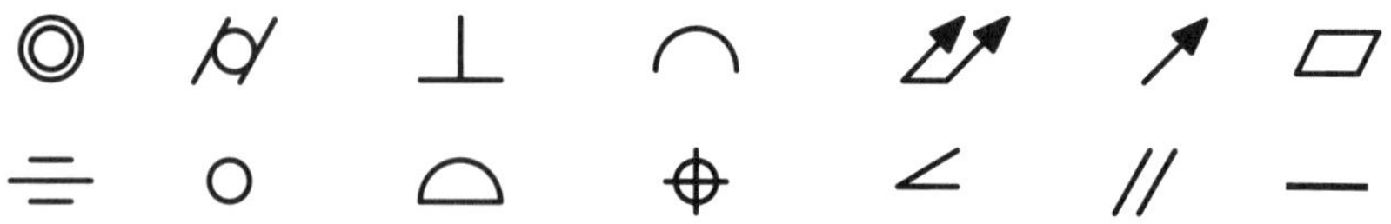

◎　⋈　⊥　⌒　⌿⌿　⌿　▱

≡　○　⌓　⊕　∠　//　—

SYMBOL	GEOMETRIC CHARACTERISTIC	TYPE OF TOLERANCE
	FLATNESS	
	STRAIGHTNESS	
	CYLINDRICITY	
	CIRCULARITY (ROUNDNESS)	
	PERPENDICULARITY	
	PARALLELISM	
	ANGULARITY	
	POSITION	
	PROFILE OF A SURFACE	
	PROFILE OF A LINE	
	TOTAL RUNOUT	
	CIRCULAR RUNOUT	
	CONCENTRICITY	
	SYMMETRY	

Workshop Exercise 1.2

1. Label the terms for the symbols shown in the table. Choose from the terms listed below.

Free State

Spherical Radius   Spherical Diameter

Projected Tolerance Zone   Dimension Origin

Tangent Plane   All Over

Least Material Condition   Controlled Radius

Unequally Disposed Tolerance   Continuous Feature

Translation   Maximum Material Condition

Term	Symbol ASME Y14.5	Symbol ISO
	S⌀	S⌀
	⌽→	⌽→
	⦶→	Proposed
	⟨CF⟩	None
	CR	None
	SR	SR
	Ⓜ	Ⓜ
	Ⓛ	Ⓛ
	Ⓕ	Ⓕ
	Ⓟ	Ⓟ
	Ⓣ	None
	▷	None
	Ⓤ	Proposed

2. Label the parts of the feature control frame. Choose from the terms below.

Shape of tolerance zone   Primary datum feature

Datum feature modifier   Geometric characteristic symbol

Tertiary datum feature   Secondary datum feature

Feature tolerance   Feature modifier

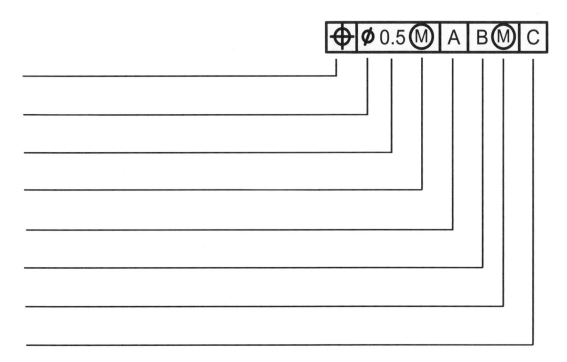

3. Next to the dimensions, label the Maximum Material Condition (MMC) and Least Material Condition (LMC) of the five features of size below.

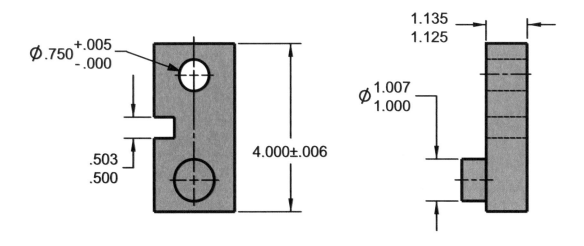

Unit 2

Limits of Size

Feature of Size Definition
Limits of Size (Rule #1, Envelope Principle, Taylor Principle)
Limits of Size also Controls Form
Limits of Size Definition per ASME Y14.5.1M-1994
Exceptions to Rule #1
Size Does Not Control Interrelationship Between Features
Directly Toleranced Dimensions Are Not Clear for Locating Surfaces
Profile Tolerance Is Used to Locate Surfaces
Dimension Origin Symbol
Definitions
Fundamental Rules
System of Limits and Fits
      Limits and Fits Tables
Workshop Exercise 2.1
Workshop Exercise 2.2

Feature of Size Definition

Feature of Size. There are two types of features of size.

Regular Feature of Size. One cylindrical or spherical surface, a circular element, a set of two opposed parallel elements or opposed parallel surfaces, each of which is associated with a directly toleranced dimension.

Irregular Feature of Size. There are two types of irregular features of size.

> 1. A directly toleranced feature or collection of features that may contain or be contained by an actual mating envelope which is a sphere, cylinder or pair of parallel planes.

> 2. A directly toleranced feature or collection of features that may contain or be contained by an actual mating envelope other than a sphere, cylinder or pair of parallel planes.

Examples of Regular Features of Size

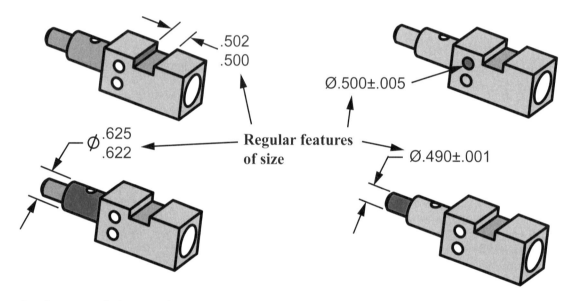

Regular features of size are the most common features of size.

Examples of Irregular Features of Size

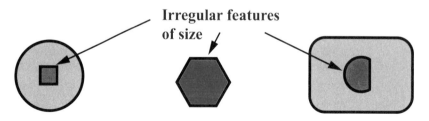

The irregular feature of size is an advanced concept and is not common. It is used in certain demanding cases where it is necessary to locate, orient or identify these type of features and apply modifiers. They are usually a collection of features that are related with a profile or position tolerance. The irregular feature of size is a new concept for the ASME Y14.5-2009 standard.

Limits of Size (Rule #1, Envelope Principle, Taylor Principle)

The limits of size (Rule #1, Envelope Principle, Taylor Principle) defines the size and form limits for regular, individual features. It requires perfect form at MMC and ensures the pin will fit in the hole.

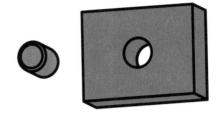

This on the drawing

The limits of size is an important concept that defines the size and form limits for an individual regular feature of size. In the ASME Y14.5, it is referred to as Rule #1 or the Envelope Principle. In the international community, it is referred to as the Envelope Principle. Historically, it has also been called the Taylor Principle because in 1905 William Taylor, an Englishman, obtained a patent on a full form go-gage for screw threads. The concept was later extended to all features of size and is the underlying principle of our go/no-go gages used in industry today.

A drawing of a pin and a hole are shown above, along with a size dimension. The limits of size ensures the pin will fit in the hole by requiring perfect form at MMC. The extreme variations of size and form allowed on an individual, regular feature of size are shown below.

Means this

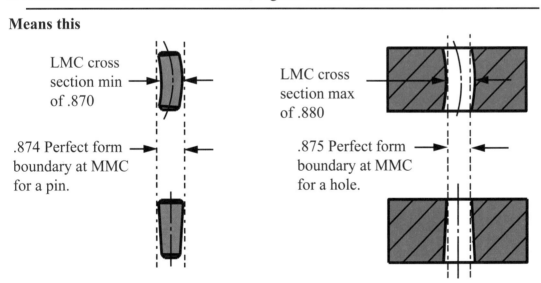

LMC cross section min of .870

.874 Perfect form boundary at MMC for a pin.

LMC cross section max of .880

.875 Perfect form boundary at MMC for a hole.

The limits of size define the size, as well as the form of an individual feature. The form of the feature may vary within the size limits. If the feature is produced at its maximum material condition, the form must be perfect. The feature may be bent, tapered or out of round as it departs from the maximum material condition.

This on the drawing

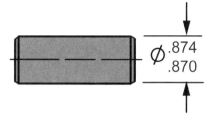

The size tolerance applied to the pin also controls its form. If it is necessary to refine the form, a form tolerance such as straightness, circularity, cylindricity or flatness is applied. Some of the possible form variations for the pin above are shown below.

Means this

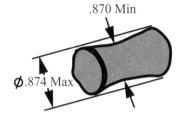

The pin can be waisted within .004. The form can be refined with straightness or cylindricity.

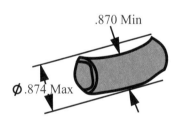

The pin can be bent within .004. The form can be refined with straightness or cylindricity.

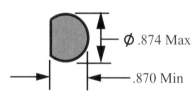

The pin can be "D" shaped or out of round within .004. The form can be refined with circularity or cylindricity.

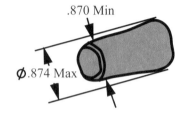

The pin can be tapered within .004. The form can be refined with cylindricity.

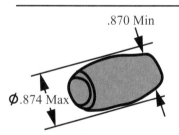

The pin can be barreled within .004. The form can be refined with straightness or cylindricity.

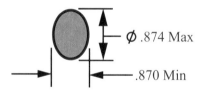

The pin can be oval or out of round within .004. The form can be refined with circularity or cylindricity.

Note: If it is necessary to relax the form control provisions of Rule #1, but still retain the size requirements, a straightness of an axis or a flatness of a median plane may be applied. See the form tolerances section (unit 10) for more information.

Limits of Size per ASME Y14.5.1M-1994, Mathematical Definitions Std.

The ASME Y14.5.1M-1994 (R2004) Mathematical Definitions standard provides a further clarification for the local size requirements of Rule #1. The actual local size is defined by a continuously expanding and contracting sphere pulled through the feature. In order to meet the local size requirements, the minimum local size sphere must never exit the material. The reason for this clarification is to protect design criteria. This definition ensures an LMC spherical cross section of the feature and limits odd-numbered lobing effects.

The mating envelope is defined as the smallest circumscribing cylinder.

This on the drawing

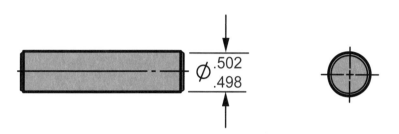

Means this

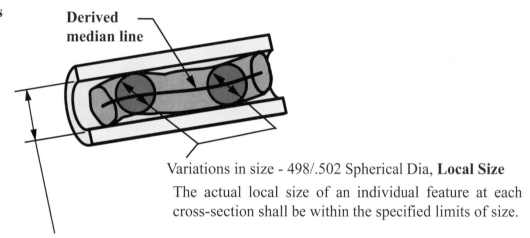

Derived median line

Variations in size - 498/.502 Spherical Dia, **Local Size**
The actual local size of an individual feature at each cross-section shall be within the specified limits of size.

Variations in Form - .502 Maximum, **Mating Envelope**

a. The surface or surfaces of a feature shall not extend beyond a boundary (envelope) of perfect form at MMC. This boundary is the true geometric form represented by the drawing. No variation in form is permitted if the regular feature of size is produced at its MMC limit of size unless a straightness or flatness tolerance is associated with the size dimension or the independency symbol is applied.

b. Where the actual local size of a regular feature of size has departed from MMC towards LMC, a local variation in form is allowed equal to the amount of such departure

c. There is no default requirement for a boundary of perfect form at LMC. Thus, a regular feature of size produced at its LMC limit of size is permitted to vary from true form to the maximum variation allowed by the boundary of perfect form at MMC.

d. In cases where a geometric tolerance is specified to apply at LMC, perfect form at LMC is required.

The definition of size is clearly defined in the ASME Y14.5-2009 standard and further clarified in the ASME Y14.5.1M-1994 (R2004) standard. The underlining theoretical definition of size makes sense based on design functional fitting requirements but can pose some challenges in verification. Size on a pin is often verified with a micrometer or calipers. This is sufficient in most cases but the quality engineer/technician should realize that in some cases, a two-point check verification will pose some associated risks.

The pin has a size tolerance of .498/.502. The pin will have two checks to verify and record values.
1. Local size is a spherical cross-sectional check.
2. Mating envelope is a perfect form envelope at MMC.

Notice that a two-point micrometer check will not verify the mating envelope. The pin could be bent beyond its perfect form at MMC envelope and still pass a micrometer check. A functional cylindrical go-gage for a shaft, or pin gage for a hole, is often used to verify perfect form at MMC.

A two-point micrometer check has some associated risk.

In some cases, pins, shafts, holes, thin-walled parts etc. can be unintentionally produced with lobes or flat spots, especially if centerless ground or diameters are clamped too tightly in a 3-jaw chuck. This odd lobed or gleichstuck shape (parts that are not round but will give equal two-point measurement, such as the impeller on a Wankel engine or Susan B Anthony coin) may yield different results when checked with a micrometer rather than using a spherical cross-sectional check. In the measuring process, it is possible for the actual local size to measure smaller and the actual local size to measure larger.

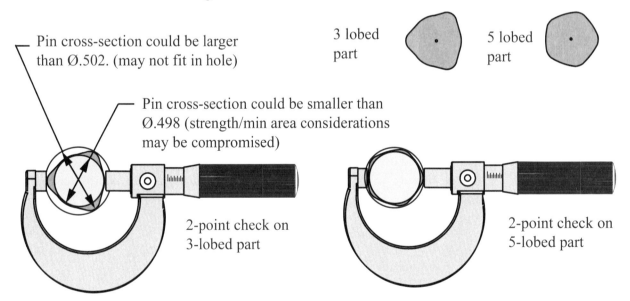

Pin cross-section could be larger than Ø.502. (may not fit in hole)

3 lobed part

5 lobed part

Pin cross-section could be smaller than Ø.498 (strength/min area considerations may be compromised)

2-point check on 3-lobed part

2-point check on 5-lobed part

The graphics above illustrate how a measuring device using two opposed points will not necessarily insure conformance to the size requirements. This does not mean calipers or a micrometer can not be used, it is just to understand that some risk is possible when demanding restrictive tolerances are used and that in some cases, a more stringent inspection procedure may be necessary. A CMM with mapping capabilities or a roundness machine may be used in these cases.

Exceptions to Rule #1

Form control does not apply (Exceptions to Rule #1)

The control of geometric form prescribed by the limits of size does not apply to the following: Stock, such as bars, sheets, tubing, structural shapes, and other items produced to established industry or government standards that prescribe limits for straightness, flatness and other geometric characteristics. Unless geometric tolerances are specified on the drawing of a part made from these items, standards for these items govern the surfaces that remain in the as-furnished condition on the finished part.

Form may also be released from the form provisions of Rule #1 by applying the Independency Principle symbol, "circle I", next to the size tolerance, or notation of the feature as shown below. Caution should be exercised when using this option as form is entirely uncontrolled. This option may be used in certain cases for example, when controlling web thickness. It is not common.

Application of Independency Principle symbol to release form control from Rule #1 requirements. (ASME Y14.5)

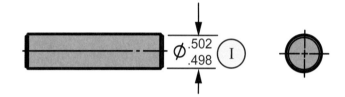

ISO 8015, 1985 -Fundamental Tolerancing Principle

In the ISO standards, the size and form requirements for features of size can be different than the ASME Y14.5-2009 requirements, depending on the specifications invoked. The ISO standards often use the Principle of Independency for size features. The Principle of Independency is defined in ISO 8015, 1985. An excerpt from that document is shown below.

Principle of Independency (ISO)
Each specified dimensional or geometrical requirement on a drawing shall be met independently, unless a particular relationship is specified. Where no relationship is specified, the geometrical tolerance applies regardless of feature size, and the two requirements are treated as being unrelated. If a particular relationship of size and form, or size and orientation, or size and location is required, it shall be specified on the drawing. Drawings to which the Principle of Independency applies are identified and marked in or close to the drawing title block as follows: Tolerancing per ISO 8015.

If the independency principle (ISO 8015) is applied to a feature dimension without individual tolerance indications, the size and form is controlled by General Tolerances and General Geometric Tolerances defined in ISO 2768 - part 1 & part 2. These standards should also be identified on the drawing, in or near the title block.

If it is necessary to invoke the Envelope Principle (Rule #1) on a drawing that is identified as being conformant to ISO 8015, then the circle E symbol is placed next to the size tolerance or as a notation on the drawing.

Size Does Not Control Interrelationship Between Individual Features

The limits of size do not control the orientation or location relationship between individual features. Features shown perpendicular, coaxial, or symmetrical to each other must be toleranced for location and orientation to avoid incomplete drawing requirements. If it is necessary to control a perfect orientation/location boundary at MMC, the boundary may be accomplished in one of the following ways.

1. Specify an applicable zero orientation or position tolerance at MMC and if applicable, the datum features at MMB.
2. Indicate the relationship between features with a note requiring perfect orientation/position at MMC for the interrelationship of features.
3. Apply profile tolerancing.

Mating parts

The drawings below have size tolerances that appear like they will fit together, but since there is no relationship shown between the features, they may not assemble. Geometric tolerancing concepts shown later in the text are necessary to assure assembly.

This on the drawing

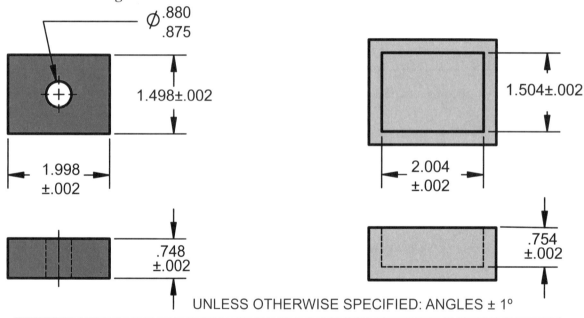

UNLESS OTHERWISE SPECIFIED: ANGLES ± 1°

Means this

The produced part may not be square and may not fit in a square box. It may be a parallelogram. The hole may not be perpendicular to any surface. Each individual feature must be within size limits but there is no implied relationship between individual features.

The only relationship between the features is the specified ±1° angle tolerance. The angle tolerance is vague, as there are many implied angles on the parts.

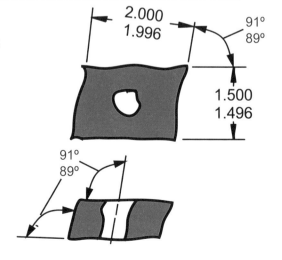

2.8

Size Does Not Control Interrelationship Between Individual Features

This on the drawing

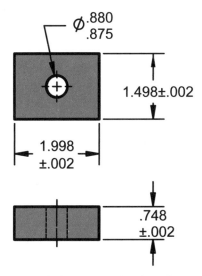

UNLESS OTHERWISE SPECIFIED:
ANGLES ± 1°

**Allows measurements to be
not square to each other**

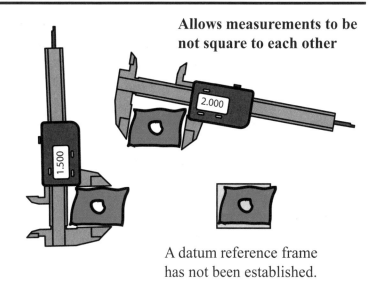

A datum reference frame
has not been established.

The ±1° angle tolerance between the features of size
allow the part to be a parallelogram and may not fit
in a 1.500 X 2.000 perfect rectangular box.

There is no implied relationship between the size toleranced features on the rectangular part above, except the specified angle tolerance. The part could be a parallelogram. Perfect form at MMC is for individual features, not the interrelationship between features.

This on the drawing

**Establishes basic 90° angles
that are square to the each
other and datum feature A.**

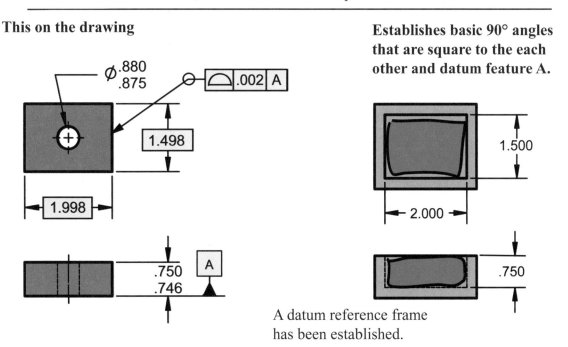

A datum reference frame
has been established.

The drawing above has the surfaces related to each other by the basic linear dimensions and the unspecified, but implied basic 90° angles. The profile tolerance is applied all around and referenced to datum feature A. This will insure the part will fit in a 1.500 X 2.000 perfectly rectangular box. Profile tolerance is explained in detail later in unit 12. A profile tolerance was shown to control the relationship between features but other geometric controls could possibly be used as well. This related feature could now be termed and used as an irregular feature of size.

Size Does Not Control Interrelationship Between Individual Features

Parts are composed of features. The shaft below displays 4 diametrical features of size. There is no relationship between features implied in the ASME Y14.5-2009 standard. The Limits of Size (Rule #1) only covers the size and form of individual features. It does not control the relationship between features.

Features shown perpendicular, coaxial, or symmetrical to each other must be controlled for location or orientation to avoid incomplete drawing requirements. There needs to be some datum reference frame relationship provided to locate the features on this shaft.

All of the diameters on the shaft below are shown on the same center, so the machinist will probably make the part with all diameters on the same center. The problem is verification, as the drawing does not define coaxiality or how much the centers are allowed to be off to each other.

This on the drawing

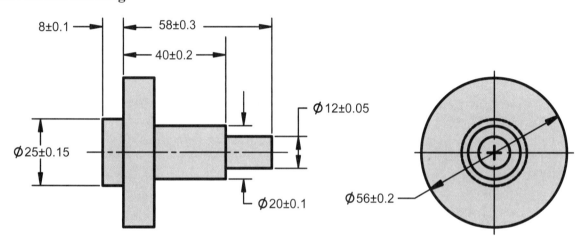

Produced part with error

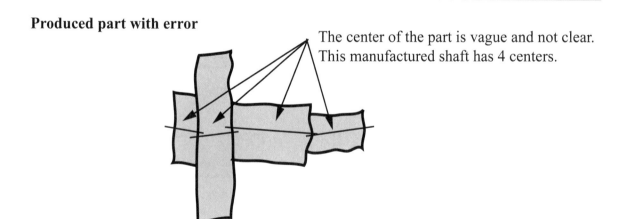

The center of the part is vague and not clear. This manufactured shaft has 4 centers.

The actual produced part has error, exaggerated here for clarity. Centerlines are only on drawings, not on parts. There is no such physical thing as the center of the part. The actual part above has four axes derived from the four cylindrical features on the part. As you can see, an inspector would get a different answer, depending on which diameter is used to set up the part. The datum reference on this part and the relationship between the features is unclear.

Size Does Not Control Interrelationship Between Individual Features

If there is a need to control the location between features, a DRF is established from specific datum features on the part. In this case, the large face is identified as the primary datum feature A and the 25 mm pilot diameter is selected as datum feature B and related to the face with a perpendicularly tolerance.

The remaining features are located to this established DRF with position tolerances. Runout and profile tolerances could be used, as well as depending on functional requirements. In some cases, the relationship between the features and the DRF can be stated with a general note that is referenced on the drawing. This makes the design intent clear.

This on the drawing

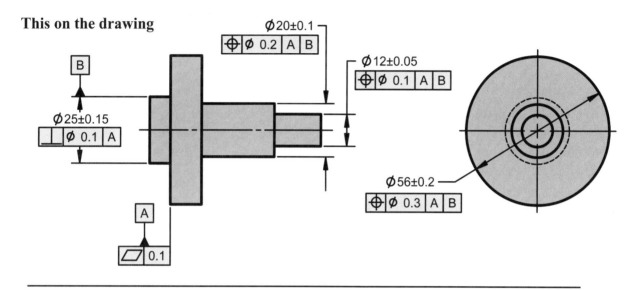

Produced part

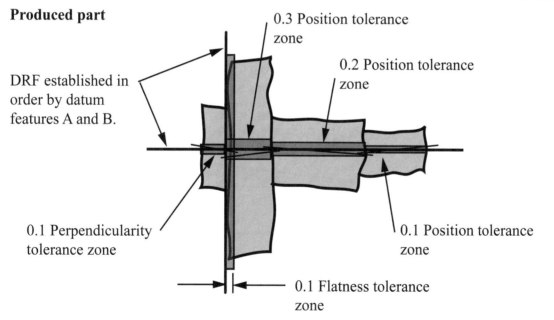

The drawing of the produced part above shows the established DRF and the position tolerance zones for each of the diametrical features. The axis of the features must fall within their respective perpendicularity and position tolerance zones. There is additional detailed information on coaxial tolerances shown later in unit 14.

Directly Toleranced Dimensions Are Not Clear for Locating Surfaces

Directly toleranced dimensions are used to define size, not location

Directly toleranced (plus/minus) dimensions that are used to locate surfaces are not clear. Directly toleranced dimensions are usually used to define size. Size dimensions require features with opposing elements. According to Rule #1, that defines size limits, the feature must have opposing elements in to order to verify the cross-sectional check.

Directly toleranced (plus/minus) dimension

As you can see from the geometry on this part, it is impossible to verify a two point micrometer or caliper check as the surfaces do not have opposing elements or points. One surface must be set as zero and the distance verified with a height gage or CMM. The distance between the 20 ±0.3 dimension on the sheet metal part is not a size dimension, it is location.

Produced part has variance. It is just a little bent. Is it produced within tolerance?

Many similar parts like this one are still toleranced with directly toleranced dimensions. This does not make it correct, as design intent may still be unclear. The manufacturing process will not make the two surfaces exactly parallel, as there will always be some variance. When the height is verified, it will depend on which surface is set up as the datum. Directly toleranced dimensions do not define an origin. Since one surface is longer than the other, different results will be obtained depending on which surface is selected as the origin or datum. There can be multiple interpretations as to what constitutes a maximum height.

A dimension origin symbol could possibly be used to help define the origin, but most parts are not this simple 2D specification and require datums to locate other additional features on the part.

Directly toleranced unopposed surfaces do not define an origin, design intent is not clear.

Leveling the short end and checking the long end will show the part bad.

Leveling the long end and checking the short end will show the part good.

Tipping the part in some orientation will show the part very good.

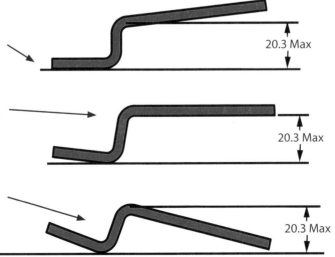

Directly toleranced dimensions that are used to locate surfaces do not provide a clear origin of measurement.

This type of part should have one surface identified as a datum feature and the other surface located with a profile tolerance.

Profile Tolerance Is Used to Locate Surfaces

Profile tolerance replaces the traditional plus/minus or direct type tolerancing for the location of surfaces. The two examples below illustrate a sheet metal part mounted in an assembly. Datums are selected based on functional mounting conditions and in each case the design intent is clear. A dimension origin symbol would work in this very simple example but considering the three dimensional requirements, profile tolerance is certainly preferred.

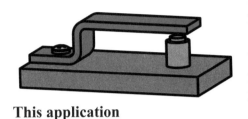

This application

In this assembly, the short face of the sheet metal part mounts to the base and the longer face must clear the pin. In this case, the short surface is identified as the datum feature and the long surface is located to the datum with a profile tolerance. The datum feature is qualified with a flatness tolerance. To inspect, mount on the short surface and indicate the longer surface.

This on the drawing

Means this

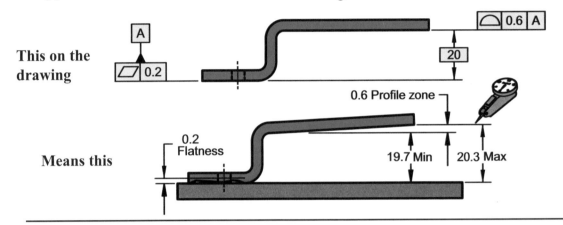

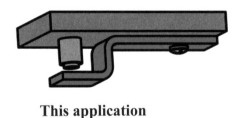

This application

In this assembly, the long face of the sheet metal part mounts to the base and the short face must clear the pin. In this case, the long surface is identified as the datum feature and the short surface is located to the datum with a flatness tolerance. The datum feature is qualified with a flatness tolerance. To inspect, mount on the long surface and indicate the short surface.

This on the drawing

Means this

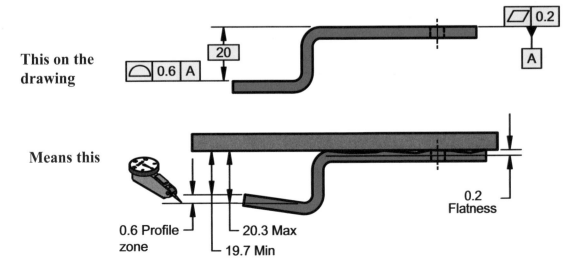

2.13

Dimension Origin Symbol

The dimension origin is a means of indicating that a directly toleranced dimension between two features originates from one of the features. A circle at the intersection of the dimension line and extension line replaces the arrowhead and indicates the origin. The high points of the surface indicated as the origin define a plane from where the dimension originates.

The example below illustrates a part where the origin is identified as the bottom surface. Without such indication, the longer surface could have been selected as the origin, thus permitting a greater linear variation between the surfaces.

The dimension origin concept is only a 2-dimensional control and is usually used in very simple 2D applications like the depth of an "O" ring groove. The dimension origin symbol does not establish a datum reference frame. If it is necessary to control more complex parts relative to a datum reference frame, a profile specification is used.

This on the drawing

Means this

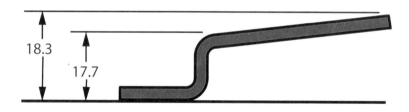

Not this

2.14

Workshop Exercise 2.1

1. According to the limits of size, also known as Rule #1, the size tolerance applied to a feature of size controls the size and _____ of the feature.

    A: Orientation   B: Location   C: Form   D: Runout

2. What is the maximum amount of bow allowed on the pin below?

    A: .003   B: .0015   C: .006   D: Not defined

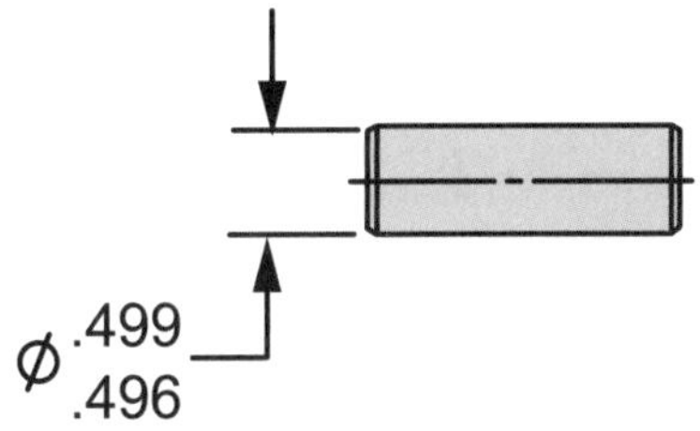

$$\phi \frac{.499}{.496}$$

3. If the pin above were produced at a size of .499, what is the maximum form variation allowed?

    A: .000   B: .003   C: .0015   D: Not defined

4. The actual size for a feature of size will have two values, the **actual mating envelope** and the **actual local size**. Match these terms to the illustrations of the produced part below.

As drawn

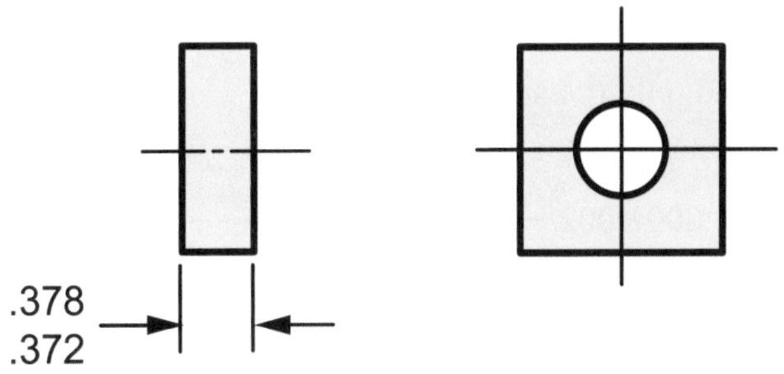

$$\frac{.378}{.372}$$

Produced part

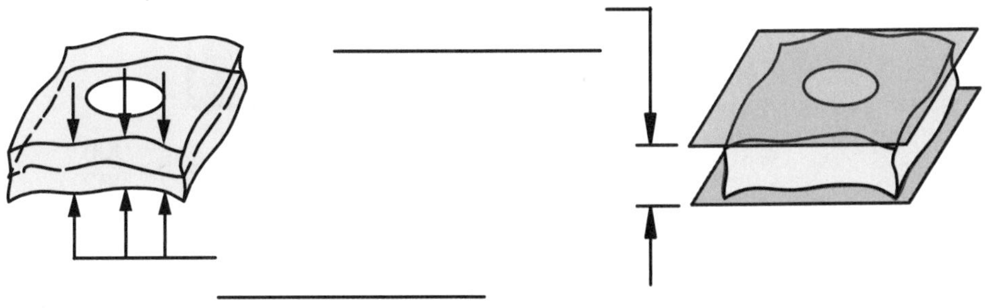

5. If a gage pin were used to verify the size of a hole, does it measure the mating envelope or the local size?

_____

6. On the shaft below, with only the limits of size defined, what coaxiality tolerance is implied among the three diameters?

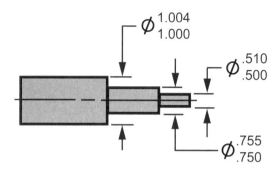

7. On the pin below, with only the limits of size defined, what perpendicularity tolerance is implied between the pin and bottom surface?

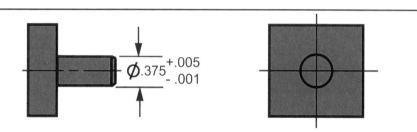

8. On the part below, circle only the size dimensions.

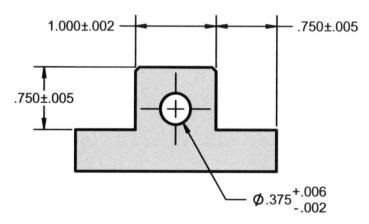

9. On the part above, you circled the size dimensions. Explain the difference between the dimensions you circled and the dimensions you did not.

_____

_____

_____

System of Limits and Fits

In order to determine mating fit tolerances between features like pins and holes or bearings and shafts etc., the engineer must define size tolerances to obtain proper functional requirements. There are standards to provide the engineer with guidelines for selecting size tolerances. The ANSI B4.1-1967 (R2004) Preferred Limits and Fits for Cylindrical Parts and ANSI B4.2-1978 (R2004) Metric Limits and Fits defines 5 main types of fits. This information can also be found in The Machinery's Handbook.

Note: This is only a brief explanation of the fits with a few sample charts and examples of how to use them. For a full description of each fit and application, consult the appropriate standard.

Running and Sliding Fits (RC) are intended to provide a similar running performance with suitable lubrication allowance, throughout the range of sizes.
   •Always have positive min clearance for lubrication
   •Applications: Translating shafts, bushings, bearings

   RC1 thru RC9   RC 1 – Close Sliding Fit   RC 9 – Loose Running Fit

Locational Clearance Fits (LC) are intended for parts which are normally stationary, but which can be freely assembled or disassembled. They range from snug fits for parts requiring accuracy of location, to the looser fastener fits where freedom of assembly is of prime importance.
   •Range from line-to-line min clearance for better alignment, to significant min clearance for ease of assembly
   •Snug fit applications: pilots, dowels, and other alignment features
   •Loose fit applications: clearance holes for fasteners

   LC 1 thru LC11   LC 1 - Snug Fit   LC11 - Loose Fit

Locational Transition Fits (LT) are a compromise between clearance and interference fits, for application where accuracy of location is important, but either a small amount of clearance or interference is permissible.
   •Applications: pilots, dowels, and other alignment features

   LT 1 thru LT6

Locational Interference Fits (LN) are used where accuracy of location is of prime importance but with no special requirements for bore pressure.
   •Applications: fits for a no "slop" alignment application
   •Can be harder to assemble and disassemble

   LN 1 thru LN 3

Force Fits (FN) constitute a special type of interference fit, normally characterized by maintenance of constant bore pressures throughout the range of sizes.
   •High bore pressures needed for high forces and/or torsion stress

   FN1 thru FN5   FN1 – Light drive fits   FN 5 – Heavy force fits

Limits and Fits Tables

Example problem:
Application: A pilot shaft and hole

Nominal Size: 1.8 in.
Fit: locational clearance, class LC3

$$H8 \quad {}^{+1.6}_{0}$$

$$h7 \quad {}^{0}_{-1}$$

Hole size 1.8016 / 1.8000

Shaft size 1.800 / 1.799

Use sample chart to find
hole and shaft tolerances.

Min clearance = .0000

Max clearance = .0026

Table 5. American National Standard Clearance Locational Fits   ANSI B4.1-1967 (R2004)

Nominal Size Range, Inches		Class LC 1			Class LC 2			Class LC 3			Class LC 4			Class LC 5		
			Standard Tolerance Limits			Standard Tolerance Limits			Standard Tolerance Limits			Standard Tolerance Limits			Standard Tolerance Limits	
Over	To	Clear-ance[1]	Hole H6	Shaft h5	Clear-ance[1]	Hole H7	Shaft h6	Clear-ance[1]	Hole H8	Shaft h7	Clear-ance[1]	Hole H10	Shaft h9	Clear-ance[1]	Hole H7	Shaft g6
							Values shown below are in thousandths of an inch								
0– 0.12		0 / 0.45	+0.25 / 0	0 / –0.2	0 / 0.65	+0.4 / 0	0 / –0.25	0 / 1	+0.6 / 0	0 / –0.4	0 / 2.6	+1.6 / 0	0 / –1.0	0.1 / 0.75	+0.4 / 0	–0.1 / –0.35
0.12– 0.24		0 / 0.5	+0.3 / 0	0 / –0.2	0 / 0.8	+0.5 / 0	0 / –0.3	0 / 1.2	+0.7 / 0	0 / –0.5	0 / 3.0	+1.8 / 0	0 / –1.2	0.15 / 0.95	+0.5 / 0	–0.15 / –0.45
0.24– 0.40		0 / 0.65	+0.4 / 0	0 / –0.25	0 / 1.0	+0.6 / 0	0 / –0.4	0 / 1.5	+0.9 / 0	0 / –0.6	0 / 3.6	+2.2 / 0	0 / –1.4	0.2 / 1.2	+0.6 / 0	–0.2 / –0.6
0.40– 0.71		0 / 0.7	+0.4 / 0	0 / –0.3	0 / 1.1	+0.7 / 0	0 / –0.4	0 / 1.7	+1.0 / 0	0 / –0.7	0 / 4.4	+2.8 / 0	0 / –1.6	0.25 / 1.35	+0.7 / 0	–0.25 / –0.65
0.71– 1.19		0 / 0.9	+0.5 / 0	0 / –0.4	0 / 1.3	+0.8 / 0	0 / –0.5	0 / 2	+1.2 / 0	0 / –0.8	0 / 5.5	+3.5 / 0	0 / –2.0	0.3 / 1.6	+0.8 / 0	–0.3 / –0.8
1.19– 1.97		0 / 1.0	+0.6 / 0	0 / –0.4	0 / 1.6	+1.0 / 0	0 / –0.6	0 / 2.6	+1.6 / 0	0 / –1	0 / 6.5	+4.0 / 0	0 / –2.5	0.4 / 2.0	+1.0 / 0	–0.4 / –1.0
1.97– 3.15		0 / 1.2	+0.7 / 0	0 / –0.5	0 / 1.9	+1.2 / 0	0 / –0.7	0 / 3	+1.8 / 0	0 / –1.2	0 / 7.5	+4.5 / 0	0 / –3	0.4 / 2.3	+1.2 / 0	–0.4 / –1.1
3.15– 4.73		0 / 1.5	+0.9 / 0	0 / –0.6	0 / 2.3	+1.4 / 0	0 / –0.9	0 / 3.6	+2.2 / 0	0 / –1.4	0 / 8.5	+5.0 / 0	0 / –3.5	0.5 / 2.8	+1.4 / 0	–0.5 / –1.4
4.73– 7.09		0 / 1.7	+1.0 / 0	0 / –0.7	0 / 2.6	+1.6 / 0	0 / –1.0	0 / 4.1	+2.5 / 0	0 / –1.6	0 / 10.0	+6.0 / 0	0 / –4	0.6 / 3.2	+1.6 / 0	–0.6 / –1.6
7.09– 9.85		0 / 2.0	+1.2 / 0	0 / –0.8	0 / 3.0	+1.8 / 0	0 / –1.2	0 / 4.6	+2.8 / 0	0 / –1.8	0 / 11.5	+7.0 / 0	0 / –4.5	0.6 / 3.6	+1.8 / 0	–0.6 / –1.8
9.85– 12.41		0 / 2.1	+1.2 / 0	0 / –0.9	0 / 3.2	+2.0 / 0	0 / –1.2	0 / 5	+3.0 / 0	0 / –2.0	0 / 13.0	+8.0 / 0	0 / –5	0.7 / 3.9	+2.0 / 0	–0.7 / –1.9
12.41– 15.75		0 / 2.4	+1.4 / 0	0 / –1.0	0 / 3.6	+2.2 / 0	0 / –1.4	0 / 5.7	+3.5 / 0	0 / –2.2	0 / 15.0	+9.0 / 0	0 / –6	0.7 / 4.3	+2.2 / 0	–0.7 / –2.1
15.75– 19.69		0 / 2.6	+1.6 / 0	0 / –1.0	0 / 4.1	+2.5 / 0	0 / –1.6	0 / 6.5	+4 / 0	0 / –2.5	0 / 16.0	+10.0 / 0	0 / –6	0.8 / 4.9	+2.5 / 0	–0.8 / –2.4

[1] Pairs of values shown represent minimum and maximum amounts of interference resulting from application of standard tolerance limits.

Table 3. American National Standard Running and Sliding Fits   ANSI B4.1-1967 (R2004)

Nominal Size Range, Inches		Class RC 1			Class RC 2			Class RC 3			Class RC 4		
			Standard Tolerance Limits			Standard Tolerance Limits			Standard Tolerance Limits			Standard Tolerance Limits	
Over	To	Clear-ance[1]	Hole H5	Shaft g4	Clearance[1]	Hole H6	Shaft g5	Clearance[1]	Hole H7	Shaft f6	Clearance[1]	Hole H8	Shaft f7
					Values shown below are in thousandths of an inch								
0– 0.12		0.1 / 0.45	+0.2 / 0	–0.1 / –0.25	0.1 / 0.55	+0.25 / 0	–0.1 / –0.3	0.3 / 0.95	+0.4 / 0	–0.3 / –0.55	0.3 / 1.3	+0.6 / 0	–0.3 / –0.7
0.12– 0.24		0.15 / 0.5	+0.2 / 0	–0.15 / –0.3	0.15 / 0.65	+0.3 / 0	–0.15 / –0.35	0.4 / 1.12	+0.5 / 0	–0.4 / –0.7	0.4 / 1.6	+0.7 / 0	–0.4 / –0.9
0.24– 0.40		0.2 / 0.6	+0.25 / 0	–0.2 / –0.35	0.2 / 0.85	+0.4 / 0	–0.2 / –0.45	0.5 / 1.5	+0.6 / 0	–0.5 / –0.9	0.5 / 2.0	+0.9 / 0	–0.5 / –1.1
0.40– 0.71		0.25 / 0.75	+0.3 / 0	–0.25 / –0.45	0.25 / 0.95	+0.4 / 0	–0.25 / –0.55	0.6 / 1.7	+0.7 / 0	–0.6 / –1.0	0.6 / 2.3	+1.0 / 0	–0.6 / –1.3
0.71– 1.19		0.3 / 0.95	+0.4 / 0	–0.3 / –0.55	0.3 / 1.2	+0.5 / 0	–0.3 / –0.7	0.8 / 2.1	+0.8 / 0	–0.8 / –1.3	0.8 / 2.8	+1.2 / 0	–0.8 / –1.6
1.19– 1.97		0.4 / 1.1	+0.4 / 0	–0.4 / –0.7	0.4 / 1.4	+0.6 / 0	–0.4 / –0.8	1.0 / 2.6	+1.0 / 0	–1.0 / –1.6	1.0 / 3.6	+1.6 / 0	–1.0 / –2.0
1.97– 3.15		0.4 / 1.2	+0.5 / 0	–0.4 / –0.7	0.4 / 1.6	+0.7 / 0	–0.4 / –0.9	1.2 / 3.1	+1.2 / 0	–1.2 / –1.9	1.2 / 4.2	+1.8 / 0	–1.2 / –2.4
3.15– 4.73		0.5 / 1.5	+0.6 / 0	–0.5 / –0.9	0.5 / 2.0	+0.9 / 0	–0.5 / –1.1	1.4 / 3.7	+1.4 / 0	–1.4 / –2.3	1.4 / 5.0	+2.2 / 0	–1.4 / –2.8
4.73– 7.09		0.6 / 1.8	+0.7 / 0	–0.6 / –1.1	0.6 / 2.3	+1.0 / 0	–0.6 / –1.3	1.6 / 4.2	+1.6 / 0	–1.6 / –2.6	1.6 / 5.7	+2.5 / 0	–1.6 / –3.2
7.09– 9.85		0.6 / 2.0	+0.8 / 0	–0.6 / –1.2	0.6 / 2.6	+1.2 / 0	–0.6 / –1.4	2.0 / 5.0	+1.8 / 0	–2.0 / –3.2	2.0 / 6.6	+2.8 / 0	–2.0 / –3.8
9.85– 12.41		0.8 / 2.3	+0.9 / 0	–0.8 / –1.4	0.8 / 2.9	+1.2 / 0	–0.8 / –1.7	2.5 / 5.7	+2.0 / 0	–2.5 / –3.7	2.5 / 7.5	+3.0 / 0	–2.5 / –4.5
12.41– 15.75		1.0 / 2.7	+1.0 / 0	–1.0 / –1.7	1.0 / 3.4	+1.4 / 0	–1.0 / –2.0	3.0 / 6.6	+2.2 / 0	–3.0 / –4.4	3.0 / 8.7	+3.5 / 0	–3.0 / –5.2
15.75– 19.69		1.2 / 3.0	+1.0 / 0	–1.2 / –2.0	1.2 / 3.8	+1.6 / 0	–1.2 / –2.2	4.0 / 8.1	+2.5 / 0	–4.0 / –5.6	4.0 / 10.5	+4.0 / 0	–4.0 / –6.5

[1] Pairs of values shown represent minimum and maximum amounts of clearance resulting from application of standard tolerance limits.

Workshop Exercise 2.2

1. What is the min and max clearance between the pin and hole above? _____

2. Name an application for a running and sliding fit.

_____

3. What type of fit would be best for an application with a pilot hole and shaft that must be freely assembled and disassembled?

_____

4. Which would provide more interference, a locational interference fit or force fit?

_____

5. Which would be a tighter fit, an LC3 or an LC9?

_____

Use sample charts earlier in the unit to find hole and shaft limits for the following nominal sizes and fits. Write your answers as min and max hole and shaft sizes. Also include the min/max clearance between the parts.

6. Application: dowel alignment pin and hole
Nominal Size: .375 in.
Fit: locational clearance, class LC3

     Min clearance = _____
     Max clearance = _____

Hole size: _____
       _____

Shaft size: _____
       _____

_____

7. Application: shaft in bushing
Nominal Size: 1 in.
Fit: close running fit, class RC4

     Min clearance = _____
     Max clearance = _____

Hole size: _____
       _____

Shaft size: _____
       _____

_____

8. Application: spool valve with translating shaft
Nominal Size: 1.2 in.
Fit: sliding fit, class RC2

     Min clearance = _____
     Max clearance = _____

Hole size: _____
       _____

Shaft size: _____
       _____

Unit 3

How the Geometric Tolerancing System Works

Plus Minus Tolerancing to Locate Surfaces Is Not Clear
Comparison - Position Tolerance vs. Plus/Minus Tolerance
How the Geometric Tolerancing System Works
Geometric Tolerancing Applied to the Hole Bar
Selecting Size and Position Tolerances along with Feature Modifiers
Effect of Maximum Material Condition Modifier
Effect of Least Material Condition Modifier
Effect of Regardless of Feature Size
Effect of Zero Tolerancing at MMC
Modifier Rules - Applicability of MMC, LMC, RFS, MMB, LMB, RMB
Modifier Rules - Former Practices
General Rules - Geo Tol Applied to Threads, Gears and Splines
When to use MMC, LMC, RFS Modifiers - Guidelines
MMC Modifier Application for Clearance
LMC Modifier Application for Location
RFS Application for Location
Workshop Exercise 3.1
Workshop Exercise 3.2
Workshop Exercise 3.3

Plus/Minus Tolerancing to Locate Surfaces Is Not Clear

The drawing below is completely defined with directly toleranced or plus/minus type dimensions. The plus/minus dimensions for the size of the two small holes, the size of the large hole and the size thickness of the bar is a clear acceptable method for defining these features. The remaining plus/minus dimensions on the part are not clear and subject to misinterpretations.

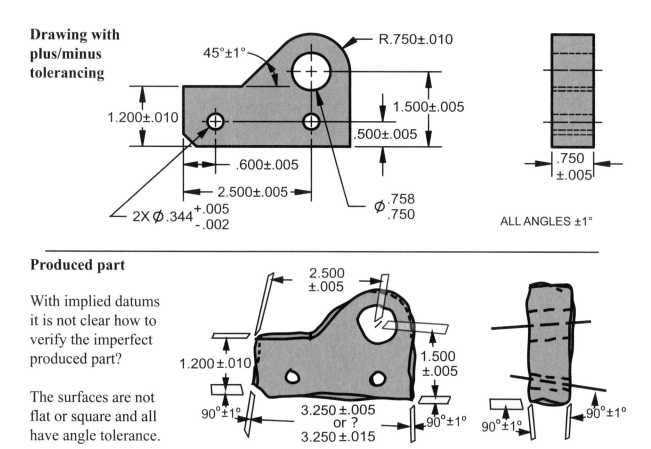

Drawing with plus/minus tolerancing

45°±1°
R.750±.010
1.200±.010
1.500±.005
.500±.005
.600±.005
2.500±.005
2X Ø.344 +.005 / -.002
Ø .758 / .750
.750 ±.005
ALL ANGLES ±1°

Produced part

With implied datums it is not clear how to verify the imperfect produced part?

The surfaces are not flat or square and all have angle tolerance.

2.500 ±.005
1.200 ±.010
1.500 ±.005
90°±1°
3.250 ±.005 or ? 3.250 ±.015
90°±1°
90°±1°
90°±1°

Manufacturing has a drawing with vague implied datums. However, manufacturing will do its best to produce the part to the mean dimension on the drawing, which allows them tolerance on either side of the line. The problem we have is in verification of the product. The drawing is not clear. There are implied datums. The features on the part are not flat, perpendicular or parallel. The holes are out of location and they are not perpendicular. All the features have implied angles to each other with a tolerance of ±1°.

How do we verify conformance? Do we measure square, like a parallelogram? What is the order of the datums? Some of the surfaces do not oppose each other. If we measure from the long surface to the short surface, we will get a different answer than if we measure from the short surface to the long surface.

The surfaces are not flat. Do we measure from the high points, low points or median points? What is the location of the .750 radius? Is it the same as the hole? What if we wanted it different? What is the tolerance on the overall length of the part? Do we add the ±.005 tolerance from the 2.500 dimension plus the ± .010 tolerance on the .750 radius? How square do the holes measure? The drawing has plus/minus 2D type tolerancing. In order to have quality parts, we need a clear 3D mathematical definition of the product. We need geometric tolerancing.

Comparison - Position Tolerancing vs Plus/Minus Tolerancing

There are two parts shown below. The large hole on the top part is located with plus/minus type tolerances and the large hole on the bottom part is located with a position tolerance. The comparison of the two methods below show position tolerancing to provide an increase in location tolerance for the hole, while establishing a datum reference frame and clearly defining design intent. Position tolerancing is a far superior method for the location of holes.

Plus/minus tolerancing

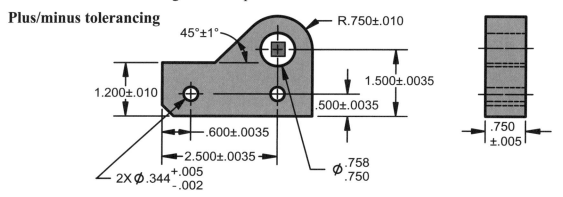

The upper example locates the large hole with a ±.0035 tolerance zone. This equates to a square zone of .007, which is approximately .010 across the corners. This allows the axis of the hole to be off in a diagonal direction more than a horizontal zone. A ±.0035 tolerance allows the hole to be off ±.005.

Position tolerancing provides an increase of location tolerance over plus/minus type tolerancing.

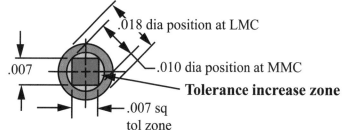

The lower example locates the large hole with a .010 diameter round tolerance zone. The axis of the hole may be off location in any direction. A position tolerance zone of .010 provides an increase in location tolerance over the plus/minus tolerancing, while still insuring the hole location is never more than .010 total. If the hole is a clearance hole, the MMC modifier specification following the position tolerance will allow additional location tolerance of up to .018 as the hole departs from MMC to LMC.

Position tolerancing

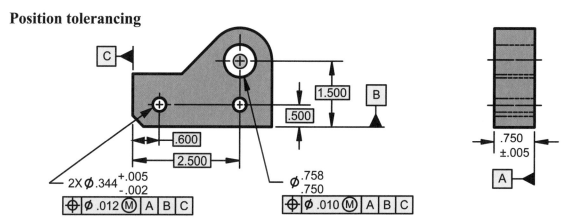

Position tolerance uses theoretically exact basic dimensions (enclosed in a box) that locate the .010 position tolerance zone. The axis of the hole must fall in this zone.

3.3

How the Geometric Tolerancing System Works

There are a few usual guidelines to consider when applying geometric tolerancing. If necessary:

1. Define the size of features such as holes, slots, tabs, pins, bosses, etc. with plus/minus dimensions. If necessary, apply form tolerances to refine the form requirements of Rule #1.

2. Identify the datum features in their order of precedence to establish a Datum Reference Frame (DRF) based on functional requirements.

3. Qualify the datum features with form, orientation and/or location tolerances as necessary.

4. Define the location of all the features with basic dimensions.

5. Apply position tolerance to locate features of size

6. Apply profile tolerances to locate surfaces

7. Apply orientation tolerances if necessary to refine the orientation provided by the location tolerances.

The example on page 3.6 displays how geometric tolerancing is applied to a part. The Hole Bar is a part that mounts into an assembly. It mounts on the back, bottom and side surfaces shown, and it is kept in place by the two mounting screws. The location of the large hole is important but its function will be kept vague at this point in time and will be discussed later. Geometric tolerancing is applied to the Hole Bar based on these design functional requirements.

On the Hole Bar, the size of two small clearance holes are defined by a plus/minus tolerance. They are oversized by 1/32 to clear a .3125 screw (standard drill size). The tolerance on the size of the hole is standard drill tolerance to provide economical production. Economical production is also an important design functional requirement.

The large hole is defined with a size and tolerance of .750/.758. The thickness of the bar is also defined with a size tolerance (could also be profile).

The mounting surfaces on the Hole Bar are selected as datum features. They are referenced in an order of precedence as stated in the feature control frames. The back face was selected as the primary datum feature A, because it was determined that the bolts will pull it up tight to this face at least 3 points of contact. The bottom surface was selected as the secondary datum feature B, because this surface is longer and provides a more reliable alignment than the shorter left-side surface. It will make contact with a minimum of 2 points. The left end surface was selected as the tertiary datum feature C and will make 1 point of contact.

The datum features are qualified with form and orientation tolerances to determine their engagement to the theoretically perfect DRF. These tolerances determine the stability or "rocking" of the part in its mounted condition.

The primary datum feature A has a flatness tolerance applied, the secondary datum feature B has an orientation tolerance (perpendicularity) referenced to A. The tertiary datum feature has a orientation tolerance (perpendicularity) applied that is referenced to A and B. See Datum Feature Controls on page 6.12 for more detailed information on relating datum features to the DRF.

All the features are located relative to the defined DRF with theoretically exact, basic dimensions that are enclosed in a box. These basic dimensions also locate the tolerance zones within which the features axes or surfaces must lie.

Profile tolerances are applied to all the feature surfaces between points X and Y. The tolerance in the profile feature control frame defines a total wide zone that is equally disposed about the true profile. The feature surfaces must fall within this zone. The profile tolerance is a relatively large tolerance at .020, so the outside surface is relatively unimportant.

Position tolerance is calculated using the fixed fastener formula and applied to the two clearance holes with a MMC modifier. This will ensure the holes will always clear the mating screws. (More on MMC later in this unit and the fastener formulas later in unit 15.)

Position tolerance is applied to the large hole based on functional mating conditions with a feature modifier MMC, LMC or RFS.

Generally, if the hole is a clearance hole, the MMC modifier is applied.

Generally, if the hole is used to locate a mating part or if wall thickness is a concern, the LMC or RFS modifier is applied.

Generally, if the hole is used to locate a mating part or it is a press fit with the mating part, the RFS modifier is applied.

For more information on modifiers see "When to apply MMC, LMC and RFS guidelines" later in this unit.

Plating and Coatings
If a part has a requirement to be plated or coated, the drawing should specify if the dimensions apply before or after the plating or coating. There is no default condition. A note on the drawing or a referenced document should state if dimensional limits apply before or after coating.

Geometric Tolerancing Applied to the Hole Bar

Geometric tolerancing is a clear and concise three dimensional, mathematical engineering language for communicating design intent based on functional requirements. A fully geometrically toleranced product drawing is shown based on assembly mounting conditions.

The datum reference frame is established by datum features A, B and C. The basic dimensions locate the center of the position and profile tolerance zone. The surfaces and hole axes must lie within the specified geometric tolerance zones.

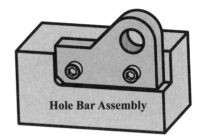

Hole Bar Assembly

This on the drawing

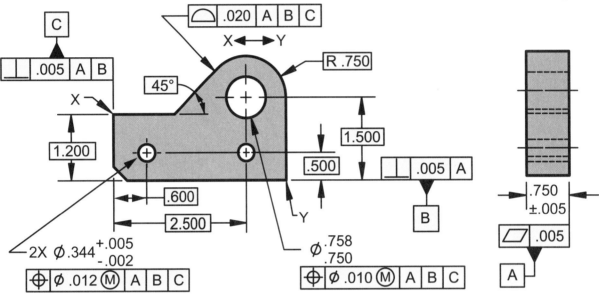

Means this on produced part

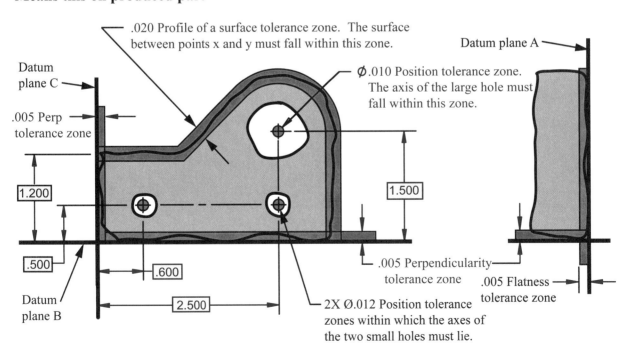

.020 Profile of a surface tolerance zone. The surface between points x and y must fall within this zone.

Datum plane A

Datum plane C

.005 Perp tolerance zone

Ø.010 Position tolerance zone. The axis of the large hole must fall within this zone.

1.200

1.500

.500

.600

2.500

Datum plane B

.005 Perpendicularity tolerance zone

.005 Flatness tolerance zone

2X Ø.012 Position tolerance zones within which the axes of the two small holes must lie.

3.6

Geometric Tolerancing Applied to the Hole Bar

This on the drawing

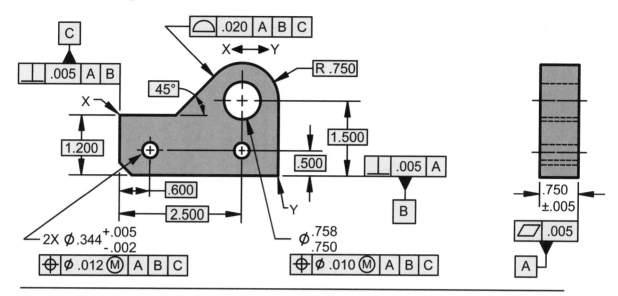

Means this

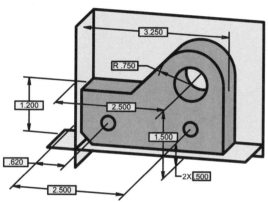

The modeled part is shown with basic dimensions that define the math data relative to the DRF.

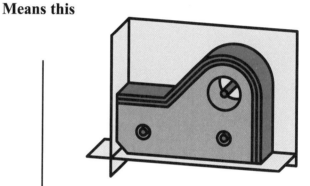

The modeled part showing the variation allowed with position and profile tolerance zones defined relative to the math data and DRF.

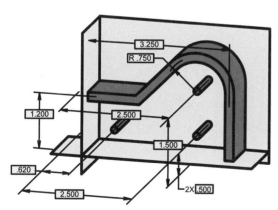

The basic dimensions on the modeled part define the theoretical exact or basic location of the position and profile tolerance zones relative to the DRF.

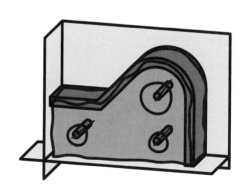

The produced part is mated, in order of precedence, to the DRF. The tolerance zones are related to the DRF. The part features must lie within the designated tolerance zones.

Selecting Size and Position Tolerances along with Feature Modifiers.

The large hole in the Hole Bar has a position tolerance applied. The value of the tolerance and the feature modifier applied is based on functional requirements of the assembly. The function of the large hole in the previous examples were purposely left vague so the general tolerancing concepts could be explained. There are two necessary items for the engineer to consider. The size tolerance on the hole and the location tolerance on the hole. The size tolerance can be determined with the Limits and Fits table shown in Unit 2. The location tolerance can be determined by the functional location requirement of the mating part or the fastener formulas shown in unit 15. The feature modifier can be determined by the feature modifier guidelines later in this unit.

Press fit application

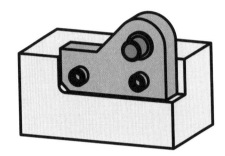

Press fit pin mounts in the large hole. The size tolerance on the hole is tight because of the required press fit. The position tolerance defines the location of the press fit pin. The feature has a RFS feature modifier applied.

**Clearance hole
application**

The large hole is used to clear a pin. The size tolerance on the hole is usually loose, because it is used for clearance and does not have a fitting requirement. The position tolerance defines the location of the hole and usually the fixed or floating fastener formula can be used. The feature has a MMC feature modifier applied. This is also a good example to consider the use of a zero position at MMC.

Locational application

The large hole is used to locate another component. The size tolerance on the hole is usually more restrictive because of the fitting requirement of the mating part. The position tolerance defines the location of the hole and mating part. The feature has a RFS or LMC feature modifier applied.

Effect of Regardless of Feature Size (RFS)

Regardless of Feature Size (RFS) is implied to apply to all geometric tolerances applied to features of size. If there is no material condition modifier following the feature tolerance, the default rule states the geometric tolerance applies at regardless of feature size. See modifier rules later in this unit. Since all geometric tolerances apply at RFS by default, there is no need for a RFS symbol, and it was eliminated in the past ASME Y14.5M-1994 (R2004) standard. In earlier editions, the symbol for RFS was circle S.

If a geometric tolerance is implied RFS, the specified geometric tolerance is independent of the actual size of the feature. The allowable geometric tolerance is limited to the specified value in the feature control frame regardless of the actual size of the feature.

The feature control frame below states that the holes must be positioned within a 0.6 diameter tolerance zone. Since there is no material condition modifier following the feature tolerance, it is implied to apply RFS. This requires the features to be positioned within a 0.6 diameter position zone regardless of the feature size. There is no additional position tolerance allowed as the holes get larger or smaller. The RFS condition is more restrictive than the MMC or LMC concept.

Position tolerance modified at RFS

The implied RFS modifier in the feature control frame invokes the RFS concept and requires the features to be positioned with a 0.6 diameter tolerance zone regardless of the feature size. See table.

Diameter Feature Size	Diameter Position Tolerance Allowed
11.7	0.6
11.8	0.6
11.9	0.6
12	0.6
12.1	0.6
12.2	0.6
12.3	0.6

Implied RFS modifier

4X Ø12±0.3

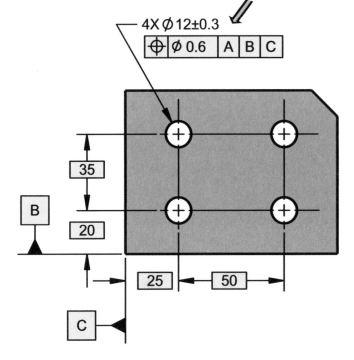

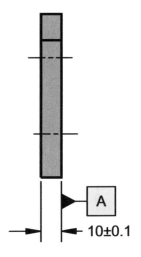

Effect of Maximum Material Condition (MMC) Modifier

The Maximum Material Condition (MMC) modifier may be applied by placing the circle M symbol in the feature control frame following the feature tolerance. This provides the feature with additional geometric tolerance as the feature's size departs from its maximum material condition. The MMC modifier may only be applied to features of size. MMC is not applicable to features without size.

The feature control frame below states that the holes must be positioned within a 0.6 diameter tolerance zone. The MMC modifier following the individual tolerance for the feature states this tolerance applies when the features are at their maximum material condition (smallest hole or largest pin).

The maximum material condition for the holes are 11.7 diameter. If the holes depart (get larger) from their maximum material condition size, they are allowed additional position tolerance equal to the amount of their departure from their MMC size of 11.7. See table.

The tolerance is always one-for-one, as the hole size is a diameter and the position tolerance is also a diameter zone. The MMC principle applies to each hole individually. Each hole may have a different position tolerance depending on the actual size of each hole. The holes could have as much as 1.2 diameter position if they are produced at their largest size of 12.3.

Position tolerance modified at MMC

The MMC modifier in the feature control frame invokes the MMC concept and allows additional position tolerance as the features depart from their MMC. See table.

Diameter Feature Size	Diameter Position Tolerance Allowed
11.7	0.6
11.8	0.7
11.9	0.8
12	0.9
12.1	1
12.2	1.1
12.3	1.2

MMC modifier

4X Ø 12±0.3

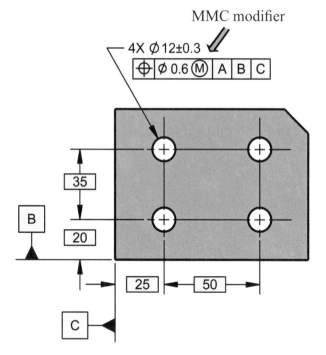

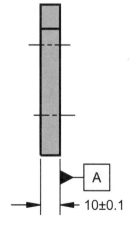

3.10

Effect of Zero Tolerancing at MMC

There are many cases where features of size may be controlled with a zero geometric tolerance. If a feature is specified with a zero geometric tolerance, it must be modified at MMC or LMC. The RFS concept is not applicable with a zero geometric tolerance.

The example below illustrates holes positioned with zero tolerance at MMC. The allowable position tolerance for the holes is dependent on the actual mating size of the holes. The MMC modifier in the feature control frame allows position tolerance equal in amount to the holes departure from their MMC size of 11.1.

The four holes below are allowed as much as 1.2 diameter position tolerance if they are produced at their maximum size of 12.3. Zero tolerancing at MMC is a little more of an advanced concept and is used to allow additional size tolerance while maintaining the same design requirement.

Zero position tolerance modified at MMC

The MMC modifier in the feature control frame invokes the MMC concept and allows additional position tolerance as the features depart from their MMC. See table.

Diameter Feature Size	Diameter Position Tolerance Allowed
11.1	0
11.2	0.1
11.3	0.2
11.4	0.3
11.5	0.4
11.6	0.5
11.7	0.6
11.8	0.7
11.9	0.8
12	0.9
12.1	1
12.2	1.1
12.3	1.2

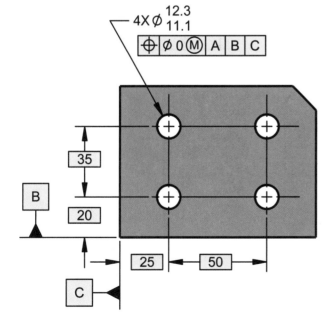

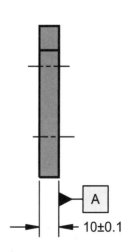

Effect of Least Material Condition (LMC) Modifier

The Least Material Condition (LMC) modifier may be applied by placing the circle L symbol in the feature control frame following the feature tolerance. This provides the feature with additional geometric tolerance as the feature's size departs from its least material condition. The LMC modifier may only be applied to features of size. LMC is not applicable to features without size.

The feature control frame below states that the holes must be positioned within a 0.6 diameter tolerance zone. The LMC modifier following the individual tolerance for the feature states this tolerance applies when the features are at their least material condition (largest hole or smallest pin).

The least material condition for the holes are 12.3 diameter. If the holes depart (get smaller) from their least material condition size, they are allowed additional position tolerance equal to the amount of their departure from their LMC size of 12.3. See table.

The tolerance is always one-for-one, as the hole size is a diameter and the position tolerance is also a diameter zone. The LMC principle applies to each hole individually. Each hole may have a different position tolerance depending on the actual size of each hole. The holes could have as much as 1.2 diameter position if they are produced at their smallest size of 11.7.

Position tolerance modified at LMC

The LMC modifier in the feature control frame invokes the LMC concept and allows additional position tolerance as the features depart from their LMC. See table.

Diameter Feature Size	Diameter Position Tolerance Allowed
12.3	0.6
12.2	0.7
12.1	0.8
12	0.9
11.9	1
11.8	1.1
11.7	1.2

LMC modifier

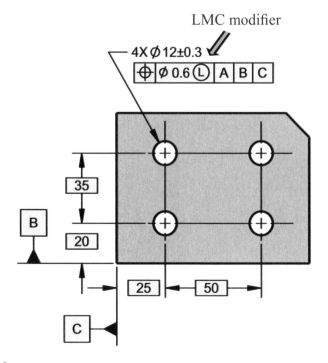

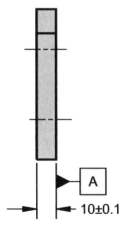

3.12

Modifier Rules - Applicability of MMC, LMC, RFS, MMB, LMB and RMB

Material condition modifiers are found in a feature control frame. They will have a different effect on the geometric specification depending on the modifier and how and where they are applied. There are two types of material condition modifiers, feature modifiers and datum feature modifiers.

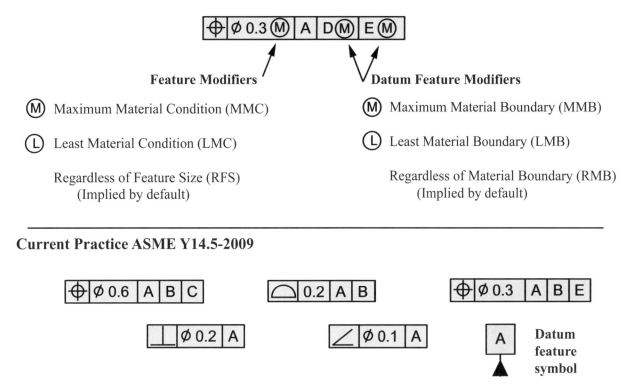

Feature Modifiers

(M) Maximum Material Condition (MMC)

(L) Least Material Condition (LMC)

Regardless of Feature Size (RFS)
(Implied by default)

Datum Feature Modifiers

(M) Maximum Material Boundary (MMB)

(L) Least Material Boundary (LMB)

Regardless of Material Boundary (RMB)
(Implied by default)

Current Practice ASME Y14.5-2009

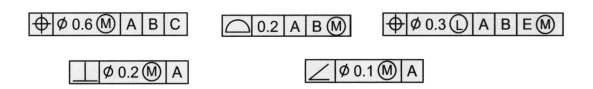

RFS and RMB apply by default for all geometric tolerances.

The applicability of these material condition modifiers are governed by the ASME Y14.5-2009 standard. According to the standard, Regardless of Feature Size (RFS) applies by default, for all feature tolerances in a feature control frame where no modifying symbol is specified. Regardless of Material Boundary (RMB) applies by default for all datum feature references in a feature control frame where no modifier symbol is specified.

MMC, LMC, MMB and LMB must be specified to apply.

If required, the feature modifiers, Maximum Material Condition (MMC) and Least Material Condition (LMC) must be specified in the feature control frame. Likewise, if the datum feature modifiers, Maximum Material Boundary (MMB) and Least Material Boundary (LMB) are required, they must be specified in the feature control frame.

Since by default, all geometric specifications apply at RFS and RMB, there are no symbols for either RFS or RMB.

Modifier Rules - Applicability of MMC, LMC, RFS, MMB, LMB and RMB

The datum feature modifiers RMB, MMB and LMB are new for the ASME Y14.5-2009 standard and replace the terms RFS, MMC and LMC used for datum features in prior versions of the standard.

The datum feature modifiers MMB, LMB and RMB were deemed to be more descriptive for datum feature boundaries than the earlier terms MMC, LMC and RFS. Applying MMC and LMC to datum features in the past was confusing. Even though these modifiers were referenced at M or L, they actually applied at their virtual condition. Virtual condition is also a boundary, but it was not applicable to RFS or profile tolerances. More on this in unit 7, datum feature modifiers.

In the ASME Y14.5-2009 standard, feature modifiers can only be applied to features of size. However, datum feature modifiers can be applied to datum features with size, and in some cases, to non-size datum features that establish orientation.

The ISO standard, by default, implies that RFS applies to both features and datum features. At this time, ISO does not distinguish a difference between feature modifiers and datum modifiers. ISO implies all material condition modifiers at RFS, and at this time, does not recognize datum feature modifiers applied to non-size features.

The geometric characteristics symmetry and concentricity always apply at RFS for the feature and at RMB for the datum feature.

The rules for the applicability of material condition modifiers have undergone many changes in past editions of the ASME Y14.5 standard. If necessary to review earlier drawings (prior to Y14.5M-1994), see Modifier Rules - Former Practices on the next page.

Former Practice ASME Y14.5M-1994 (R2004)

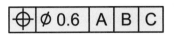

The ASME Y14.5M-1994 standard stated that RFS applied for all geometric tolerances and datum references.

If MMC or LMC are to apply for geometric tolerances or datum references, they must be specified.

Datum feature symbol

Former Practice ANSI Y14.5M-1982

The ANSI Y14.5M-1982 standard stated that MMC, RFS and LMC must be specified when applying position tolerances for both feature tolerance and datum feature references.

Datum feature symbol

RFS is implied for all feature tolerances and datum features on all other geometric tolerances. If MMC or LMC is required, it must be specified.

Former Practice ANSI Y14.5M-1973

The ANSI Y14.5M-1973 standard stated that MMC was implied for position tolerances for both the feature tolerances and datum feature references. If LMC or RFS is required, it must be specified.

 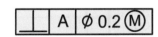 

Datum feature symbol

RFS is implied for all feature tolerances and datum features on all other geometric tolerances. If MMC or LMC is required, it must be specified. Notice the datum feature references preceed the tolerance in the ANSI Y14.5M-1973 standard.

Screw Thread Rule

A screw thread has three diameters, the major diameter, minor diameter and pitch diameter. The pitch diameter is usually the theoretical functional mating diameter. According to the ASME Y14.5-2009 standard, the pitch diameter applies by default to the geometric tolerance and/or the datum reference. Thread wires or a thread gage is often used to establish the axis in inspection.

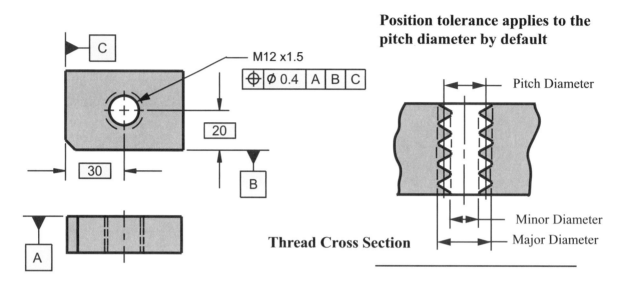

Position tolerance applies to the pitch diameter by default

Thread Cross Section

If the major diameter or minor diameter is required to apply, it must be specified. The term MAJOR or MINOR diameter is stated under the feature control frame or under the datum feature symbol, as applicable. The adjacent example illustrates a position tolerance applied to the major diameter and a datum feature reference applied to the pitch diameter.

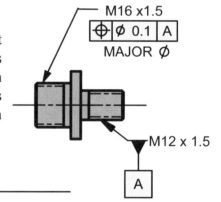

Gear and Spline Rule

A gear or spline is composed of a variety of diameters. Generally, these diameters are the major, minor, and pitch diameter. According to the ASME Y14.5-2009 standard and ISO standard it must be specifically stated which diameter is desired. There is no default condition. The required diameter, MAJOR DIA, PITCH DIA, MINOR DIA is to be stated directly under the feature control frame.

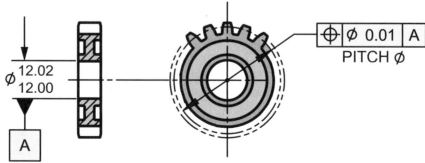

When to Use MMC, LMC and RFS Modifiers - Guidelines

Feature modifiers are shown in the feature control frame following the feature tolerance. They maybe applied at MMC, LMC or implied to be RFS. In the past, it seemed that all holes were positioned at MMC. This felt good, and we generally understood that MMC provided more tolerance and allowed the use of functional gages. This allowed engineers to get by without really thinking about feature modifiers. In reality, the engineers should carefully think about the application before just blindly applying MMC to all features.

It is important to properly apply the correct feature modifier in order to achieve design functional requirements as well as allowing maximum manufacturing tolerance. Feature modifier selection can be painful to think about at first. Once applied a few times, it becomes evident and can be spotted intuitively. As you might imagine, modifier selection is based on the purpose of the hole. It is based on design intent of mating parts. If the hole is being used to clear a mating part, then MMC is generally applied. If the hole is being used to locate a mating part, then RFS or LMC is generally applied.

On the next page, there are three examples of a rectangular top plate mounting in the bottom left corner and forming an assembly with the base plate. The datums applied in the feature control frame represent the mounting condition of the part in each of the three applications. (Note: The mounting condition and the value of the size and position tolerances are only shown as representative for illustrative and comparison purposes.)

The three different applications will each signal the use of different feature modifiers based on functional requirements. It is important to study the functional requirements of the hole in the three examples to understand the correct use of each modifier.

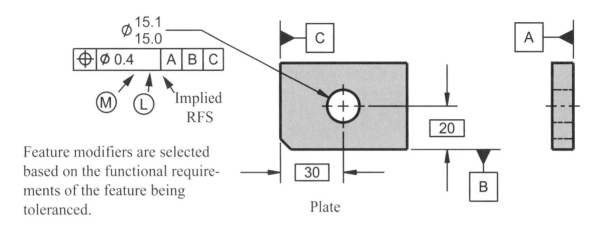

Feature modifiers are selected based on the functional requirements of the feature being toleranced.

Plate

MMC is applied when the feature is used for clearance applications.

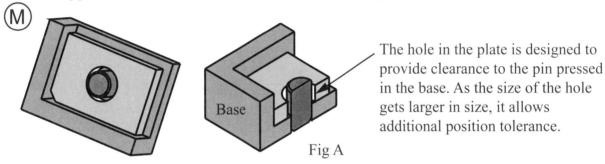

The hole in the plate is designed to provide clearance to the pin pressed in the base. As the size of the hole gets larger in size, it allows additional position tolerance.

Fig A

LMC is used for location requirements when the effect of the size tolerance on the hole also contributes to the location.

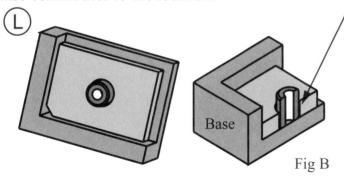

The hole in the plate is designed to locate a bushing. The bushing fits with some clearance, it is not pressed. The size tolerance on the hole also contributes to the location ("slop"). LMC counteracts this contribution and provides additional position tolerance as the hole gets smaller. RFS may be used as well.

Fig B

RFS is applied when the feature is used for location applications or when the size tolerance of the feature has no relation to the location.

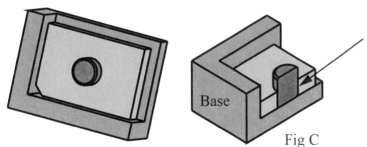

The hole in the plate is designed to locate and accept a press fit pin. The size of the hole has no effect on the location of the pin. The hole is located regardless of feature size.

Fig C

3.18

MMC Modifier Application for Clearance

The assembly below illustrates an MMC application.

With the part properly mounted in the DRF, the Ø15.1/15 hole in the top plate is designed to clear a fixed pin or object that is mounted in the base plate. The amount of clearance available is based both on the size of the hole and the position tolerance placed on the hole. In order to clear the pin, the size of the hole is also allowed to contribute to the position tolerance.

MMC Application

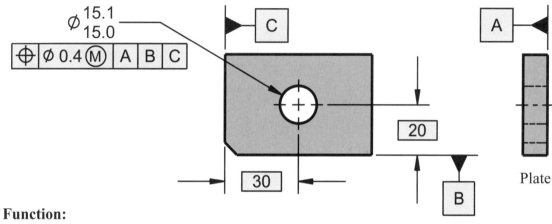

Function:
The hole in the plate is designed to provide clearance to the pin pressed in the base. As the size of the hole gets larger, it allows additional position tolerance.

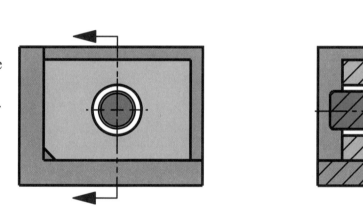

This MMC specification states that the size of the hole and the location of the hole are interrelated to the amount of clearance available. As the hole gets larger in size (and provides more clearance) its center is allowed to be off position by an equal amount. This, in effect, creates an Ø14.6 MMC inner boundary of which no portion of the surface of the feature may enter. This allows manufacturing to vary both the size of the feature and the position of the feature proportionately, based on gaining maximum manufacturing tolerance. The feature must be within the limits of size, as well as lie within the MMC boundary.

As an option; If it is truly a clearance application, and function dictates that the size and location are integral and are considered as only one variable, then zero tolerancing should be considered. In zero tolerancing, the size tolerance of the feature is increased by the amount of position tolerance applied to the feature. The Ø14.6/15.1 feature is then positioned zero at MMC. This provides manufacturing all of the advantages explained in the previous example plus, it relaxes size requirements of the holes. It provides manufacturing the maximum tolerance for a clearance application. See zero tolerancing.

The assembly below illustrates an LMC application.

With the part properly mounted in the DRF, the Ø15/15.1 hole in the top plate is designed to locate a fixed pin or object that mounts in the hole. The amount of movement available for the mating pin is based on both the size and position tolerance placed on the hole. In order to properly locate the pin, the size of the hole as well as the position tolerance must be taken into account.

LMC Application

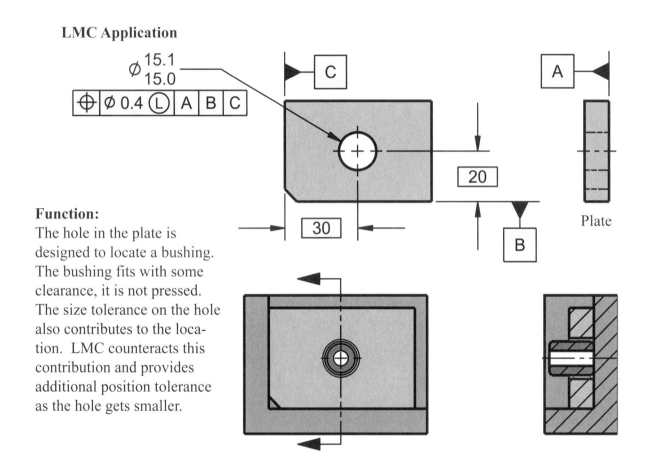

Function:
The hole in the plate is designed to locate a bushing. The bushing fits with some clearance, it is not pressed. The size tolerance on the hole also contributes to the location. LMC counteracts this contribution and provides additional position tolerance as the hole gets smaller.

This LMC specification states that the size of the hole and the location of the hole are interrelated and both will have an effect on the location of the mating part. As the hole gets smaller in size (and provides a better fit with the mating pin), its center is allowed to be off position by an equal amount. In effect, this creates a Ø15.5 LMB outer boundary of which no portion of the surface may extend beyond. This allows manufacturing to vary both the size and position of the feature based on gaining maximum manufacturing tolerance.

As an option: If it is truly a locational application, and function dictates that the size and location are integral and considered as only one variable, then zero tolerancing at LMC may be considered. If it is properly applied, it provides manufacturing the maximum tolerance for a locational application.

LMC Modifier Application for Location

Note: Use caution in the application of LMC modifiers, as they can cause serious interference problems if they are improperly used for clearance applications. Be sure and know what you are doing before applying LMC modifiers. If in doubt, use RFS rather than LMC.

Holes for self piloting weld nuts in a sheet metal part might be a good application for a LMC modifier. Weld nuts mount in the three holes shown below. The holes locate the pilot of the weld nut and have a small clearance but are still intended to fixture the weld nut in place for welding.

In determining the final location of the weld nut, the engineer must consider the size of the holes that accepts the pilot of the weld nut, as well as the position tolerance on the hole. The size of the hole is important because as the hole increases in size, the pilot on the weld nut will rattle around more in the hole and in effect, increases the location variation of the weld nut. The LMC modifier would require the holes to be at a closer position when they are at LMC (because of the rattle problem) but allow more position as the holes get smaller and move towards MMC.

LMC Weld Nut Application

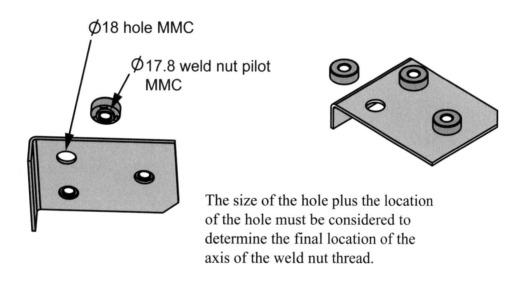

Ø18 hole MMC

Ø17.8 weld nut pilot MMC

The size of the hole plus the location of the hole must be considered to determine the final location of the axis of the weld nut thread.

LMC can also be used effectively in controlling wall thickness conditions. LMC is also used for locating holes on a casting that will later be machined. LMC provides less location when the hole is at its largest size, but more location when the hole is smaller, thus preserving a boundary of guaranteed machine stock.

Note: According to the ASME Y14.5-2009 standard, when applying a position tolerance at LMC, perfect form at LMC is required, but perfect form at MMC is not required for the toleranced feature(s). In extreme cases of form deviation, this could pose a problem for fit of a mating part (no Rule #1 applied). If the feature depth is minimal, as in sheet metal or other thin material, it is not a problem. If the feature depth is significant, and it requires a fit with a mating part, it might be prudent to add a note to the feature positioned at LMC that perfect form at MMC is required. An additional feature control frame, perpendicularity of zero at MMC, would also ensure the perfect form at MMC.

The assembly below illustrates an RFS application.

With the part properly mounted in the DRF, the Ø15/15.1 hole in the top plate is designed to accept a fixed pin or object that is pressed/interference fit with the hole. The press fit pin will always center itself in the hole (no "slop"). Since the pin is centered in the hole, the amount of movement available for the pin is based solely on the position tolerance. The size tolerance on the hole will not contribute to the location. In order to properly locate the pin, only the position tolerance is taken into account.

RFS Application

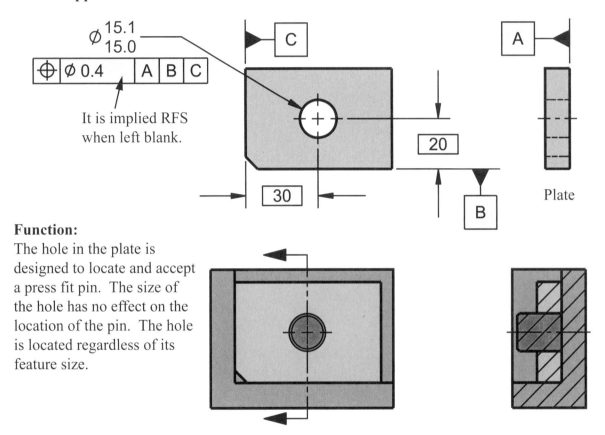

Function:
The hole in the plate is designed to locate and accept a press fit pin. The size of the hole has no effect on the location of the pin. The hole is located regardless of its feature size.

This RFS specification states that the size of the hole has no relation to the location of the hole. The size of the hole and location of the hole are separate and unrelated. If the hole gets larger or smaller in size it allows no additional positional tolerance. This requires manufacturing to process the size of the feature and the position of the feature as individual requirements. Zero tolerancing is not an option with RFS applications.

In an RFS application, if the hole were produced larger or smaller, it would just have a heavier or lighter press fit. The fit of the pin has nothing to do with the location of the hole. The fit of the pin in the hole is one consideration and the location of the hole is another consideration.

Study the drawing below and answer the following questions.

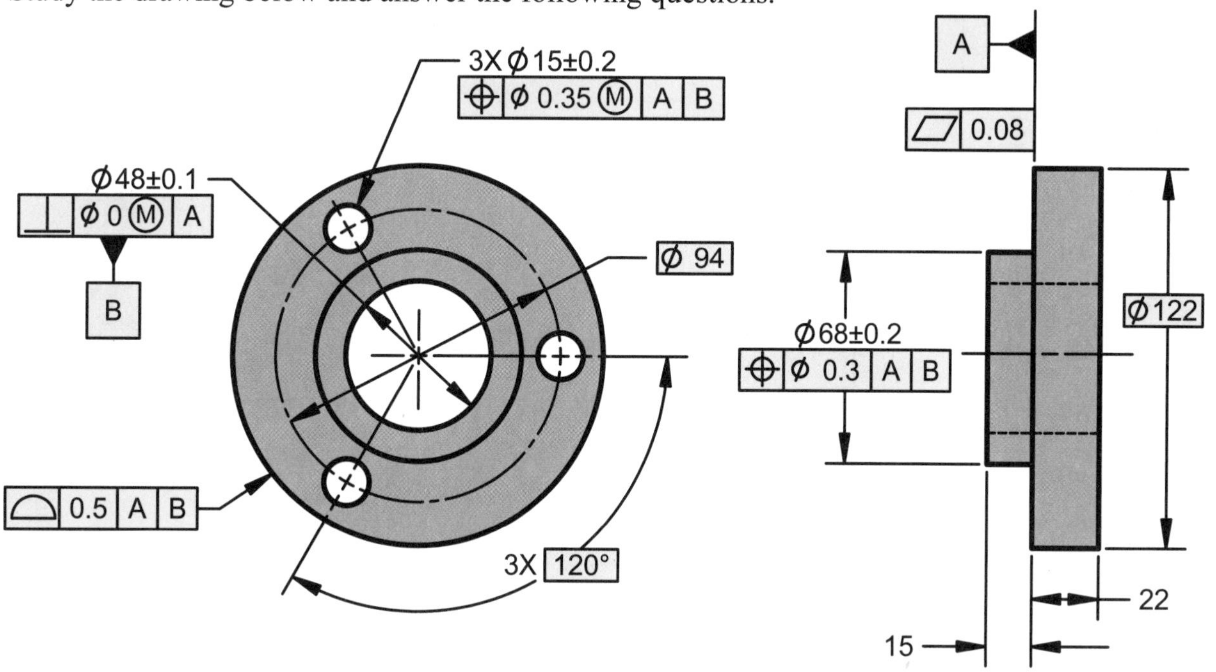

3X Ø15±0.2

⊕ | Ø 0.35 Ⓜ | A | B

Ø48±0.1

⊥ | Ø0 Ⓜ | A

B

Ø 94

A ◄

▱ | 0.08

Ø122

Ø68±0.2

⊕ | Ø 0.3 | A | B

⌒ | 0.5 | A | B

3X 120°

22

15

UNLESS OTHERWISE SPECIFIED:
ALL DIMENSIONS ±0.3
ALL ANGLES ±1°

1. What is the MMC of the Ø68 mm boss?

2. Is the position tolerance on the boss referenced at MMC, LMC, or RFS?

3. How much position tolerance is allowed on the boss if it is produced at a size of 68.1 mm?

4. What is the MMC of the 3 holes?

5. How much position tolerance is allowed if the 3 holes are produced at a size of 14.8 mm?

6. How much position tolerance is allowed if the 3 holes are produced at a size of 15 mm?

7. What is the perpendicularity tolerance to datum A for the 3 holes?

8. What could be the maximum perpendicularity tolerance on the center hole?

9. Does the 94 mm bolt circle dimension have tolerance? If so, what is it?

This drawing, using the direct tolerancing method, allows the design intent to be unclear. Study the assembly below and convert this drawing to geometric tolerancing. Add datum feature symbols to create a datum reference frame. Make necessary dimensions basic. Apply geometric tolerancing to define your design intent for this assembly. At this point, apply just enough geometric tolerancing to meet your own specific needs, and estimate the tolerances. It may help to review the Hole Bar problem again. Keep it simple, keep it neat.

Conversion: Plus/Minus to Geo Tol

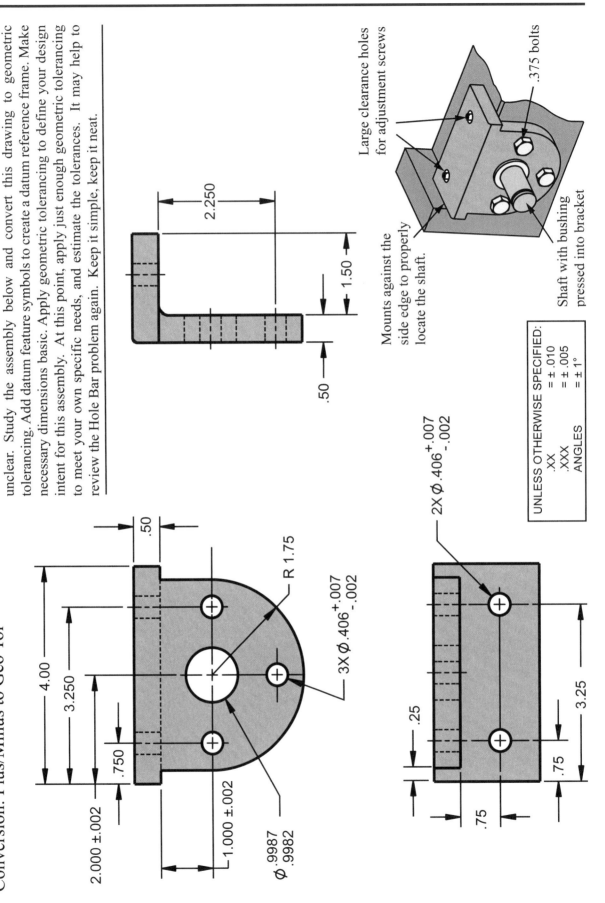

Large clearance holes for adjustment screws

.375 bolts

Mounts against the side edge to properly locate the shaft.

Shaft with bushing pressed into bracket

2.250

1.50

.50

2X ϕ .406 $^{+.007}_{-.002}$

UNLESS OTHERWISE SPECIFIED:	
.XX	= ± .010
.XXX	= ± .005
ANGLES	= ± 1°

.50

R 1.75

4.00

3.250

.750

2.000 ±.002

1.000 ±.002

ϕ .9987 / .9982

3X ϕ .406 $^{+.007}_{-.002}$

.25

3.25

.75

.75

Workshop Exercise 3.3

Apply geometric tolerancing to part according to the instructions below. This exercise will provide you with experience in correctly applying and interpreting geometric tolerancing symbology. If you have trouble applying the symbols, page through this unit looking for similar examples. The Hole Bar in this unit is a good reference. Draw the symbols and feature control frames clearly and neatly.

1. Establish the right face in the side view as datum feature A. Qualify this feature with a flatness of .005.

2. Establish the top surface in the front view as datum feature B. Qualify this feature with a perpendicularity of .005 to datum feature A.

3. Establish the left surface in the front view as datum feature C. Qualify this feature with a perpendicularity of .005 to datum features A and B.

4. Establish all necessary dimensions as basic.

5. Position the .250 hole within a diameter of .005 RFS relative to datum features A, B, C.

6. Position the three holes within a diameter of .012 at MMC relative to datum features A,B,C.

7. In the front view, identify the upper right corner as point "Y" and the lower left corner as point "X". Apply a profile tolerance of .020 between points X and Y relative to datum features A, B, C.

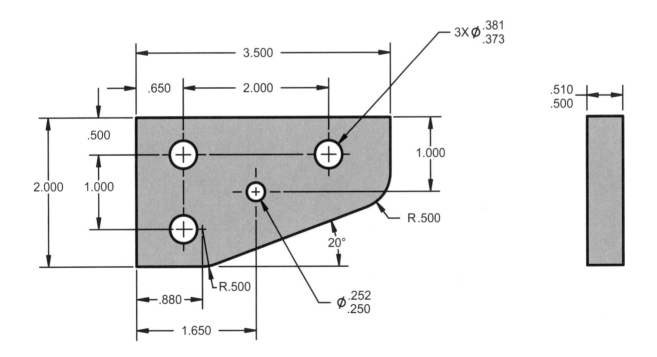

Front View   Side View

8. The produced part made from the drawing on the previous page is shown below, mated to the datum reference frame. Draw and shade in the geometric tolerance zones that were applied on the previous problem. Show the flatness, perpendicularity, profile and position zones. Illustrate these zones clearly and neatly. If you need help, look at the examples from previous pages in this unit.

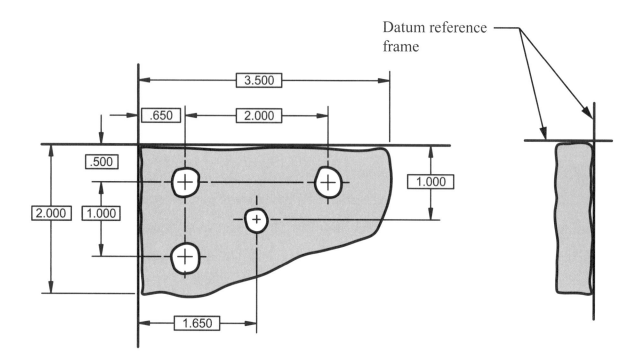

Datum reference frame

9. On the drawing below, a position tolerance with an MMC modifier has been applied to the four holes. The MMC modifier allows additional position tolerance as the features depart from their MMC. Complete the chart below to show the allowed position tolerance for the range of hole sizes.

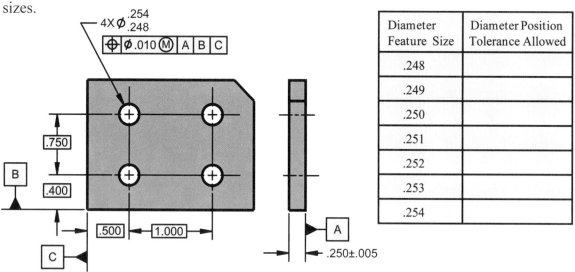

Diameter Feature Size	Diameter Position Tolerance Allowed
.248	
.249	
.250	
.251	
.252	
.253	
.254	

10. According to the current ASME Y14.5-2009 standard, if a feature control frame has no feature modifier present, is the tolerance for the features implied to be at MMC, LMC, or RFS?

Unit 4

Position Verification

Introduction to Position Verification
Position - Hole Verification at MMC
Position - HoleVerification at LMC
Class Exercise - Position Verification, Inch
Class Exercise - Position Verification, Metric
Position Conversion Table - Metric
Position Conversion Table - Inch
Workshop Exercise 4.1 - Metric
Workshop Exercise 4.2 - Metric
Workshop Exercise 4.3 - Metric
Workshop Exercise 4.4 - Metric
Workshop Exercise 4.5 - Metric
Workshop Exercise 4.6 - Inch
Workshop Exercise 4.7 - Inch
Workshop Exercise 4.8 - Inch
Workshop Exercise 4.9 - Inch
Paper Gage Concept
        Paper Gage Evaluation of 4.1
        Paper Gage Cartesian and Polar Coordinate

Introduction to Position Verification

This unit covers the verification of position tolerances. It is not intended to make everyone an expert on inspecting parts but rather provide an overview on the verification process.

The exercises in this unit contain both metric and inch values and allow the user to choose example problems to meet specific needs. These exercises are a valuable tool to ensure and solidify the understanding of the concepts covered in unit 3, also defining the effect of the MMC, LMC and RFS feature modifiers.

If you have been working with plus/minus tolerances in the past, there are some significant changes in verifying position tolerances. In plus/minus tolerancing, the dimensions had an implied origin that was subject to many interpretations, and the deviation value of the plus/minus dimension was recorded on an inspection report. Position tolerancing is different.

In position tolerancing, a clear datum reference frame is established and the value of the dimension is basic (theoretically exact) without tolerance. Tolerances do not accumulate on basic dimensions. The basic dimensions locate the center of the tolerance zone from the datum reference frame that is defined in the position feature control frame. The feature control frame contains the size of the tolerance zone, the applicable feature modifier, and specified datum feature references. It is important to follow the directions stated in the feature control frame for each of the features. The origin of the basic dimension is unimportant and could be defined in a CAD model data set.

The location of the positioned holes are reported on an inspection report as deviations from basic in the x and y direction. Positive and negative numbers are used in the inspection report to identify the direction of the deviation, so processing adjustments can be made. Conversion tables in metric and inch are shown to convert the x and y deviations to the actual position.

There is also a section in the rear of this unit on the paper gage concept. This graphical procedure will allow a visual picture of the hole deviations. The paper gage concept provides information on how to adjust and center the processes and is a valuable tool in the material review process. Additional advanced paper gage concepts are illustrated in unit 7 for datum feature modifiers and again in unit 13 for composite position tolerancing.

Understanding the measurement procedure is important and is covered extensively in this text. William Thomson, 1st Baron Kelvin (also known as Lord Kelvin), was a Irish-born British mathematical physicist and engineer. At Glasgow University, he did much to unify the emerging discipline of physics in its modern form.

Lord Kelvin is widely known for developing the Kelvin scale of absolute temperature measurement. He also has some famous quotes attributed to him about design specifications. Below is a famous quote from Lord Kelvin regarding the importance of understanding measurement.

> *"I often say that when you can measure what you are speaking about, and express it in numbers, you know something about it; but when you cannot measure it, when you cannot express it in numbers, your knowledge is of a meagre and unsatisfactory kind;"*

Position - Hole Verification at MMC

The example below illustrates the procedure to verify a hole with a position tolerance at MMC.

As drawn

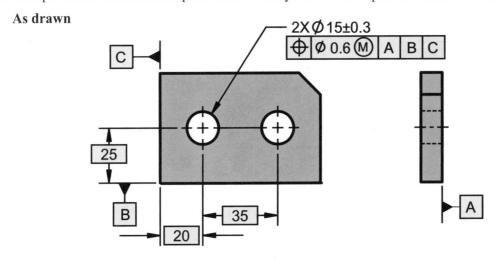

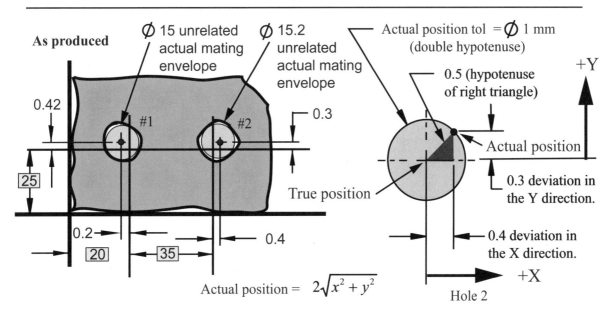

$$\text{Actual position} = 2\sqrt{x^2 + y^2}$$

Actual position tolerance is calculated by using "X" and "Y" deviations in the position formula or position table. In order to pass inspection, the hole's actual position tolerance must be less than the allowed position.

Allowed position tolerance is calculated by taking the hole size departure from MMC and adding it to the position tolerance stated in the feature control frame.

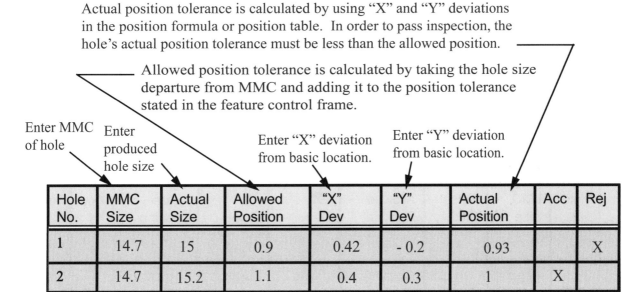

Hole No.	MMC Size	Actual Size	Allowed Position	"X" Dev	"Y" Dev	Actual Position	Acc	Rej
1	14.7	15	0.9	0.42	- 0.2	0.93		X
2	14.7	15.2	1.1	0.4	0.3	1	X	

4.3

The example below illustrates the procedure to verify a hole with a position tolerance at LMC.

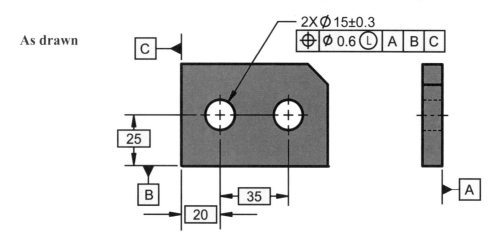

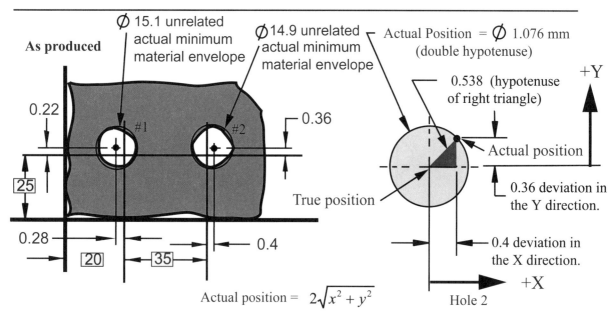

Actual position tolerance is calculated by using "X" and "Y" deviations in the position formula or position table. In order to pass inspection, the hole's actual position tolerance must be less than the allowed position.

Allowed position tolerance is calculated by taking the departure of the hole from LMC and adding it to the position tolerance stated in the feature control frame.

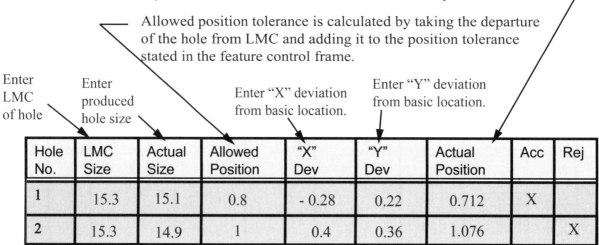

Hole No.	LMC Size	Actual Size	Allowed Position	"X" Dev	"Y" Dev	Actual Position	Acc	Rej
1	15.3	15.1	0.8	- 0.28	0.22	0.712	X	
2	15.3	14.9	1	0.4	0.36	1.076		X

Class Exercise - Position Verification at MMC, Inch

The top drawing is a part with position tolerance applied. The lower drawing is the produced part. Evaluate the dimensions on the produced part to verify conformance to the engineering drawing. Use the table below to record your calculations. Review the previous examples if you have trouble.

As drawn

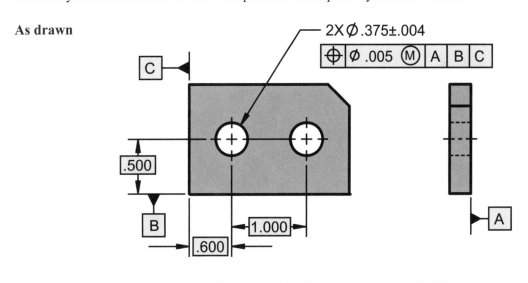

As produced

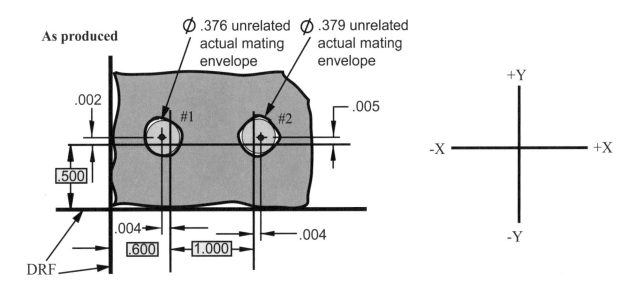

Hole No.	MMC Size	Actual Size	Allowed Position	"X" Dev	"Y" Dev	Actual Position	Acc	Rej
1								
2								

Class Exercise - Position Verification at MMC, Metric

The top drawing is a part with position tolerance applied. The lower drawing is the produced part. Evaluate the dimensions on the produced part to verify conformance to the engineering drawing. Use the table below to record your calculations. Review the previous examples if you have trouble.

As drawn

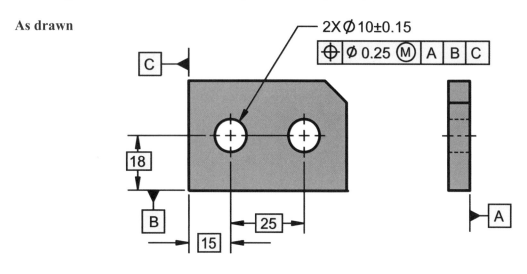

As produced

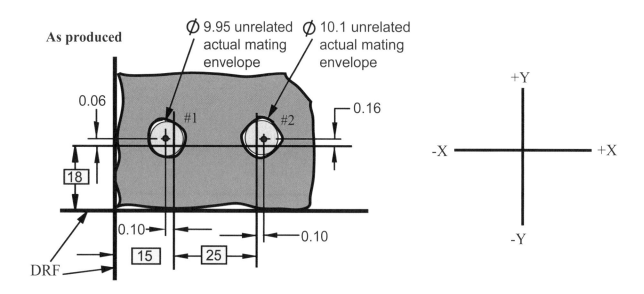

Hole No.	MMC Size	Actual Size	Allowed Position	"X" Dev	"Y" Dev	Actual Position	Acc	Rej
1								
2								

Conversion of Coordinate Measurement to Position Tolerance Zones - Inch

The chart below will convert coordinate (x and y) inch measurements to position diameter tolerance zones. It is simply, double the Pythagorean Theorem. The formula is shown below.

Position Zone (Diameter)																				
Y (DEV) \ X (DEV)	**.001**	**.002**	**.003**	**.004**	**.005**	**.006**	**.007**	**.008**	**.009**	**.010**	**.011**	**.012**	**.013**	**.014**	**.015**	**.016**	**.017**	**.018**	**.019**	**.020**
.025	.0500	.0502	.0504	.0506	.0510	.0514	.0519	.0525	.0531	.0539	.0546	.0555	.0564	.0573	.0583	.0594	.0605	.0616	.0628	.0640
.024	.0480	.0482	.0484	.0487	.0490	.0495	.0500	.0506	.0513	.0520	.0528	.0537	.0546	.0556	.0566	.0577	.0588	.0600	.0612	.0625
.023	.0460	.0462	.0464	.0467	.0471	.0475	.0481	.0487	.0494	.0502	.0510	.0519	.0528	.0539	.0549	.0560	.0572	.0584	.0597	.0610
.022	.0440	.0442	.0444	.0447	.0451	.0456	.0462	.0468	.0475	.0483	.0492	.0501	.0511	.0522	.0533	.0544	.0556	.0569	.0581	.0595
.021	.0420	.0422	.0424	.0428	.0432	.0437	.0443	.0449	.0457	.0465	.0474	.0484	.0494	.0505	.0516	.0528	.0540	.0553	.0566	.0580
.020	.0400	.0402	.0404	.0408	.0412	.0418	.0424	.0431	.0439	.0447	.0457	.0466	.0477	.0488	.0500	.0512	.0525	.0538	.0552	.0566
.019	.0381	.0382	.0385	.0388	.0393	.0398	.0405	.0412	.0420	.0429	.0439	.0449	.0460	.0472	.0484	.0497	.0510	.0523	.0537	.0552
.018	.0361	.0362	.0365	.0369	.0374	.0379	.0386	.0394	.0402	.0412	.0422	.0433	.0444	.0456	.0469	.0482	.0495	.0509	.0523	.0538
.017	.0341	.0342	.0345	.0349	.0354	.0361	.0368	.0376	.0385	.0394	.0405	.0416	.0428	.0440	.0453	.0467	.0481	.0495	.0510	.0525
.016	.0321	.0322	.0326	.0330	.0335	.0342	.0349	.0358	.0367	.0377	.0388	.0400	.0412	.0425	.0439	.0453	.0467	.0482	.0497	.0512
.015	.0301	.0303	.0306	.0310	.0316	.0323	.0331	.0340	.0350	.0361	.0372	.0384	.0397	.0410	.0424	.0439	.0453	.0469	.0484	.0500
.014	.0281	.0283	.0286	.0291	.0297	.0305	.0313	.0322	.0333	.0344	.0356	.0369	.0382	.0396	.0410	.0425	.0440	.0456	.0472	.0488
.013	.0261	.0263	.0267	.0272	.0279	.0286	.0295	.0305	.0316	.0328	.0341	.0354	.0368	.0382	.0397	.0412	.0428	.0444	.0460	.0477
.012	.0241	.0243	.0247	.0253	.0260	.0268	.0278	.0288	.0300	.0312	.0326	.0339	.0354	.0369	.0384	.0400	.0416	.0433	.0449	.0466
.011	.0221	.0224	.0228	.0234	.0242	.0251	.0261	.0272	.0284	.0297	.0311	.0326	.0341	.0356	.0372	.0388	.0405	.0422	.0439	.0457
.010	.0201	.0204	.0209	.0215	.0224	.0233	.0244	.0256	.0269	.0283	.0297	.0312	.0328	.0344	.0361	.0377	.0394	.0412	.0429	.0447
.009	.0181	.0184	.0190	.0197	.0206	.0216	.0228	.0241	.0255	.0269	.0284	.0300	.0316	.0333	.0350	.0367	.0385	.0402	.0420	.0439
.008	.0161	.0165	.0171	.0179	.0189	.0200	.0213	.0226	.0241	.0256	.0272	.0288	.0305	.0322	.0340	.0358	.0376	.0394	.0412	.0431
.007	.0141	.0146	.0152	.0161	.0172	.0184	.0198	.0213	.0228	.0244	.0261	.0278	.0295	.0313	.0331	.0349	.0368	.0386	.0405	.0424
.006	.0122	.0126	.0134	.0144	.0156	.0170	.0184	.0200	.0216	.0233	.0251	.0268	.0286	.0305	.0323	.0342	.0361	.0379	.0398	.0418
.005	.0102	.0108	.0117	.0128	.0141	.0156	.0172	.0189	.0206	.0224	.0242	.0260	.0279	.0297	.0316	.0335	.0354	.0374	.0393	.0412
.004	.0082	.0089	.0100	.0113	.0128	.0144	.0161	.0179	.0197	.0215	.0234	.0253	.0272	.0291	.0310	.0330	.0349	.0369	.0388	.0408
.003	.0063	.0072	.0085	.0100	.0117	.0134	.0152	.0171	.0190	.0209	.0228	.0247	.0267	.0286	.0306	.0326	.0345	.0365	.0385	.0404
.002	.0045	.0057	.0072	.0089	.0108	.0126	.0146	.0165	.0184	.0204	.0224	.0243	.0263	.0283	.0303	.0322	.0342	.0362	.0382	.0402
.001	.0028	.0045	.0063	.0082	.0102	.0122	.0141	.0161	.0181	.0201	.0221	.0241	.0261	.0281	.0301	.0321	.0341	.0361	.0381	.0400

Y (DEV) ← → (vertical axis)

X → (DEV) (horizontal axis)

Actual position = $2\sqrt{x^2 + y^2}$

Conversion of Coordinate Measurement to Position Tolerance Zones - Metric

The chart below will convert coordinate (x and y) metric measurements to position diameter tolerance zones. It is simply, double the Pythagorean Theorem. The formula is shown below.

Position Zone (Diameter)

Y (DEV) \ X (DEV)	0.02	0.04	0.06	0.08	0.10	0.12	0.14	0.16	0.18	0.20	0.22	0.24	0.26	0.28	0.30	0.32	0.34	0.36	0.38	0.40	0.42	0.44	0.46	0.48	0.50
0.60	1.201	1.203	1.206	1.211	1.217	1.224	1.232	1.242	1.253	1.265	1.278	1.292	1.308	1.324	1.342	1.360	1.379	1.399	1.420	1.442	1.465	1.488	1.512	1.537	1.562
0.58	1.161	1.163	1.166	1.171	1.177	1.185	1.193	1.203	1.215	1.227	1.241	1.255	1.271	1.288	1.306	1.325	1.345	1.365	1.387	1.409	1.432	1.456	1.481	1.506	1.532
0.56	1.121	1.123	1.126	1.131	1.138	1.145	1.154	1.165	1.176	1.189	1.203	1.219	1.235	1.252	1.271	1.290	1.310	1.331	1.354	1.376	1.400	1.424	1.449	1.475	1.501
0.54	1.081	1.083	1.087	1.092	1.098	1.106	1.116	1.126	1.138	1.152	1.166	1.182	1.199	1.217	1.235	1.255	1.276	1.298	1.321	1.344	1.368	1.393	1.419	1.445	1.472
0.52	1.041	1.043	1.047	1.052	1.059	1.067	1.077	1.088	1.101	1.114	1.129	1.145	1.163	1.181	1.201	1.221	1.243	1.265	1.288	1.312	1.337	1.362	1.389	1.415	1.443
0.50	1.001	1.003	1.007	1.013	1.020	1.028	1.038	1.050	1.063	1.077	1.093	1.109	1.127	1.146	1.166	1.187	1.209	1.232	1.256	1.281	1.306	1.332	1.359	1.386	1.414
0.48	0.961	0.963	0.967	0.973	0.981	0.990	1.000	1.012	1.025	1.040	1.056	1.073	1.092	1.111	1.132	1.154	1.176	1.200	1.224	1.250	1.276	1.302	1.330	1.358	1.386
0.46	0.921	0.923	0.928	0.934	0.941	0.951	0.962	0.974	0.988	1.003	1.020	1.038	1.057	1.077	1.098	1.121	1.144	1.168	1.193	1.219	1.246	1.273	1.301	1.330	1.359
0.44	0.881	0.884	0.888	0.894	0.902	0.912	0.923	0.936	0.951	0.967	0.984	1.002	1.022	1.043	1.065	1.088	1.112	1.137	1.163	1.189	1.217	1.245	1.273	1.302	1.332
0.42	0.841	0.844	0.849	0.855	0.863	0.874	0.885	0.899	0.914	0.930	0.948	0.967	0.988	1.010	1.032	1.056	1.081	1.106	1.133	1.160	1.188	1.217	1.246	1.276	1.306
0.40	0.801	0.804	0.809	0.816	0.825	0.835	0.848	0.862	0.877	0.894	0.913	0.933	0.954	0.977	1.000	1.024	1.050	1.076	1.103	1.131	1.160	1.189	1.219	1.250	1.281
0.38	0.761	0.764	0.769	0.777	0.786	0.797	0.810	0.825	0.841	0.859	0.878	0.899	0.921	0.944	0.968	0.994	1.020	1.047	1.075	1.103	1.133	1.163	1.193	1.224	1.256
0.36	0.721	0.724	0.730	0.738	0.747	0.759	0.773	0.788	0.805	0.824	0.844	0.865	0.888	0.912	0.937	0.963	0.990	1.018	1.047	1.076	1.106	1.137	1.168	1.200	1.232
0.34	0.681	0.685	0.691	0.699	0.709	0.721	0.735	0.752	0.769	0.789	0.810	0.832	0.856	0.881	0.907	0.934	0.962	0.990	1.020	1.050	1.081	1.112	1.144	1.176	1.209
0.32	0.641	0.645	0.651	0.660	0.671	0.684	0.699	0.716	0.734	0.755	0.777	0.800	0.825	0.850	0.877	0.905	0.934	0.963	0.994	1.024	1.056	1.088	1.121	1.154	1.187
0.30	0.601	0.605	0.612	0.621	0.632	0.646	0.662	0.680	0.700	0.721	0.744	0.768	0.794	0.821	0.849	0.877	0.907	0.937	0.968	1.000	1.032	1.065	1.098	1.132	1.166
0.28	0.561	0.566	0.573	0.582	0.595	0.609	0.626	0.645	0.666	0.688	0.712	0.738	0.764	0.792	0.821	0.850	0.881	0.912	0.944	0.977	1.010	1.043	1.077	1.111	1.146
0.26	0.522	0.526	0.534	0.544	0.557	0.573	0.591	0.611	0.632	0.656	0.681	0.708	0.735	0.764	0.794	0.825	0.856	0.888	0.921	0.954	0.988	1.022	1.057	1.092	1.127
0.24	0.482	0.487	0.495	0.506	0.520	0.537	0.556	0.577	0.600	0.625	0.651	0.679	0.708	0.738	0.768	0.800	0.832	0.865	0.899	0.933	0.967	1.002	1.038	1.073	1.109
0.22	0.442	0.447	0.456	0.468	0.483	0.501	0.522	0.544	0.569	0.595	0.622	0.651	0.681	0.712	0.744	0.777	0.810	0.844	0.878	0.913	0.948	0.984	1.020	1.056	1.093
0.20	0.402	0.408	0.418	0.431	0.447	0.466	0.488	0.512	0.538	0.566	0.595	0.625	0.656	0.688	0.721	0.755	0.789	0.824	0.859	0.894	0.930	0.967	1.003	1.040	1.077
0.18	0.362	0.369	0.379	0.394	0.412	0.433	0.456	0.482	0.509	0.538	0.569	0.600	0.632	0.666	0.700	0.734	0.769	0.805	0.841	0.877	0.914	0.951	0.988	1.025	1.063
0.16	0.322	0.330	0.342	0.358	0.377	0.400	0.425	0.453	0.482	0.512	0.544	0.577	0.611	0.645	0.680	0.716	0.752	0.788	0.825	0.862	0.899	0.936	0.974	1.012	1.050
0.14	0.283	0.291	0.305	0.322	0.344	0.369	0.396	0.425	0.456	0.488	0.522	0.556	0.591	0.626	0.662	0.699	0.735	0.773	0.810	0.848	0.885	0.923	0.962	1.000	1.038
0.12	0.243	0.253	0.268	0.288	0.312	0.339	0.369	0.400	0.433	0.466	0.501	0.537	0.573	0.609	0.646	0.684	0.721	0.759	0.797	0.835	0.874	0.912	0.951	0.990	1.028
0.10	0.204	0.215	0.233	0.256	0.283	0.312	0.344	0.377	0.412	0.447	0.483	0.520	0.557	0.595	0.632	0.671	0.709	0.747	0.786	0.825	0.863	0.902	0.941	0.981	1.020
0.08	0.165	0.179	0.200	0.226	0.256	0.288	0.322	0.358	0.394	0.431	0.468	0.506	0.544	0.582	0.621	0.660	0.699	0.738	0.777	0.816	0.855	0.894	0.934	0.973	1.013
0.06	0.126	0.144	0.170	0.200	0.233	0.268	0.305	0.342	0.379	0.418	0.456	0.495	0.534	0.573	0.612	0.651	0.691	0.730	0.769	0.809	0.849	0.888	0.928	0.967	1.007
0.04	0.089	0.113	0.144	0.179	0.215	0.253	0.291	0.330	0.369	0.408	0.447	0.487	0.526	0.566	0.605	0.645	0.685	0.724	0.764	0.804	0.844	0.884	0.923	0.963	1.003
0.02	0.057	0.089	0.126	0.165	0.204	0.243	0.283	0.322	0.362	0.402	0.442	0.482	0.522	0.561	0.601	0.641	0.681	0.721	0.761	0.801	0.841	0.881	0.921	0.961	1.001

X ——▶ (DEV)

◀—— Y ——▶ (DEV)

Actual position = $2\sqrt{x^2 + y^2}$

4.8

The top drawing is a plate with position tolerancing applied to the three holes. The lower drawing is the produced part. Evaluate the dimensions on the produced part to verify conformance to the position tolerances. Use the table below to record your calculations.

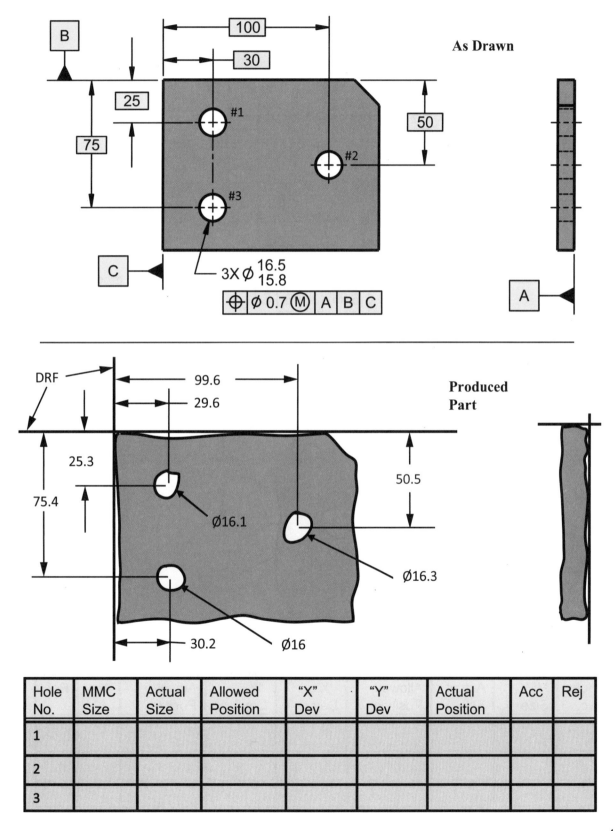

Hole No.	MMC Size	Actual Size	Allowed Position	"X" Dev	"Y" Dev	Actual Position	Acc	Rej
1								
2								
3								

The top drawing is a plate with position tolerancing applied to the three holes. The lower drawing is the produced part. Evaluate the dimensions on the produced part to verify conformance to the position tolerances. Use the table below to record your calculations.

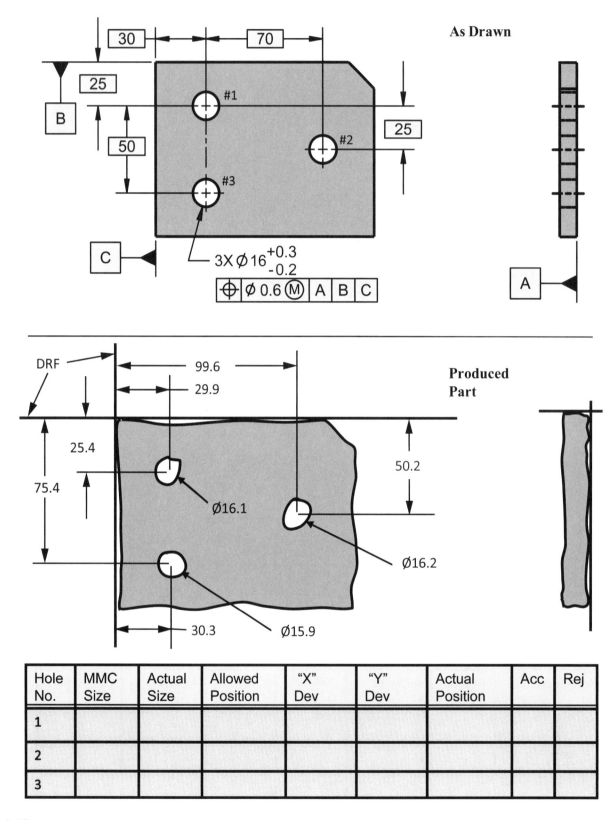

Hole No.	MMC Size	Actual Size	Allowed Position	"X" Dev	"Y" Dev	Actual Position	Acc	Rej
1								
2								
3								

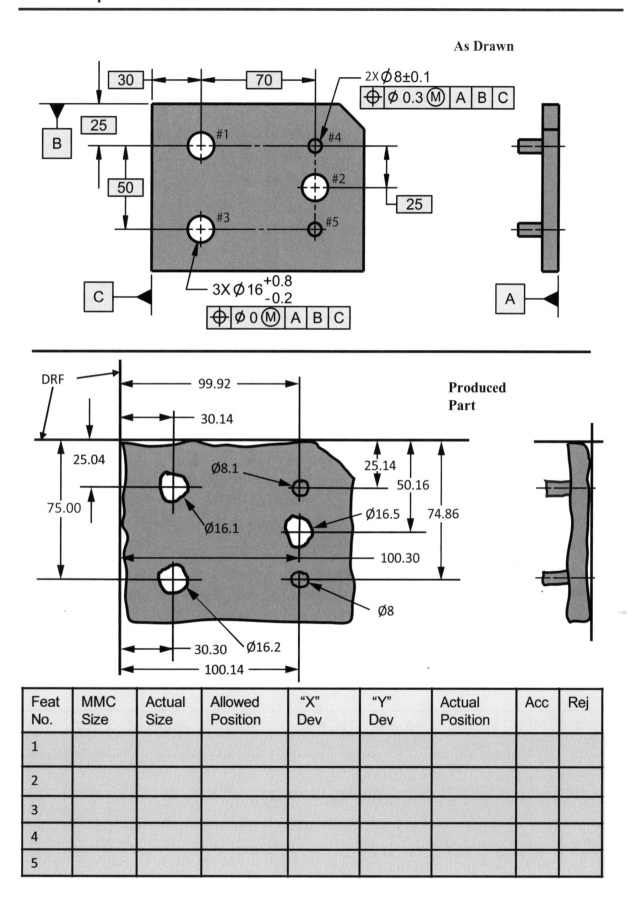

As Drawn

Produced Part

Feat No.	MMC Size	Actual Size	Allowed Position	"X" Dev	"Y" Dev	Actual Position	Acc	Rej
1								
2								
3								
4								
5								

4.11

As Drawn

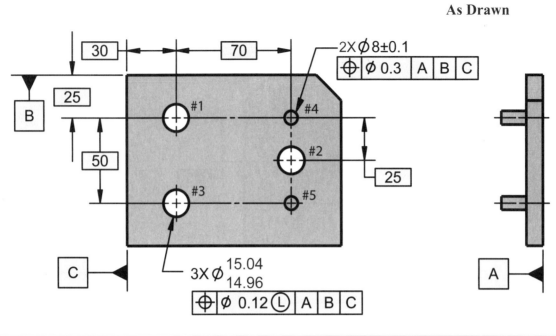

Produced Part

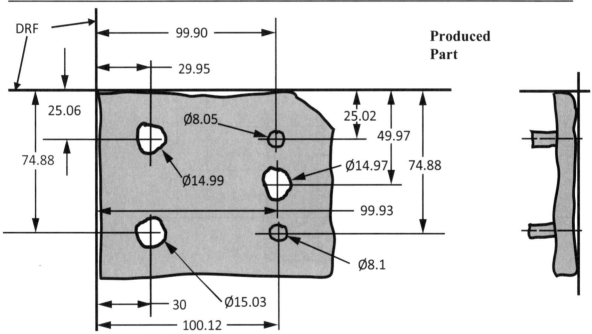

Feat No.	LMC Size	Actual Size	Allowed Position	"X" Dev	"Y" Dev	Actual Position	Acc	Rej
1								
2								
3								
4	N/A							
5	N/A							

The top drawing is a part with position tolerance applied. The lower drawing is the produced part set up to the datum reference frame as referenced in the feature control frame for the 3 holes. Evaluate the dimensions on the produced part for the 3 holes to verify conformance to the position tolerances. Use the table below to record your calculations. The datum references are implied to apply at RMB. Later in unit 6, this part will be evaluated again with the datum references at MMB.

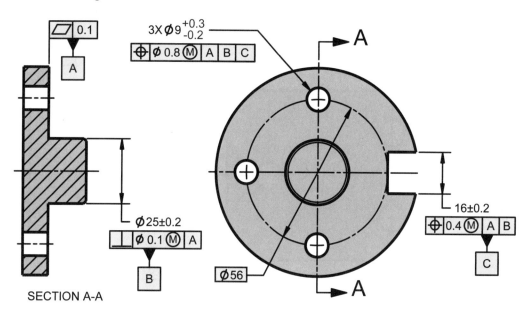

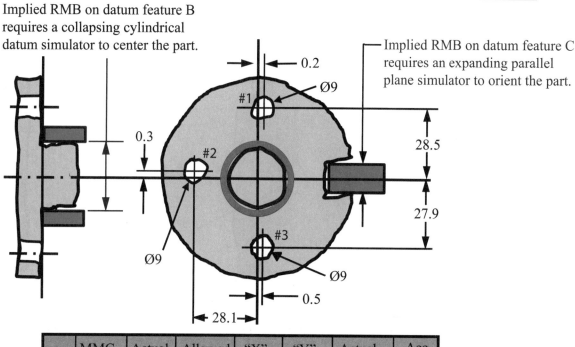

Implied RMB on datum feature B requires a collapsing cylindrical datum simulator to center the part.

Implied RMB on datum feature C requires an expanding parallel plane simulator to orient the part.

No.	MMC Size	Actual Size	Allowed Position	"X" Dev	"Y" Dev	Actual Position	Acc Rej
1							
2							
3							

4.13

The top drawing is a plate with position tolerancing applied to the three holes. The lower drawing is the produced part. Evaluate the dimensions on the produced part to verify conformance to the position tolerances. Use the table below to record your calculations.

As Drawn

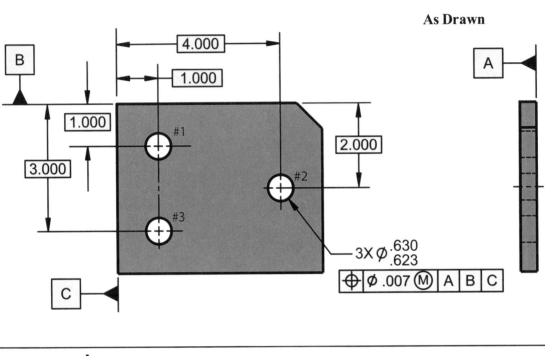

Produced Part

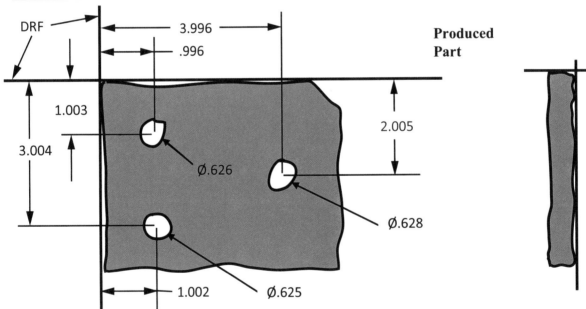

Hole No.	MMC Size	Actual Size	Allowed Position	"X" Dev	"Y" Dev	Actual Position	Acc	Rej
1								
2								
3								

Workshop Exercise 4.7 - Inch

The top drawing is a plate with position tolerancing applied to the three holes. The lower drawing is the produced part. Evaluate the dimensions on the produced part to verify conformance to the position tolerances. Use the table below to record your calculations.

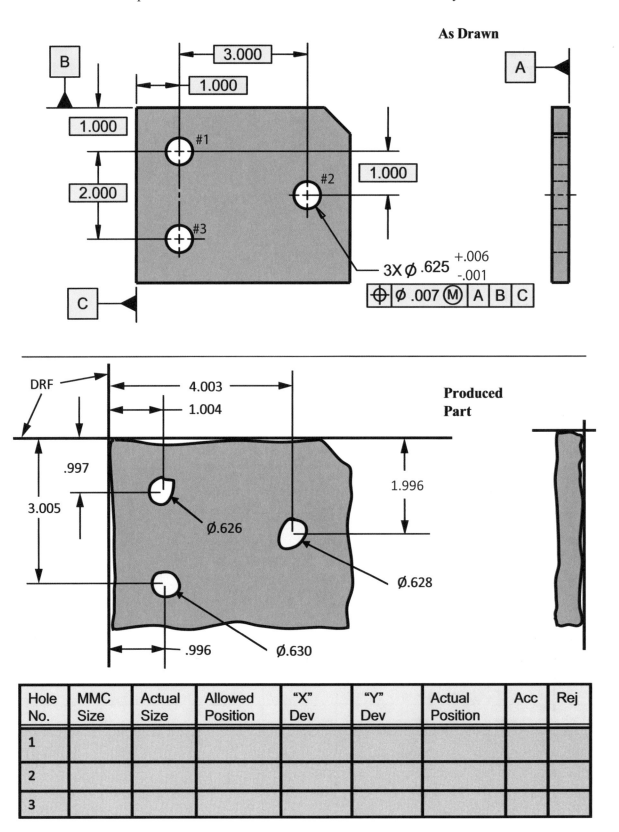

As Drawn

3.000

1.000

B

1.000

#1

1.000

#2

2.000

#3

3X⌀.625 +.006 / -.001

⌖ ⌀.007 Ⓜ A B C

A

C

Produced Part

DRF

4.003

1.004

.997

1.996

3.005

⌀.626

⌀.628

.996 ⌀.630

Hole No.	MMC Size	Actual Size	Allowed Position	"X" Dev	"Y" Dev	Actual Position	Acc	Rej
1								
2								
3								

4.15

Workshop Exercise 4.8 - Inch

As Drawn

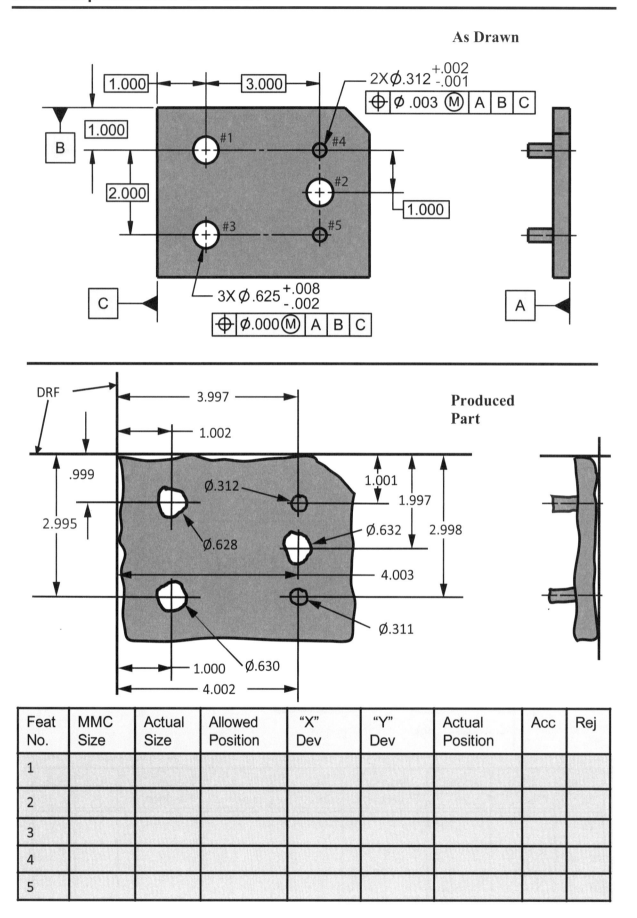

2X⌀.312 +.002/-.001

| ⌖ | ⌀ .003 Ⓜ | A | B | C |

3X⌀.625 +.008/-.002

| ⌖ | ⌀.000 Ⓜ | A | B | C |

1.000 3.000

1.000

B

2.000

#1   #4

#2

1.000

#3   #5

C

A

Produced Part

DRF

3.997

1.002

.999

2.995

⌀.312

⌀.628

1.001

1.997

⌀.632

2.998

4.003

⌀.311

1.000 ⌀.630

4.002

Feat No.	MMC Size	Actual Size	Allowed Position	"X" Dev	"Y" Dev	Actual Position	Acc	Rej
1								
2								
3								
4								
5								

As Drawn

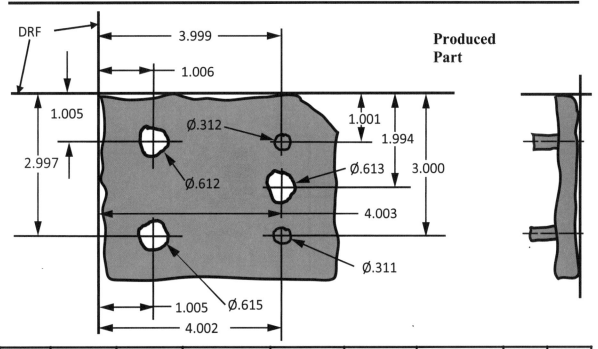

Feat No.	LMC Size	Actual Size	Allowed Position	"X" Dev	"Y" Dev	Actual Position	Acc	Rej
1								
2								
3								
4	N/A							
5	N/A							

Paper Gage Concept

The paper gage concept is a good tool for understanding and evaluating geometric tolerancing. Paper gage concepts are used throughout this book to help illustrate many of the geometric concepts.

The paper gage concept is also a handy tool for evaluating data that is collected from coordinate measuring machines (CMM) or surface plate set-ups. Many coordinate measuring machines have software that is capable of evaluating datum feature modifiers and hole patterns as datum features. This simple exercise will show you the basic math behind these concepts.

The paper gage concept illustrated in this workbook uses a Cartesian coordinate and polar coordinate system. In order to understand the concepts in the workbook, it is important to understand how to plot the deviations of the feature from its basic location.

When plotting points, it is important that all users plot consistently and in the same method. In this book, all points are plotted relative to the Cartesian coordinate system. In other words, each hole or feature has a basic location. Its address, in theory, is X = 0, Y = 0.

The features should be plotted on the graph relative to their theoretical or basic address in the view in which the holes are specified on the drawing. When plotting in the X direction, if a feature falls to the right of its basic address, it will have a plus X value. If it falls to the left of its basic location, it will have a minus X value. Additionally, when plotting in the Y direction, if a feature falls above its basic address, it will have a plus Y value. If it falls below its basic location, it will have a minus Y value.

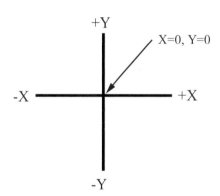

The paper gage concept explained in this book is a simple rendering. Another more precise method to calculate position tolerances involves inputting the information in a CAD system to evaluate the data using separate tolerance zones for each hole on basic locations. The method will allow the evaluation of holes in a rotated position, as well as a linearly shifted position. The simple method shown here is sufficient for most examples as long as inspection has correctly aligned to the datum features or features when collecting the data.

Paper Gage Evaluation of 3 Hole Part (Problem 4.1- Metric)

The part below was evaluated in exercise 4.1. Hole number 2 was found to be out of position. The paper gage example on the following page evaluates the manufacturing problem and suggests possible solutions.

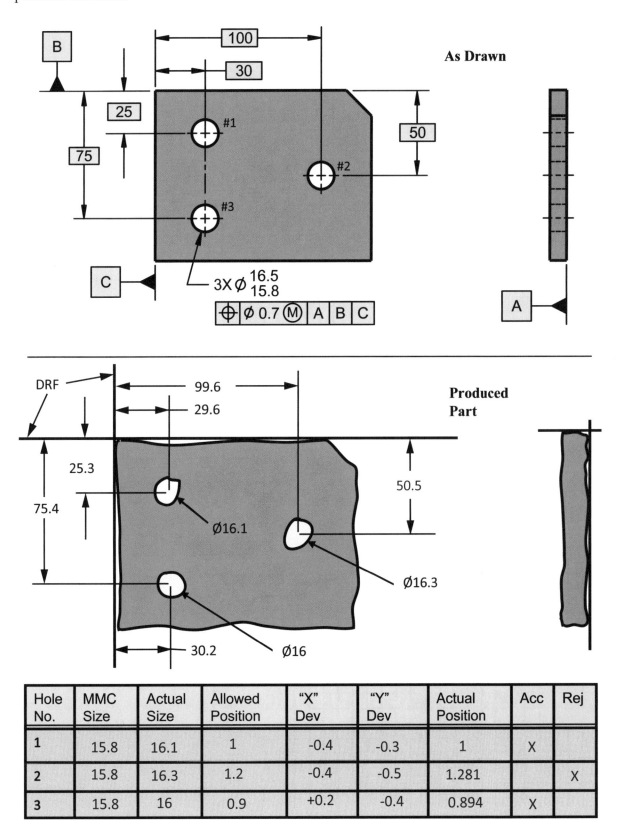

Hole No.	MMC Size	Actual Size	Allowed Position	"X" Dev	"Y" Dev	Actual Position	Acc	Rej
1	15.8	16.1	1	-0.4	-0.3	1	X	
2	15.8	16.3	1.2	-0.4	-0.5	1.281		X
3	15.8	16	0.9	+0.2	-0.4	0.894	X	

The paper gage example below is an evaluation of the three hole part in workshop exercise 4.1-Metric. On the produced part, hole number 2 was reported out of position tolerance. We can easily evaluate the problem after plotting the three holes on the graph below.

At first glance, it appears that hole #2 must be relocated in the manufacturing process. After a paper gage evaluation, it can be seen that the location of the 3 holes are good to each other. The group of 3 holes is just displaced to the DRF.

The example below shows the 3 holes positioned to each other within a group of 0.7 diameter zone. In order to center the process, the origin of the holes as a group (0.7 ring) must be relocated by plus 0.1 in the X direction and plus 0.4 in the Y direction. It seems there is an elevation and windage problem relative to the DRF.

The paper gage concept is a good tool for evaluating and adjusting the manufacturing process. Of course, an appropriate sampling of parts should be considered for proper representation.

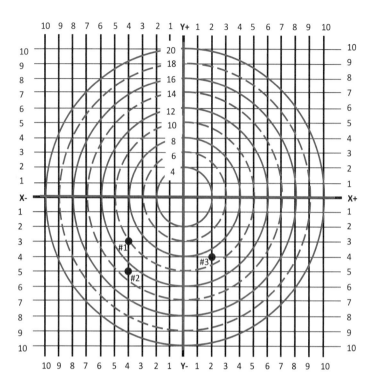

**Pattern of 3 holes to the DRF
shows hole 2 is out of location.**

In this example, it seems the manufacturing process is capable of locating the three holes to each other within a diameter of 0.7, but there is a problem locating the holes relative to the DRF.

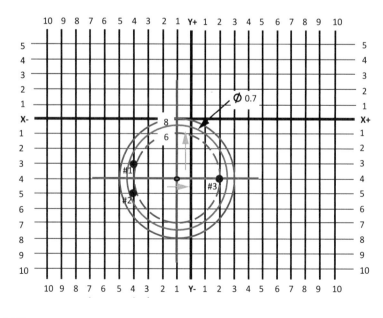

Grouping of 3 holes to each other shows within 0.7 dia. but they are displaced to the DRF by -0.1 in X and -0.4 in Y.

Paper Gage - Cartesian and Polar Coordinate System

Paper copies of this Cartesian coordinate system can be used for the paper gage exercises in this chapter and other exercises later in the book. In using the coordinate system graphs for paper gage calculations, it is important that the circles on the polar coordinate system and the squares line up. The Cartesian coordinate and the polar coordinate must be of the same scale. They can be any scale as long as the circles and squares correspond and line up. The graphs can be used for metric or inch problems.

Copies produced from an original on office copiers tend to stretch or shrink the copies somewhat. If using the paper gage concept for work related evaluations, it might be more reliable if new coordinates were drawn and printed from a CAD system. The Cartesian coordinate can be graph paper. The polar coordinate is a transparency.

The paper gage calculations can also be completed with a CAD system or appropriate CMM software.

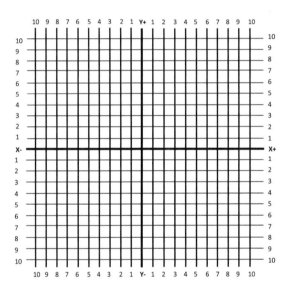

Cartesian coordinate

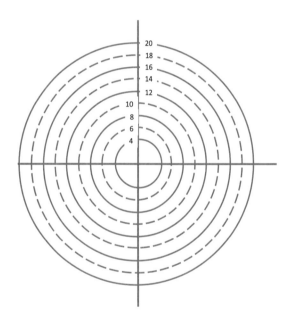

Polar coordinate

Unit 5

Product Plans and Virtual Condition

The Hub - Design Product Drawing

The drawing below has a lot of geometric tolerancing applied. The datum features are defined and qualified. There are applications of form, orientation, position, runout and profile tolerances.

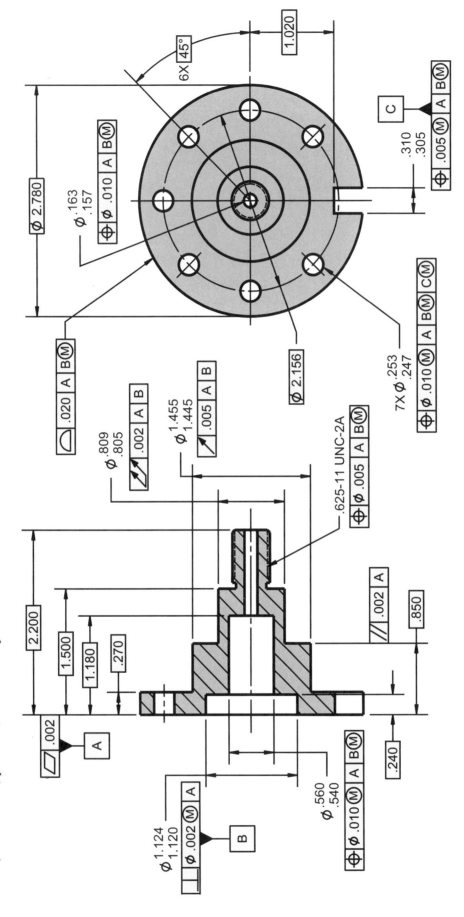

ALL UNTOLERANCED SURFACES:

INTERPRET THIS DRAWING
PER ASME Y14.5-2009

5.2

Three Plans for a Quality Product

In order to build quality parts, there must be a plan. In fact, we need three plans. These three documents or plans require input from the three main players: design, manufacturing and quality.

The three documents or plans are:

1. **The Design Product Drawing**

   -Defines the functional requirements of the product.

2. **The Manufacturing Process Plan**

   -Defines how to make the product.

3. **The Quality Dimensional Measurement Plan**

   -Defines how to verify the product.

These three documents can vary in complexity. In some cases, these steps can be very detailed written documents. In other cases, these steps can be informal verbal instructions. The formality of these documents depends on the complexity of the part, the amount of parts to be produced, and the particular organization.

Design Product Drawing: This document defines the product and contains detailed information about the functional requirements of the part. The drawing is the most valuable document that we have. Without this document we do not have a product. This document is usually controlled by design engineering and kept locked in a vault or master database where no one can change this drawing without the approval of design engineering.

The selected datums and specified requirements on this drawing define the functional requirements of the part. It explains what the part is supposed to look like before it goes out the door or is assembled with mating parts. The datum reference frame established on the product drawing is not for manufacturing. This drawing does not define how to build the part. For example, a hole call out lists the size of a hole but does not list whether it is drilled, reamed or punched, unless that information is critical to the function of the part. Datum features are selected on the basis of part function. These datum features define how the part mounts or how it is set up for functional requirements.

Manufacturing Process Plan: This document defines how to make the product. Every organization has manufacturing process plans. They may be formal or informal. They are made up of a collection of documents called by different names, such as routing, process or operation sheets. This document is sometimes very formal, as in the manufacturing of a cylinder block. Or, it can be informal, as in the manufacturing of an initial prototype part.

The process plan defines how to build the product and is usually drawn up by a manufacturing or process engineer. The process engineer designs the process to meet the requirements of the product drawing. This process plan can be very detailed, such as listing the processes, procedures, sequence of operations and the equipment necessary to build the product. In other cases, the process plan may be as simple as a CNC tape and the drawing of a fixture that holds the part.

These process plans are flexible and can be changed at will by the manufacturing engineer. Sometimes, a single part may be made by different processes in different locations. In this case, the part has multiple process plans to reflect the operations of the multiple processes.

Quality Dimensional Measurement Plan: This document establishes the methods of measurement. It tells how to inspect or verify the part. Just as with the manufacturing process plan, this dimensional measurement plan can be a formal written document or it can be informal, and left up to the individual inspectors discretion.

This inspection plan is usually drawn up by a quality engineer and lists step-by-step which characteristics to check, how often they should be checked and the method and type of equipment that should be used. The plan formulated by the quality engineer is usually developed after a careful evaluation of the functional requirements specified on the product drawing. Once the quality engineer understands the functional requirements, he/she then examines the manufacturing process sheet to understand how the part is made. The first article inspection may be a complete verification of all the characteristics. Once we know something about the process variation and what could go wrong, it helps define what characteristics must be checked and how often. It also helps determine the inspection tools, procedures and methods to be used.

Different inspection methods or devices yield different results during the verification of the products. This is evident by the different results obtained from a CMM vs. open setup surface plate techniques. This is often referred to as methods divergence. The dimensional measurement plan should define the acceptable measurement method and the associated uncertainty of results.

As you can see, it is very important to have some plan or organization to build interchangeable parts. This is especially important as we buy and sell parts all over the world and strive to achieve world class quality. The three plan system allows our parts to be followed and tracked from beginning to end, no matter who works with the product. This makes it easy to identify and solve any problems or differences we may have along the way. In order to have quality parts, you must have communication and documentation throughout the design, manufacturing and verification process.

Sample Operation Sheet

Part Name : Hub	**Machine Type** : CNC Lathe
Part Number: 38342831	**Machine Number**: 28L

Instructions:
Clamp datum feature X face as primary and datum feature Y diameter 2.780 shown. Turn .809 and 1.455 diameters. Verification chart required.

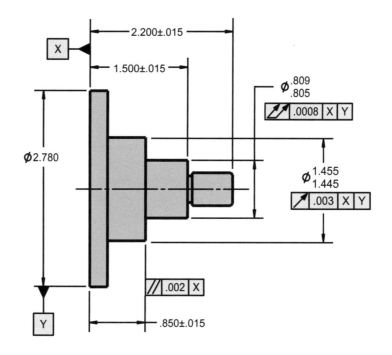

Tools Required	Tool Number
Gage ring .8085	GL8085
Gage ring .8055	GL8055
Height gage w/ .0001 indicator	6987
Calipers	---
CNC Tape	---
Turn Tool	TL1532

Dimensional Measurement Plan

Part Name : Hub

Part Number: 38342831

Special Instructions:
Watch for Burrs
First article 100% check, 3 parts per shift

① ▱ .002 — A

② ⌀ 1.124 / 1.120

③ ▯ ⌀ .002 Ⓜ A — B

④ ⌀ .560 / .540

⑤ ⊕ ⌀ .010 Ⓜ A B Ⓜ

⑥ ∥ .002 A

⑦ 7X ⌀ .253 / .247

⑧ ⊕ ⌀ .010 Ⓜ A B Ⓜ C Ⓜ

⑨ ⌒ .020 A BⓂ

⑩ ⌀ .809 / .805

⑪ ⌰ .002 A B

⌀ 2.780

6X 45°

⌀ 2.156

C

ALL UNTOLERANCED SURFACES:
⑫ ⌒ .030 A B Ⓜ

Item Number	Inspection Requirement	Reporting Method	Equipment or gage
1	Check with height gage min 6 locations	B	height gage
2	go/ no-go	A	gage pin
3	functional gage	A	gage #1599
4	go/ no-go	A	gage pin
5	functional gage	A	gage #1599
6	Check with height gage min 10 locations	B	height gage
7	go/ no-go	B	gage pin
8	Height gage to find X, Y locations	C	height gage
9	Check with height gage, set up in vee block using gage pin, rotate 360 degrees	C	height gage, gage pin, 90° vee block
10	two point check, min 4 locations	B	calipers
11	Check with height gage, set up in vee block using gage pin, rotate 360 degrees	B	height gage, gage pin, 90° vee block
12	Check min 5 locations, must be within total 60% of tolerance range. If any outside 60%, check min 10 locations.	A	height gage

Reporting Methods:
A: Attribute data (pass/fail)
B: A single reported value or zone (actual profile zone, actual position zone)
C: Variable data for each sampled location or feature checked (plus/minus material for profile or x,y locations for position)

Virtual Condition

Depending on function, a feature may be controlled with a size tolerance and geometric control. The geometric control may also include an MMC modifier.

Consideration must be given to the collective effects of MMC and applicable tolerances in determining the clearance between parts and in establishing gage feature sizes.

When a MMC modifier is applied to the feature, it creates a constant virtual boundary in which the feature must be contained. This calculated boundary is used to find the worst case clearance for two parts fitting together. The following shows how to calculate these virtual size boundaries.

The two parts below will fit together with a minimum of Ø.004 clearance.

Pins appear larger at virtual

To calculate virtual size for external features, start with the largest feature and add the geometric tolerance.

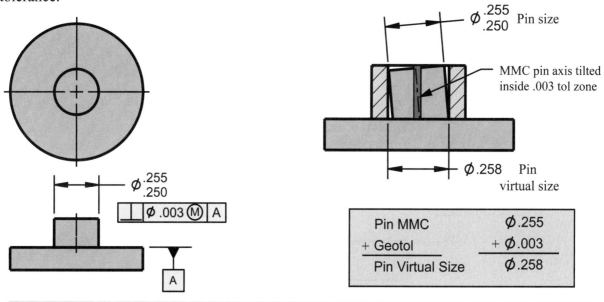

Pin MMC	Ø.255
+ Geotol	+ Ø.003
Pin Virtual Size	Ø.258

Holes appear smaller at virtual

To calculate virtual size for internal features, start with the smallest feature and subtract the geometric tolerance.

Hole MMC	Ø.265
- Geotol	- Ø.003
Hole Virtual Size	Ø.262

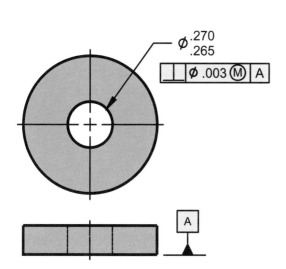

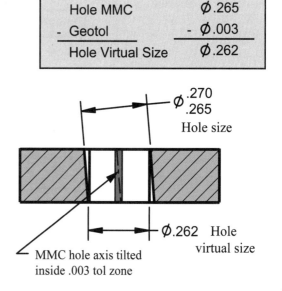

Virtual Condition - Multiple Feature Control Frames

Virtual condition is a constant boundary generated by the collective effects of size and the geometric tolerance applied. Sometimes, features may have multiple geometric controls. This will create multiple virtual sizes.

The pin below has a size tolerance and two feature control frames applied, and therefore two virtual size boundaries are created. The .255 diameter virtual size relative to datum A is a result of the perpendicularity tolerance. The .265 diameter virtual size relative to datums A, B, and C is a result of the position tolerance.

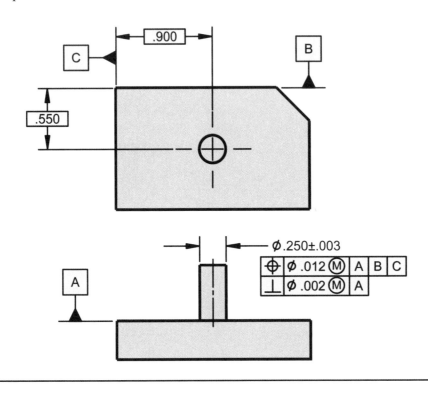

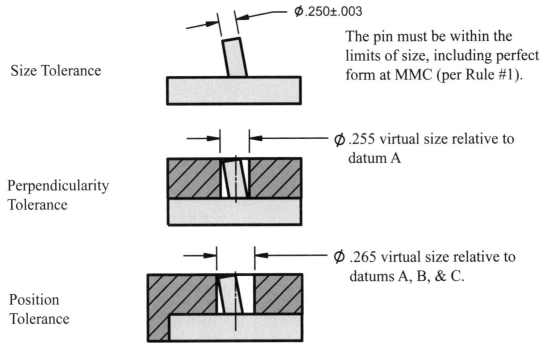

Size Tolerance

The pin must be within the limits of size, including perfect form at MMC (per Rule #1).

Perpendicularity Tolerance

ϕ.255 virtual size relative to datum A

Position Tolerance

ϕ.265 virtual size relative to datums A, B, & C.

Workshop Exercise 5.1

On the drawing below, calculate the virtual sizes for the identified features.
Remember, when calculating virtual with MMC, holes appear smaller and pins appear larger.

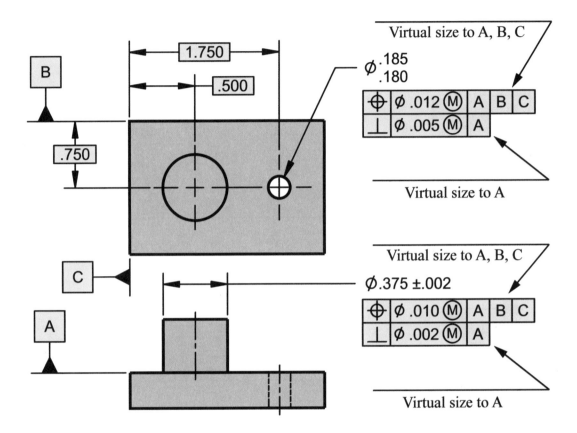

Workshop Exercise 5.2

Below are three assemblies for the part on the previous page. Use the drawing on the previous page to calculate the smallest virtual or mating hole size possible on the three rings. Pay close attention to how the rings mount in the assemblies.

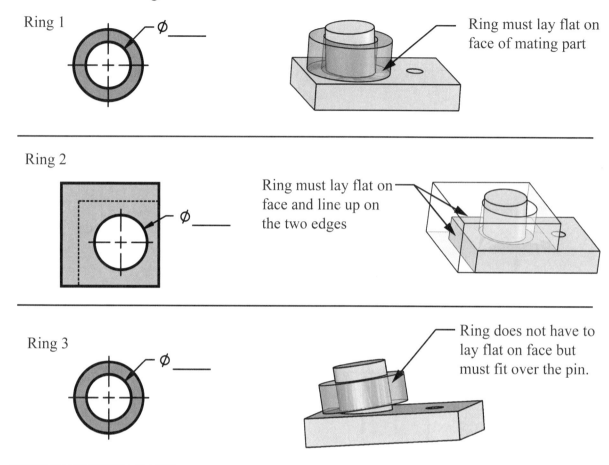

Ring 1 — ∅____ — Ring must lay flat on face of mating part

Ring 2 — ∅____ — Ring must lay flat on face and line up on the two edges

Ring 3 — ∅____ — Ring does not have to lay flat on face but must fit over the pin.

Workshop Exercise 5.3

Fill in the two feature control frames below for rings 1 and 2. Use the virtual sizes from above, and size tolerance below, to calculate the position or perpendicularity tolerances. Make sure to include the correct feature modifier.

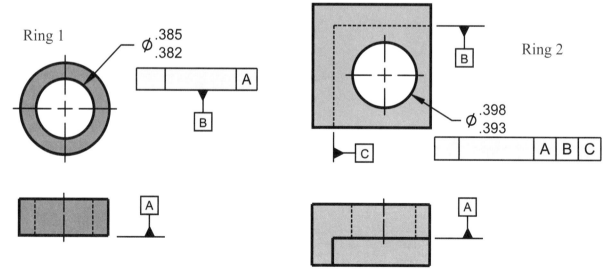

Ring 1 — ∅ .385 / .382

Ring 2 — ∅ .398 / .393

Unit 6

The Datum Reference Frame

Datum Reference Frame (DRF) and Degrees of Freedom (DOF)
Datum Reference Frame Simulators
Right Hand Rule
Positioning Parts in a Datum Reference Frame
Implied Datums are not Clear
Datum Reference Frame Used in Engineering
Datum Reference Frame Used in Manufacturing and Inspection
Theoretical Datum Feature Simulators and Datum Feature Controls
Datum Feature Simulator Requirements
Physical Datum Feature Simulators
Establishing Datums from Primary Planar Surfaces
Planar Surfaces as Primary Datum Features
Workshop Exercise 6.1
Establishing Datums from Primary Datum Features of Size at RMB
Datum Feature Symbol Placement is Important
Partial Datum Features and Multiple Datum Features
Workshop Exercise 6.2
Degrees of Freedom - Primary Datum Feature at RMB
Degrees of Freedom Table - Primary Datum Feature at RMB
Workshop Exercise 6.3
Establishing a DRF - Planar Surfaces
Establishing a DRF - Plane, Axis, Plane
Establishing a DRF - Axis, Plane, Center Plane
Establishing a DRF - Plane, Plane, Axis
Datum Feature Precedence
Workshop Exercise 6.4

Datum Reference Frame (DRF) and Degrees of Freedom (DOF)

The Datum Reference Frame (DRF) is the most important concept in geometric tolerancing. The DRF is the skeleton of the geometric system. It is the "frame of reference" to which all referenced geometric specifications are related.

The DRF consists of three mutually perpendicular datum planes and three mutually perpendicular datum axes that occur at the intersection of the datum planes. The intersection of three planes establishes a point. The datum reference frame is composed of a point, axes and planes. The datum reference frame is the origin of all the dimensions and geometric specifications that are related to it.

A datum reference frame constraints or arrests 6 degrees of freedom

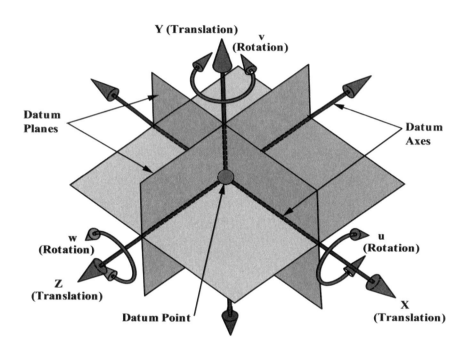

Three Translation Controls
X = Translation in "X" direction
Y = Translation in "Y" direction
Z = Translation in "Z" direction

Three Rotation Controls
u = Rotation in "u" direction
v = Rotation in "v" direction
w = Rotation in "w" direction

A datum reference frame establishes 6 Degrees of Freedom (DOF), three translations and three rotations. The three translational degrees of freedom are termed X, Y and Z. The three rotational degrees of freedom are termed u (rotation about the X axis), v (rotation about the Y axis) and w (rotation about the Z axis).

In order to design, manufacture, and verify parts, the necessary DOF must be constrained. Parts are mated to the DRF so measurements, processing and calculations can be made.

In defining a part, an engineer will identify datum features on a part that are most important to the functional requirements of the design. These datum features are usually the features that mount the part in the assembly. There are leaders and followers on a part. The datum features are the leaders.

Datum features referenced in the rear compartments of a feature control frame, in an order of precedence, will mate the part to the datum reference frame. The mating of a part and DRF process takes place by a series of events.

The datum features on a part are mated with datum feature simulators, which are usually the inverse of the datum feature. Datums are established from the datum feature simulators. Datums which are points, lines, and planes, are the building components used in an order to establish the datum reference frame. The details of establishing a DRF and mating it to the part are shown on the following pages of this unit.

Datum Reference Frame Simulators

Design engineering, manufacturing and inspection all share a common three plane concept. It is a communication tool used between these groups. These three mutually perpendicular planes are perfect and are 90 degrees to each other.

In geometric tolerancing, this is called the Datum Reference Frame (DRF). In mathematics, it is called the Cartesian coordinate system. The Cartesian coordinate system was invented by a French mathematician, René Descartes (1596-1650) The DRF and Cartesian coordinate system are identical.

Design engineers use the DRF in geometric tolerancing to mate this coordinate system to the produced part from which they locate features. The 3 plane DRF concept is fundamentally imbedded in all our engineering, manufacturing and inspection equipment. The equipment is not perfect but is of such quality that it will adequately simulate the DRF. The equipment shown below are common datum reference frame simulators that use the 3 plane concept.

Datum Reference Frame Simulators

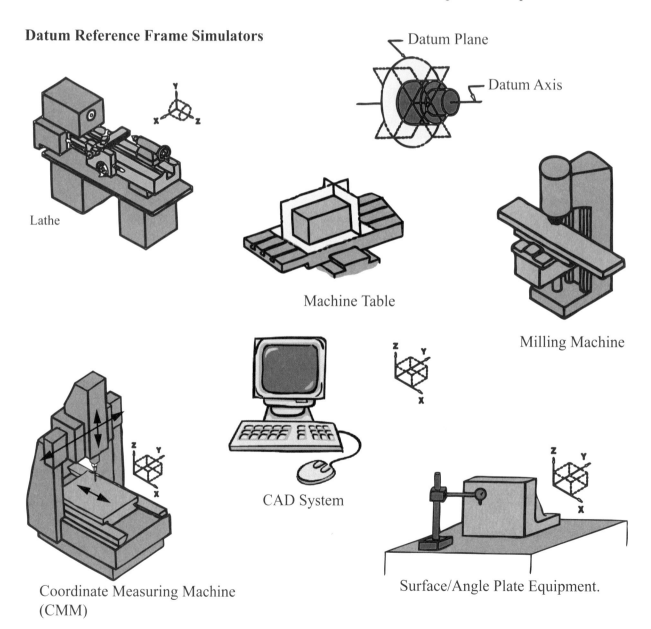

Lathe

Machine Table

Milling Machine

CAD System

Coordinate Measuring Machine
(CMM)

Surface/Angle Plate Equipment.

Using the right hand to visualize a right hand coordinate system

A method to understand the "right hand rule" is to use the right hand to establish the Cartesian coordinate system. The thumb, index finger and middle finger of the right hand are held so they form three right angles. The thumb indicates the x axis, the index finger the y axis, and the middle finger the z axis. The thumb and two figures always point in the plus direction. Right hand rotation about the "X" axis is in the pointing direction of the ring and little finger.

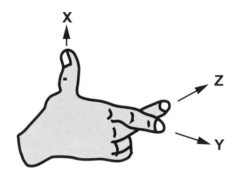

The right-handed Cartesian coordinate system is the standard coordinate systems for most engineering, manufacturing and quality systems. Left-handed Cartesian coordinate systems follow the same rule, just using the left hand.

Using a folded paper to visualize a right-hand coordinate system

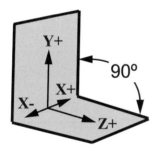

Another way to visualize the right-hand coordinate system is to draw and label the Cartesian coordinate system on a small rectangular piece of paper, as shown in the view to the left.

Crease and fold the paper at a 90° angle along the "X" axis. The paper can be rotated around as necessary to view the spatial orientation of a right hand coordinate system, along with its associated translations and rotations.

Positioning Parts in a Datum Reference Frame

The part shown below has four holes that are located relative to a datum reference frame. The detailed requirements for the 4 holes can be found in the feature control frame, along with the relationship to the datums. The datum features referenced in the rear compartments of the feature control frame specify how the part is mated to the datum reference frame.

The primary datum feature is shown in the first compartment. In this case, it mates with the DRF on the highest 3 points of the surface. The secondary datum feature is shown in the second compartment and it mates with the highest 2 points of contact on the surface. The tertiary datum feature is shown in the third compartment and it mates with one point of contact.

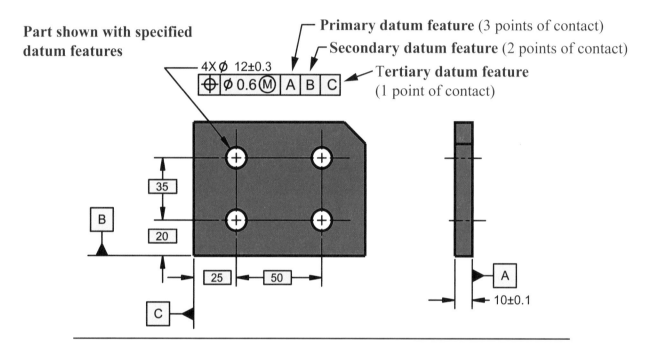

Part shown with specified datum features

Primary datum feature (3 points of contact)

Secondary datum feature (2 points of contact)

Tertiary datum feature (1 point of contact)

4X Ø 12±0.3

10±0.1

The order of the datum features in the rear compartment of the feature control frame mate the DRF to the part.

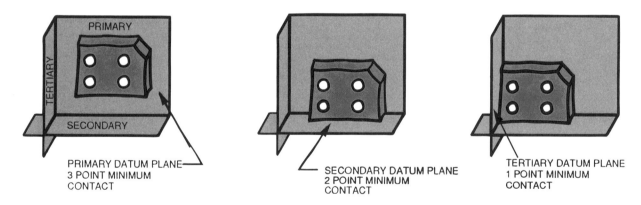

PRIMARY DATUM PLANE
3 POINT MINIMUM
CONTACT

SECONDARY DATUM PLANE
2 POINT MINIMUM
CONTACT

TERTIARY DATUM PLANE
1 POINT MINIMUM
CONTACT

In order to have interchangeable parts, it is mandatory to have a datum reference frame established on the parts. The DRF is a 3D coordinate system where all dimensions and measurements originate. If a part does not have a well-defined DRF, the product will not be well-defined, which leads to misunderstandings, confusion and poor quality.

Note: Not all parts mate to the DRF with the 3, 2, 1 concept, as explained later in this unit.

The produced part shown below (exaggerated for clarity) is not perfect. It will never be perfect due to inherent errors in the manufacturing process. We cannot build perfect parts. None of the surfaces or edges are flat or square. It is an acceptable part however, as it is within acceptable limits for size, flatness and squareness.

The imperfect produced part is mated to the (perfect) datum reference frame according to the order the datum features are referenced in the rear compartments of the feature control frame. The alphabetical order in the compartments is not important. It reads from left to right. All dimensions and measurements originate from the datum reference frame and not the part. If the order of the datum features referenced in the feature control frame is changed, it will change the order the part is mated to the datum reference frame.

The graphics below illustrate how changing the order of the datum features in the feature control frame will change the order in which the part is mated to the datum reference frame. Engineers usually select the order of these datum features based on how the part mates in an assembly. Quality control will set up the part for inspection based on the order of the datum features referenced in the feature control frame.

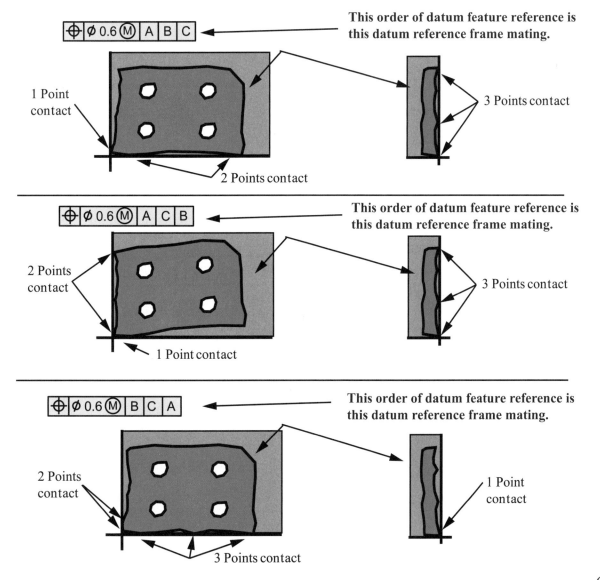

6.7

Implied Datums are not Clear

The part below has 4 holes that has dimensions to the edges. This part has implied datums. We call them implied datums because it is not very clear. It is assumed that the dimensions originate from the bottom edge and left hand edge. This is vague and not clear enough if we want to mate this part to a coordinate system.

Part shown with unclear definition and implied datums

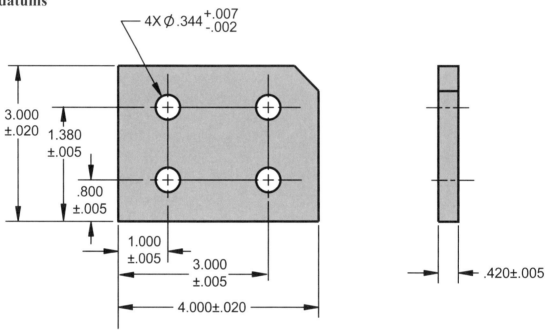

The actual produced part shown below (exaggerated for clarity) is not perfect and has inherent manufacturing error. It will never be perfect, as we cannot manufacture perfect parts. Notice the surfaces are not flat or square. The part is within tolerance however, as it is within the accepted limits for size and squareness.

Produced part

Since the produced part is imperfect, we need to place the part in a perfect datum reference frame to define geometry boundaries or make measurements. When the imperfect part is mated to a perfect datum reference frame, it does not fit exactly and will rock back and forth because of error in the implied datum surfaces. The implied datums do not have a specified order in which to mate the part to the datum reference frame.

The graphics below illustrate the problem of mating an imperfect, un-square part to a perfect datum reference frame. The are a number of combinations in which to the unspecified datum features to the datum reference frame. Implied datums will result in different interpretations on how the imperfect part is mated to a perfect datum reference frame.

3 possible DRF mating conditions with implied datums and no clear idea on the condition of the features.

The datum reference frame is perfect, the parts are not perfect with implied datums.

The parts rattle around, it is not clear how to measure conformance.

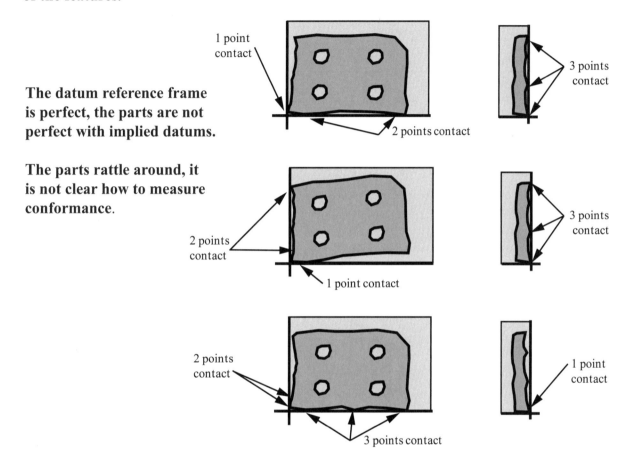

Implied datums are not clear. It is impossible to have interchangeable parts with implied datums. Engineering, manufacturing, and inspection may all have different interpretations as to how the part is to be mated to the datum reference frame. This will result in confusion as to where the holes or other features must lie on the part. Everyone must interpret the drawings and understand design intent in exactly the same way to achieve a high quality interchangeable product.

This four hole part is a simple example. As the parts become more complex with more holes and surfaces, the order of datums becomes even more important. If the order of the datums is not specified, it is possible to misinterpret design intent and reject good parts or even worse, accept bad parts.

Engineering, manufacturing, and quality control must have a common DRF. Creating a DRF is mandatory in order to produce interchangeable parts. Our modern methods of engineering, manufacturing and quality control demand a clear concise DRF along with geometric tolerancing to provide a clear definition of the product.

The **Datum Reference Frame (DRF)** in design engineering is a three dimensional, Cartesian coordinate system. The engineers work in this theoretically perfect coordinate system to mathematically define their product and make necessary calculations. Dimensions and tolerances originate from this DRF. It is an integral part of the CAD system.

Mathematics
Cartesian coordinate system

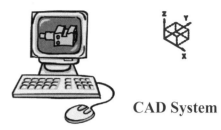

CAD System

The DRF consists of three mutually perpendicular datum planes and 3 mutually perpendicular datum axes that occur at the intersection of the datum planes and a datum point that occurs at the intersection of the three axes. This DRF is established on the part by engaging the datum features with simulators and establishing datums in an order of precedence.

Datum Planes

Datum Axes

Datum Point

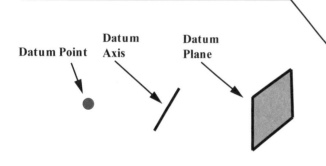

Datum Point   **Datum Axis**   **Datum Plane**

Datums are points, axes (lines), and planes or some combination of these components. They are established from theoretical datum feature simulators.

Theoretical datum feature simulator is the perfect inverse of a datum feature. (cylinders, widths etc.) They engage with datum features and establish datums. *Formerly called True Geometric Counterpart (TGC).*

Datum simulator

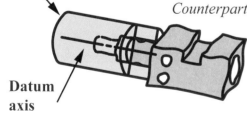

Datum axis

Datum plane

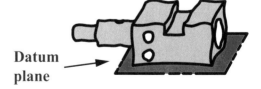

Datum features are the actual, physical features on the part. They are not perfect, they have variation.

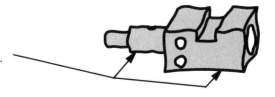

Simulated Datum Reference Frame used in Manufacturing

The **Simulated Datum Reference Frame** in manufacturing and verification is imbedded in the processing, tooling, and inspection equipment, such as a machine center, lathe and coordinate measuring machine etc. Since all physical equipment is not perfect, the simulated datum reference frame is not perfect, but it is of sufficient quality that it is used for measurement and verification in manufacturing and quality.

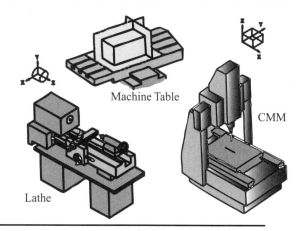

Machine Table

CMM

Lathe

The **Simulated Datum Reference Frame** used in manufacturing is a similar coordinate system used in design engineering. It also has datum planes, datum axes and datum points. This DRF is mated in an order of precedence to datum features through the physical datum simulators and establish simulated datums.

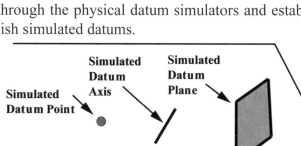

Simulated Datum Point

Simulated Datum Axis

Simulated Datum Plane

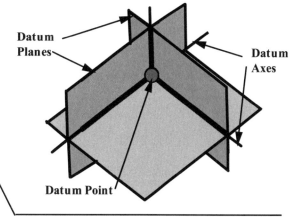

Datum Planes

Datum Axes

Datum Point

Simulated datums are points, axes (lines), and planes or some combination of these components that are established from physical datum feature simulators, such as a plane established from a surface plate or the axis of a collet or chuck. Simulated datums are not perfect but are of sufficient quality to be used in manufacturing and inspection.

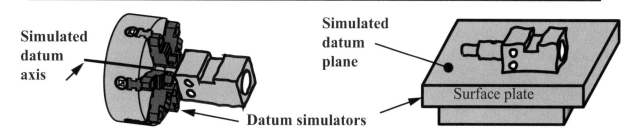

Simulated datum axis

Datum simulators

Simulated datum plane

Surface plate

Physical datum feature simulators are tooling and inspection equipment such as a collet, surface plate, gage blocks etc. These simulators establish simulated datums. Simulated datums are the practical representation of the theoretical datums used in design.

Datum features are the actual, physical features on the part. They are not perfect, they have variation.

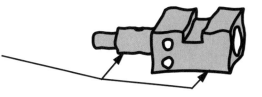

Theoretical Datum Feature Simulators and Datum Feature Controls

Datums are established from datum feature simulators. There are two types of datum feature simulators used in geometric tolerancing. They both have the same name, so the concept can be a bit confusing at first. The **Theoretical** datum feature simulators are used by design to define the product and are perfect and free of any variation or tolerance. The **Physical** datum feature feature simulators are used in tooling, manufacturing and verification, and shop tolerances will apply.

Datum Feature Simulator (Theoretical) *Formerly called True Geometric Counterpart (TGC) in the ASME Y14.5M-1994 standard.*

A theoretically perfect boundary such as a plane, cylinder, width etc. that is used to establish a datum from a specified datum feature. A (theoretical) datum feature simulator is perfect and exists only in theory. This datum feature simulator (theoretical) is defined in the ASME Y14.5-2009 standard and can be one of the following:

A maximum material boundary (MMB)
A least material boundary (LMB)
An actual mating envelope
A minimum material envelope
A tangent plane
A datum target
A mathematically defined contour

Unless otherwise specified, the theoretical datum feature simulator is the perfect inverse geometry of the datum feature. How the datum features mates with the datum feature simulators depends on the physical condition of the datum feature, any qualifying control, its order of precedence and the datum feature modifier specified. Datum feature modifiers are usually only applied to features that contain, or can be contained by, an actual mating envelope (a feature of size). In some cases, datum feature modifiers can be applied to datum features with established boundaries that constrain rotation about a higher precedence datum point or axis. See pages 9.10/9.14.

DATUM FEATURE CONTROLS

Geometric tolerances related to a datum reference frame do not take into account any variations in form, orientation, or location of the datum features. Datum features shall be controlled directly by applying geometric tolerances or indirectly as in a size tolerance applied to a primary datum feature. This makes it possible to calculate the datum feature simulator boundaries for each datum feature in a datum reference frame. Relationships between datum features to be considered are;

1. The qualification of the primary datum feature(s) form and/or the location between features in a pattern used to establish the primary datum.
2. The secondary datum feature(s) orientation and/or location as applicable, to the primary datum.
3. The tertiary datum feature(s) orientation and/or location as applicable to the primary and secondary datum features.

See pages 6.28 thru 6.32 for examples of how the primary datum is qualified, the secondary datum feature is related to the primary and the tertiary datum feature is related to the primary and secondary datum.

Datum feature simulator requirements for establishing a
Datum Reference Frame (DRF)

Referencing datum features in an order of precedence in a feature control frame establishes a series of datum feature simulators. The size, form, orientation, location, growth or movement of these simulators are based on the datum feature modifier specified according to the datum feature simulator requirements shown below.

Datums are established from these datum feature simulators. These datums in their referenced order of precedence, establish the datum reference frame. The datum feature simulator requirements shown below are the foundation principles for establishing a DRF from datum features referenced in a feature control frame. Know them well, as there are critical to your understanding of datum reference frame development.

Datum Feature Simulators

1. They shall have perfect form. They are perfectly round, cylindrical, straight, flat and with perfect geometry.

2. They have basic orientation relative to one another as referenced in a feature control frame. The orientation (tilt), like angularity, perpendicularity or parallelism, is perfect.

3. They have basic location relative to one another as referenced in a feature control frame, unless a translation modifier or moveable datum target symbol is specified.

4. They have movable location when the translation modifier or the movable datum target symbol is specified.

5. They expand, compress or otherwise progress in their order of precedence, to make maximum contact with the datum feature when specified at RMB.

6. They are fixed at the designated size, when specified at MMB or LMB.

A lower precedence datum feature can not override, constrain or eliminate any datums or degrees of freedom already established by any higher precedence datum feature(s). Each subsequent referenced datum feature in the feature control frame will constrain all the degrees of freedom that is possible within its ability.

Note: If it is necessary to override these foundational datum feature simulator requirements it may be accomplished with a customized datum reference frame shown later in unit 9.

Datum Feature Simulator (Physical) *Formerly called Datum Simulator in the ASME Y14.5M-1994 (R2004) standard*

A physical boundary, such as a surface plate, collet, chuck, gage block etc., is used to establish a simulated datum from a specified datum feature. These datum simulators are not perfect, but they are of such quality that their surfaces, center planes, and axes are used for measurement and verification in manufacturing and inspection.

The design engineer works in the theoretical CAD world, where everything is perfect. The datum reference frame, datums, datum simulators and basic dimensions are all perfect and free of any variation or tolerance. This "perfect concept" allows an engineer to define requirements and make necessary calculations based on a solid mathematical foundation. This CAD world also allows the engineer to convey this "perfect design intent" with allowable geometric tolerances, to manufacturing and inspection.

Manufacturing and inspection systems, produced parts, datum features, datum simulators and features have errors and imperfections. In practice, it is impossible to manufacture or measure from a theoretical datum reference frame, because perfect planes do not exist in the "real world." To accommodate this disparity in manufacturing and inspection, physical datum feature simulators are used to establish simulated datums.

The physical datum feature simulator is the bridge from the engineering theoretical mathematical world to the practical world of manufacturing and quality control. These physical datum feature simulators mimic the theoretical datum feature simulators as close as possible within reasonable gage and tool tolerances.

Error derived from the gage or inspection equipment, software etc., must be accounted for within the measurement values to insure parts still lie within engineering design tolerance limits. Physical datum simulators are not perfect and have associated gage and tool tolerances. Gage tolerances are usually 10% to 15% of part tolerances. There is additional information on tool and gage tolerances in ASME Y14.43-2003 (R2008).

All dimensions and engineering specifications are relative to the theoretical datums established from theoretical datum feature simulators. Theoretical datums do not take into account any tolerances or error in the inspection or processing equipment. Tolerances on inspection or processing equipment must fall within the engineering specifications and are not additive.

Note: Whenever the term datum feature simulator is used in this text or the ASME Y14.5-2009 standard, it refers to the datum feature simulator (theoretical) unless otherwise specified.

Planar datum features on the drawing

The detail drawing shows datum feature symbols applied to planar features. The datum feature symbol is attached to, or associated with, the surface of the feature. A datum feature simulator, which is the inverse of the datum feature, is mated with the datum feature. A datum plane is then derived from the simulator. The examples below show the sequence of events to establish a datum in theory, and a simulated datum in practice, from a datum feature.

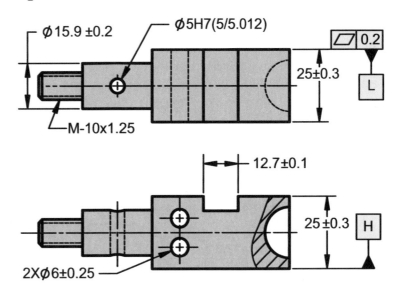

Establishing datums from planar datum features

Theory

Datum features H and L are mated with a theoretical perfect plane contacting the high points of the datum feature.

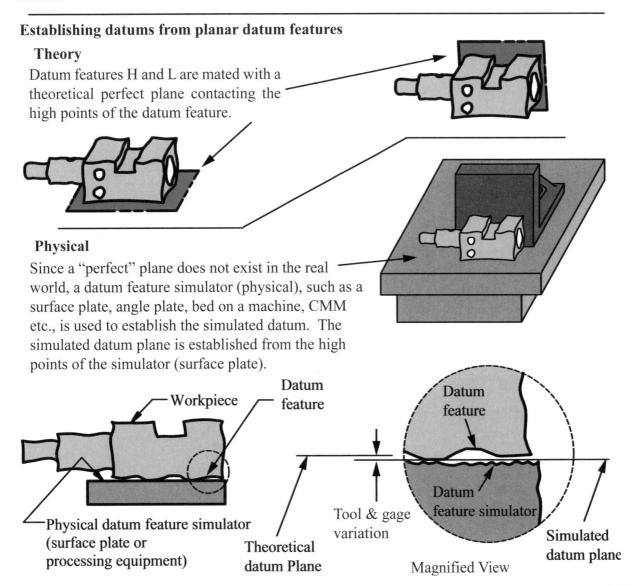

Physical

Since a "perfect" plane does not exist in the real world, a datum feature simulator (physical), such as a surface plate, angle plate, bed on a machine, CMM etc., is used to establish the simulated datum. The simulated datum plane is established from the high points of the simulator (surface plate).

Planar Surfaces as Primary Datum Features

The drawing below shows examples of the datum feature symbol applied to planar surfaces. These datum features establish a datum plane which contacts the high points of the surface.

The datum feature symbol may be attached to the surface or to an extension line of a surface. The datum feature symbol can also be placed on a feature control frame that is attached to the extension line as shown with datum feature L. The datum feature symbol can also be placed on a leader line that is directed to the surface or extension line as shown with datum feature N. If a datum plane is to be derived from multiple surfaces, the number of surfaces should be noted, as shown with datum feature K.

Planar datum features

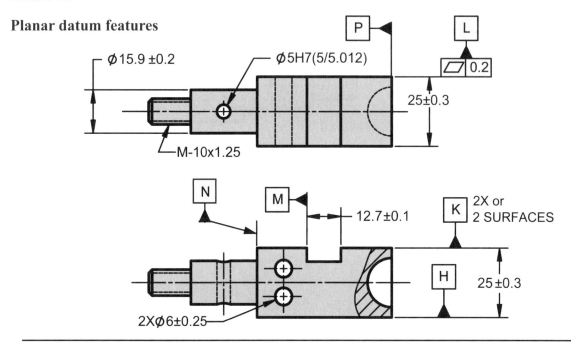

Datum planes established from planar datum features

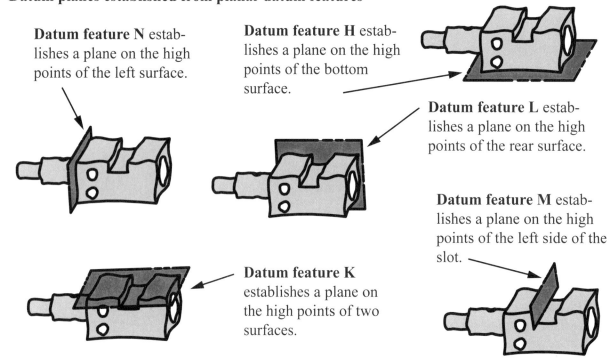

Datum feature N establishes a plane on the high points of the left surface.

Datum feature H establishes a plane on the high points of the bottom surface.

Datum feature L establishes a plane on the high points of the rear surface.

Datum feature M establishes a plane on the high points of the left side of the slot.

Datum feature K establishes a plane on the high points of two surfaces.

1. The datum reference frame (DRF) consists of points, lines and planes. Label these components on the illustration below.

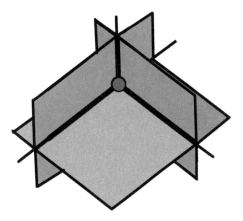

2. The datum reference frame will arrest six degrees of freedom, 3 (x, y, z) translations and 3 (u,v,w) rotations. Label these six degrees of freedom on the illustration below. Translation X has already been shown.

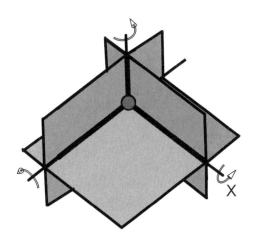

3. On the drawing and produced part below, label the following terms:

Datum feature symbol | Datum | Datum feature

Datum feature simulator | Simulated datum

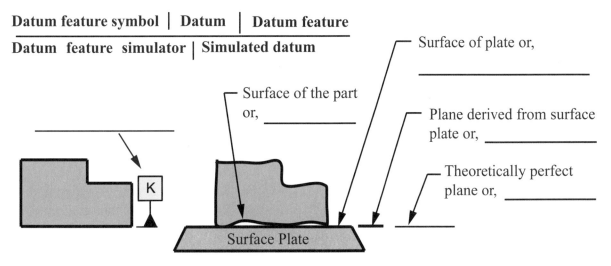

Surface of plate or, _____

Surface of the part or, _____

Plane derived from surface plate or, _____

Theoretically perfect plane or, _____

The detail drawing shows datum feature symbols applied to features of size that establish a datum axis, center plane or point. The datum feature symbol is never applied directly to the center line of a feature.

The datum feature symbol is attached to, or associated with the size dimension of a feature. A datum feature simulator, which is the inverse of the feature, which is mated with the feature and the datum axis, center plane or point (or combination of), is then derived from the datum feature simulator. The examples below and on the next two pages show the sequence of events to establish a datum in theory, and a simulated datum in practice, from a primary datum feature at Regardless of Material Boundary (RMB).

Datum features of size

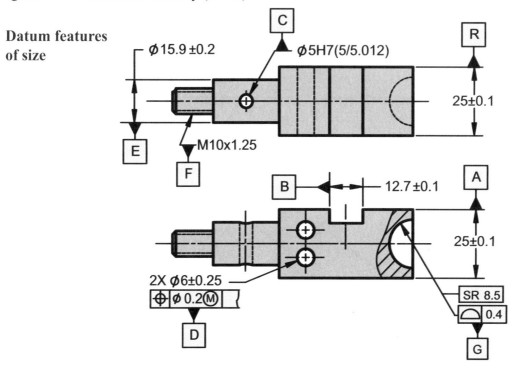

Establishing a datum axis from an external cylinder, datum feature E at RMB

Theory

Datum axis E is the axis of the theoretical datum feature simulator.

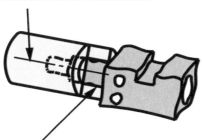

The cylindrical datum feature is mated with a theoretical datum feature simulator that is a circumscribing cylinder collapsing on the high points of the feature. The datum axis is the axis of the theoretical datum feature simulator.

Physical

Simulated datum axis E is the axis of a collet, chuck, CMM etc.

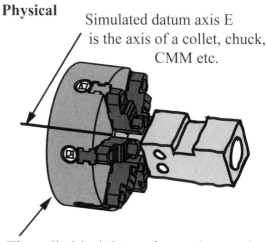

The cylindrical datum feature is mated with a physical datum feature simulator, like a collapsing collet. The simulated datum axis is the axis of the physical simulator or collet.

Establishing a datum center plane from an external width, datum feature A at RMB

Theory

Datum center plane A is the center plane of the datum feature simulator.

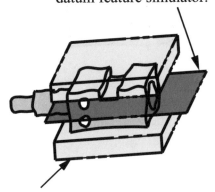

The external width datum feature is mated with a theoretical datum feature simulator, that is two parallel planes at minmum separation contacting the high points of the feature. The datum is the center plane of the datum feature simulator.

Physical

Simulated datum center plane A is the center of the vice jaws.

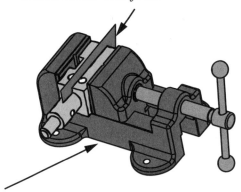

The external width datum feature is mated with a physical datum feature simulator, such as the jaws on a machinist vice or CMM, etc. The simulated datum is the center plane of this inspection or processing equipment.

Establishing a datum center plane from an internal width, datum feature B at RMB

Theory

Datum center plane B is the center plane of the datum feature simulator.

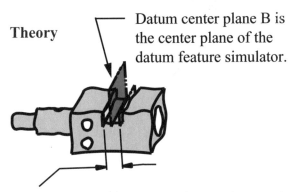

The internal width datum feature is mated with a theoretical datum feature simulator, that is two parallel planes at maximum separation contacting the high points of the feature. The datum is the center plane of the datum feature simulator.

Physical

Simulated datum center plane B is the center plane of the gage block.

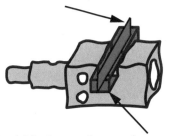

The internal width datum feature is mated with a physical datum feature simulator, such as a gage block, CMM, etc., contacting the high points of the feature. The simulated datum is the center plane of this inspection or processing equipment.

Note: All dimensions and engineering specifications are relative to the theoretical datums established from theoretical datum simulators. It does not take into account any tolerances or error in the inspection or processing equipment. Tolerances on inspection or processing equipment must fall within the engineering specifications and are not additive.

Establishing a datum axis from a hole, datum feature C at RMB

Theory

Datum axis C is the axis of the datum feature simulator.

Physical

Simulated datum axis C is the axis of a gage pin.

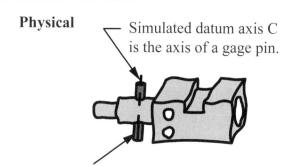

The hole is mated with a theoretical datum feature simulator that is an inscribing cylinder expanding to contact the high points of the feature. The datum is the axis of the datum feature simulator.

The hole is mated with a physical datum simulator, such as a gage pin, mandrel, CMM etc. The simulated datum is the axis established in this inspection or processing equipment.

Establishing datum axis and 2 planes from a pattern of two holes, datum feature D at RMB

Theory

Datum D, axis and two planes are established from a pattern of two holes.

Physical

Simulated datum feature D axis and two intersecting planes established from gage.

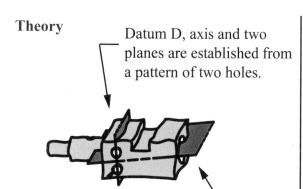

The two holes are mated with a theoretical datum feature simulator that is two inscribing cylinders fixed at basic location expanding simultaneously and making contact with the high points of the features. The datum axis and 2 planes are established from the datum feature simulator.

The two holes are mated with a physical datum feature simulator, like a hard or soft gage, CMM etc. The axis and two planes are established in this inspection or processing equipment.

Establishing a datum point from a spherical datum feature G at RMB

Datum point

Theory- Datum feature simulator is an expanding sphere
Physical- CMM, gage ball, etc.

Note: This SR is not technically a feature of size, but it does establish a boundary

Establishing a datum axis from the pitch dia of a thread, datum feature F at RMB

Datum axis

Theory- Datum feature simulator is a collapsing pitch cylinder
Physical- Thread gage, thread wires etc.

6.20

Datum Feature Symbol Placement is Important

The placement of the datum feature symbol is very important. Datum features M and K are datum features without size and establish datum planes that are derived from the high points of the planar surface of the datum feature. The datum feature symbol is placed on the extension line of the feature, clearly removed from the size dimension line.

Datum features B and R are datum features with size, and establish the center plane of the features as datums. The datum feature symbol is placed in line with the size dimension line.

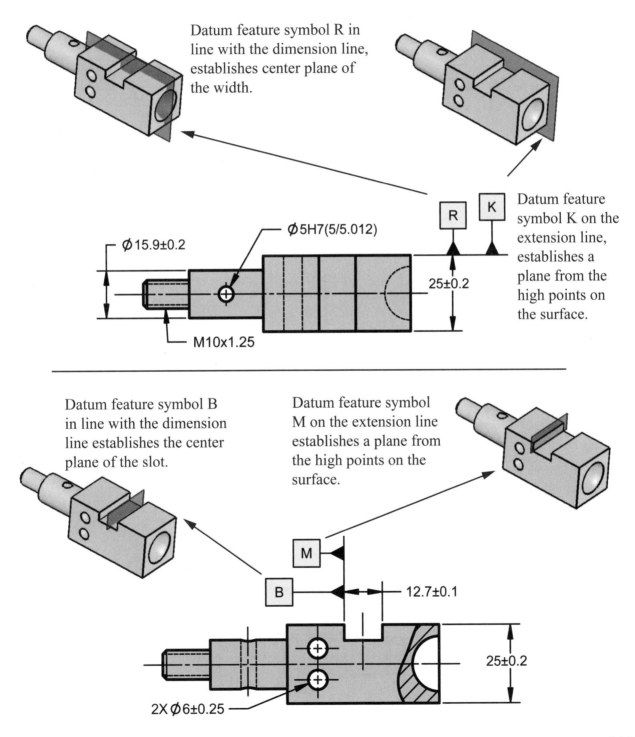

Datum feature symbol R in line with the dimension line, establishes center plane of the width.

Datum feature symbol K on the extension line, establishes a plane from the high points on the surface.

Datum feature symbol B in line with the dimension line establishes the center plane of the slot.

Datum feature symbol M on the extension line establishes a plane from the high points on the surface.

Partial Datum Features and Multiple Datum Features

If a feature or group of features are identified as a datum feature, the entire surface of the feature shall be used to establish the datum. In some cases, an engineer might find it desirable to identify only a portion of a surface as a datum feature. In these cases, a partial datum feature may be identified on the drawing.

Examples of partial datum features are shown below. A leader line terminating with a dot within the outline of a surface can be used to identify a partial datum within the outline of a surface. If necessary, the partial area is identified with basic dimensions and may be cross-hatched for clarity. Datum targets are another method to establish partial datum features. See datum target section.

Partial datum featues

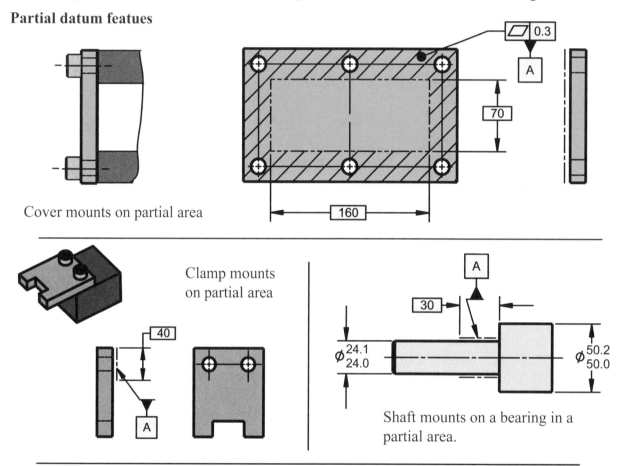

Cover mounts on partial area

Clamp mounts on partial area

Shaft mounts on a bearing in a partial area.

The shaft mounts on two bearing surfaces simultaneously. The datum feature letters are entered in the primary compartment of the feature control frame, separated by a dash. Both datum features together establish a single primary datum axis.

Multiple datum featues

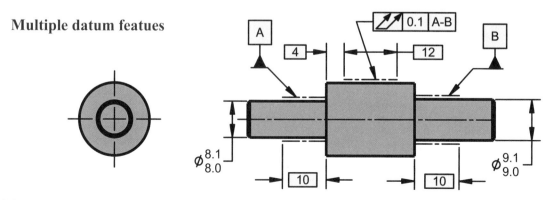

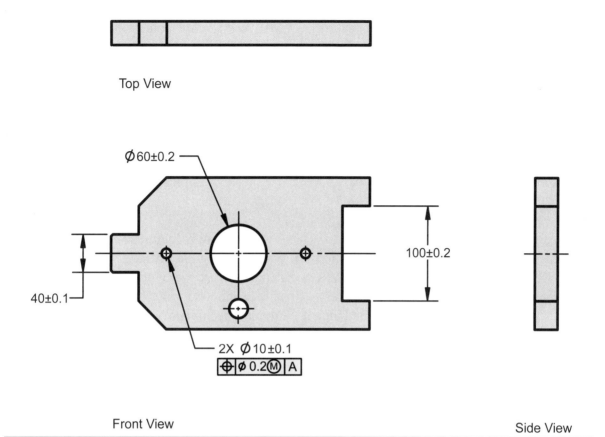

Top View

Front View   Side View

This exercise will provide practice on properly applying datum feature symbols. On the drawing above, apply datum feature symbols to create the following theoretical datums. If you need help, see application examples on pages 6.6 through 6.22.

1. Create datum plane A on the right hand face in the side view.

2. Create datum plane B on the bottom surface in the front view.

3. Create datum plane C on the top surface of the 40 mm tab.

4. Create datum center plane D in the center of the tab.

5. Create datum center plane E in the center of the 100 mm slot.

6. Create datum plane F on the left side of the slot.

7. Create a partial datum plane G on the top of the part in the front view. The length is 120 mm from the right side.

8. Create datum H, axis and two planes, through the two 10 mm holes.

9. Create datum axis J from the 60 mm hole.

10. Datum features can be placed in two categories - datum features of size and datum features without size. Place an "X" next to all the datum features of size.

A theoretical datum feature simulator is the mathematically perfect inverse of a datum feature. When referenced as a primary datum feature at RMB in a feature control frame, these datum features are mated with a collapsing/expanding inverse simulator. A datum, such as point, line, plane or some combination of these datums, are established from the datum feature simulators.

For example, a datum feature simulator for a hole is a perfect inscribing cylinder, and the datum feature simulator for a pin is a circumscribing cylinder. The datum feature simulator for a slot is two parallel planes at maximum separation, and the simulator for a tab is two parallel planes at minimum separation. The simulator for a planar surface is a plane. These are the usual or classic type features that can be found on a part. Many of these examples were shown on previous pages.

There are cases where more complex parts may have more complex features identified as a datum features, such as a conical surface, linear cross section or irregular surface. Since any feature on a part can be identified as a datum feature, the datum feature simulator, which is the inverse of the datum feature, can also have a variety of shapes such as a cone, linear cross section, or complex. Examples of datum features and their datum feature simulators are shown on the degrees of freedom table on page 6.26.

A primary datum feature may establish a single datum such as plane, axis, point or a combination of these datums, depending on the geometrical shape of the datum feature. The datum feature and the datum feature simulator are engaged according to the order of precedence and modified conditions applied to the datum feature in the feature control frame. The datum feature simulator restricts the movement of the feature and establishes the datum.

Degrees of Freedom Table - Primary Datum Feature at RMB

The Degrees of Freedom (DOF) table on the next page describes the theoretical succession of events, beginning with the identification of a datum feature on a drawing, its related primary simulator at RMB, the datum it establishes, and the degrees of freedom it constrains.

Collections of features may also be used to establish a single datum, but for simplicity, the Degrees of Freedom (DOF) table illustrates only single datum features.

The table has 5 columns.

The first column identifies the type of feature. Labeled a through g.

The second column illustrates how the datum feature symbol is applied on the drawing.

The third column identifies the datum feature.

The fourth column identifies the datum and the datum feature simulator.

The fifth column identifies the datum and the degrees of freedom it constrains.

a) A nominally flat planar feature establishes a plane that limits three degrees of freedom, one translation and two rotations.

b) A width of two opposed parallel surfaces establishes a center plane that limits three degrees of freedom, one translation and two rotations.

c) A cylindrical feature establishes an axis (line) that limits four degrees of freedom, two translations and two rotations.

d) A spherical feature establishes a center point that limits three translations.

e) A conical feature establishes an axis and a point that limits five degrees of freedom, three translations and two rotations.

f) A bounded linear extruded feature (as illustrated) establishes a plane and an axis that limits five degrees of freedom, two translations and three rotations.

g) A complex feature composed of a compound curve or collection of features establishes a plane, point and an axis that limits six degrees of freedom, three translations and three rotations.

The DOF table illustrates the datum feature, the datum, the datum feature simulator and the arrested degrees of freedom for primary datum features modified at RMB only. The degrees of freedom constrained for a lower precedence datum feature depends on its geometry and order of precedence as a secondary or tertiary datum feature.

A lower precedence datum feature can not override, constrain or eliminate any datums or degrees of freedom already established by any higher precedence datum features. Each subsequent referenced datum feature in the feature control frame will constrain all the degrees of freedom that is possible, within its ability.

Degrees of Freedom Table - Datum features referenced in a feature control frame as primary datum features at RMB.

Feature Type	On the Drawing	Datum Feature	Datum and Datum Feature Simulator	Datum and Constraining DOF
Planar (a)			Plane	1 Trans 2 Rotate
Width (b)			Plane	1 Trans 2 Rotate
Cylindrical (c)			Axis	2 Trans 2 Rotate
Spherical (d)			Point	3 Trans 0 Rotate
Conical (e)			Axis & Plane	3 Trans 2 Rotate
Linear Extrusion (f)			2 Planes	2 Trans 3 Rotate
Complex (g)			3 Planes	3 Trans 3 Rotate

Workshop Exercise 6.3 - Datum Feature Exercise

The part below has a variety of features that are labeled as datum features. Assume all are primary datum features and the datum features of size apply at RMB.

Next to the datum feature in the table below, identify if it establishes a point, axis, plane, or some combination of these datums. Also list the names of translational (x,y,z) and rotational (u,v,w) degrees of freedom constrained by the datum feature. See the isometric view to the right for the labeled degrees of freedom.

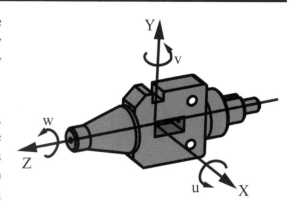

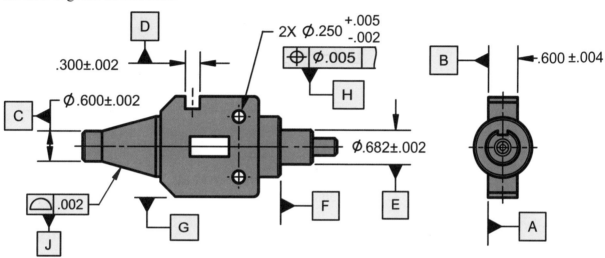

1. Point   **2.** Axis   **3.** Plane   **4.** Center Plane   **5.** Axis & 2 Planes   **6.** Axis & Plane

	Primary Datum Feature	Established Datum	Constrained Degrees of Freedom Translations (x,y,z)	Constrained Degrees of Freedom Rotations (u,v,w)
1.	A	3	x	v, w
2.	B			
3.	C-E			
4.	D			
5.	F			
6.	G			
7.	H			
8.	J			

Discussion: Some of the above datum features may be too unstable (short in length) to be a primary datum feature and constrain all of the listed degrees of freedom. Which ones are they?

The H, L, P - DRF on the drawing

The figures below illustrate the sequence of events to establish the H, L, P datum reference frame to constrain the 6 degrees of freedom on the part.

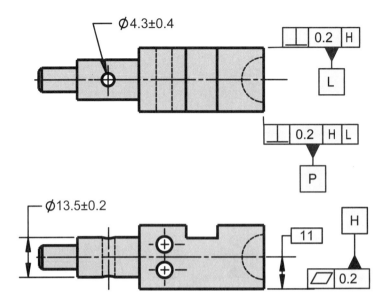

A lower precedence datum can not override the degrees of freedom already established by a previous datum. The qualification of the datum features with flatness and perpendicularity are necessary to relate the imperfect datum features to the perfect DRF.

Establishing the H, L, P - DRF on the part

① | ⊕ | ⌀ 0.4 Ⓜ | H | |

Datum feature H is entered in the first compartment as the primary datum feature. It establishes a plane and constrains 1 translation and 2 rotations. The surface is qualified with a flatness tolerance.

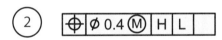

3 point min contact to establish primary plane

2 point min contact to establish secondary datum plane

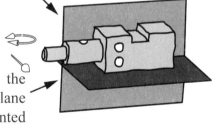

② | ⊕ | ⌀ 0.4 Ⓜ | H | L |

Datum feature L is entered in the second compartment as the secondary datum feature. The feature surface establishes a plane and constrains 1 translation and 1 rotation. The feature is oriented with a perpendicularity tolerance.

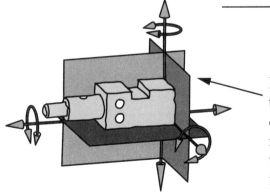

③ | ⊕ | ⌀ 0.4 Ⓜ | H | L | P |

Datum feature P is entered in the third compartment as the tertiary datum feature. It establishes a plane and constrains 1 translation. The datum feature is oriented relative to datums H and L with a perpendicularity tolerance. The DRF is complete and all 6 degrees of freedom are constrained on the part.

The N, E, H - DRF on the drawing

The figures below illustrate the sequence of events to establish the N, E, H datum reference frame to constrain the 6 degrees of freedom on the part.

| ⊕ | ⌀ 0.4 Ⓜ | N | E | H |

A lower precedence datum can not override the degrees of freedom already established by a previous datum. The qualification of the datum features with flatness, perpendicularity and profile are necessary to relate the imperfect datum features to the perfect DRF.

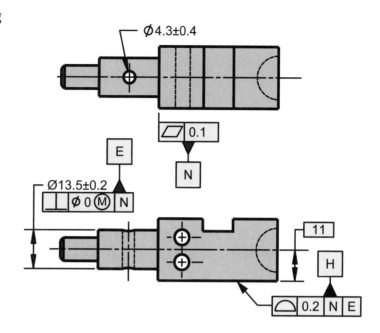

Establishing the N, E, H - DRF on the part

(1) | ⊕ | ⌀ 0.4 Ⓜ | N | | |

Datum feature N is entered in the first compartment as the primary datum feature. It establishes a plane and constrains 1 translation and 2 rotations. The surface is qualified with a flatness tolerance.

3 point min contact to establish primary plane

(2) | ⊕ | ⌀ 0.4 Ⓜ | N | E |

Datum feature E is entered in the second compartment as the secondary datum feature. The feature establishes an axis and constrains 2 translations. The intersection of the axis and plane establish a point. The feature is oriented with a perpendicularity tolerance.

3 point min contact to establish secondary datum axis

(3) | ⊕ | ⌀ 0.4 Ⓜ | N | E | H |

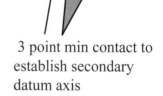

Datum feature H is entered in the third compartment as the tertiary datum feature. It establishes a plane and constrains 1 rotation. The datum feature is located relative to datums N and E with a profile tolerance. The DRF is complete and all 6 degrees of freedom are constrained on the part.

The E, N, S - DRF on the drawing

The figures below illustrate the sequence of events to establish the E, N, S datum reference frame to constrain the 6 degrees of freedom on the part.

⊕ | Ø 0.4 Ⓜ | E | N | S

A lower precedence datum can not override the degrees of freedom already established by a previous datum. The qualification of the datum features with size, perpendicularity and position are necessary to relate the imperfect datum features to the perfect DRF.

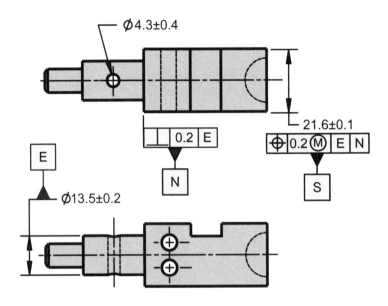

Establishing the E, N, S - DRF on the part

① ⊕ | Ø 0.4 Ⓜ | E | |

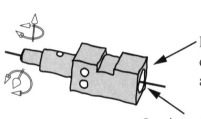

Datum feature E is entered in the first compartment as the primary datum feature. It establishes an axis and constrains 2 translations and 2 rotations. The diameter is qualified with a size tolerance.

5 point min contact to establish an axis

---

② ⊕ | Ø 0.4 Ⓜ | E | N |

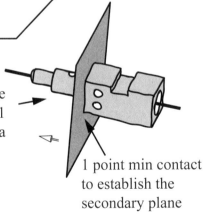

Datum feature N is entered in the second compartment as the secondary datum feature. It establishes a plane and constrains 1 translation. The intersection of the axis and plane establish a point. The feature is oriented with a perpendicularity tolerance.

1 point min contact to establish the secondary plane

③ ⊕ | Ø 0.4 Ⓜ | E | N | S

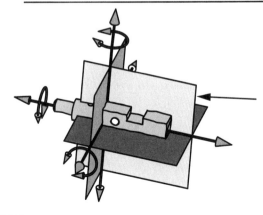

Datum feature S is entered in the third compartment as the tertiary datum feature. The feature establishes a center plane and constrains 1 rotation. The datum feature is located relative to datums E and S with a position tolerance. The DRF is complete and all 6 degrees of freedom are constrained on the part.

The H, N, E - DRF on the drawing

The figures below illustrate the sequence of events to establish the H, N, E datum reference frame to constrain the 6 degrees of freedom on the part.

| ⊕ | ⌀ 0.4 Ⓜ | H | N | E |

A lower precedence datum can not override the degrees of freedom already established by a previous datum. The qualification of the datum features with flatness, perpendicularity and position are necessary to relate the imperfect datum features to the perfect DRF.

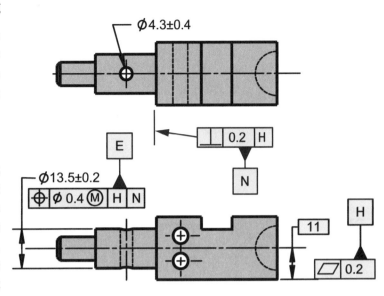

Establishing the H, N, E datum reference frame on the part

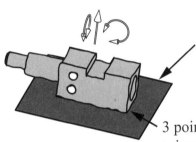

(1) | ⊕ | ⌀ 0.4 Ⓜ | H | | |

Datum feature H is entered in the first compartment as the primary datum feature. It establishes a plane and constrains 1 translation and 2 rotations. The surface is qualified with a flatness tolerance.

3 point min contact to establish primary plane

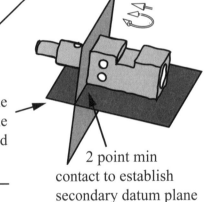

---

(2) | ⊕ | ⌀ 0.4 Ⓜ | H | N |

Datum feature N is entered in the second compartment as the secondary datum feature. The feature surface establishes a plane and arrests 1 translation and 1 rotation. The feature is oriented with a perpendicularity tolerance.

2 point min contact to establish secondary datum plane

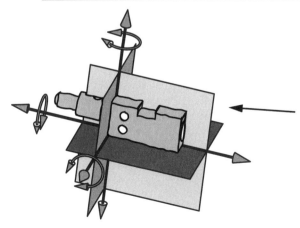

(3) | ⊕ | ⌀ 0.4 Ⓜ | H | N | E |

Datum feature E is entered in the third compartment as the tertiary datum feature. The feature axis establishes a plane and constrains 1 translation. The datum feature is located relative to datums H and N with a position tolerance. The DRF is complete and all 6 degrees of freedom are constrained on the part.

6.31

Datum Feature Precedence and Modifiers

The order of datum features referenced in a feature control frame are selected based on functional assembly requirements. In the three examples below, the feature control frame for the four holes represents the established DRF for the mounting condition shown.

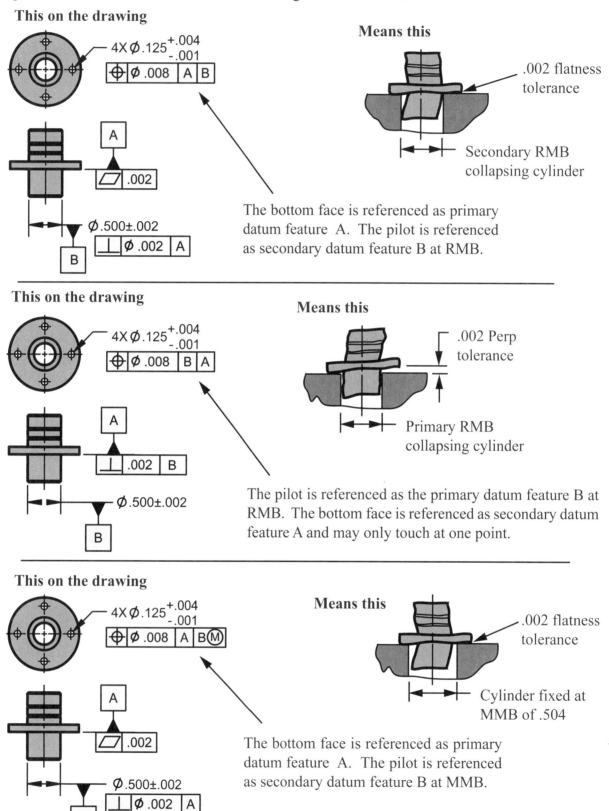

This on the drawing

$4X \phi.125^{+.004}_{-.001}$

Means this

.002 flatness tolerance

Secondary RMB collapsing cylinder

The bottom face is referenced as primary datum feature A. The pilot is referenced as secondary datum feature B at RMB.

This on the drawing

$4X \phi.125^{+.004}_{-.001}$

Means this

.002 Perp tolerance

Primary RMB collapsing cylinder

The pilot is referenced as the primary datum feature B at RMB. The bottom face is referenced as secondary datum feature A and may only touch at one point.

This on the drawing

$4X \phi.125^{+.004}_{-.001}$

Means this

.002 flatness tolerance

Cylinder fixed at MMB of .504

The bottom face is referenced as primary datum feature A. The pilot is referenced as secondary datum feature B at MMB.

Constrained Degrees of Freedom Example

Datum features referenced in a feature control frame establish a datum reference frame based on an order of precedence. These datum features constrain the six degrees of freedom on a part.

See the datum feature simulator requirements on page 6.13. "A lower precedence datum feature can not override, constrain or eliminate any datums or degrees of freedom already established by any higher precedence datum features. Each subsequent referenced datum feature in the feature control frame will constrain all the degrees of freedom that is possible within its ability."

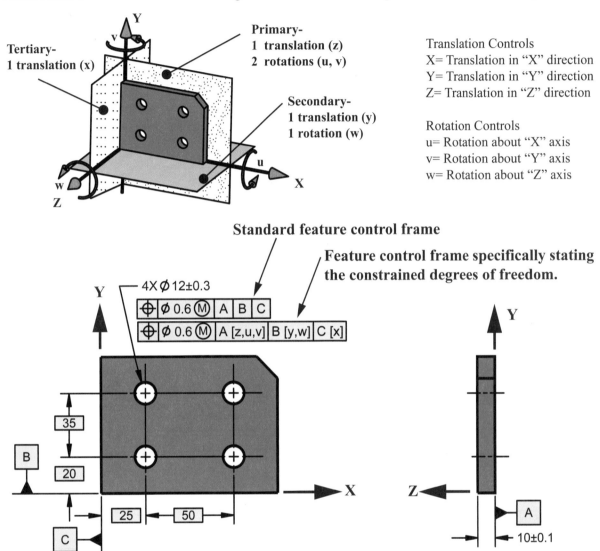

The mating of the datum features to the datum reference frame can be expressed mathematically as shown by the lower feature control frame for the 4 holes. The lower feature control frame is an example that explicitly states the degrees of freedom constrained by each datum feature shown in the upper feature control frame.

In a standard feature control frame it is not necessary to state the degrees of freedom constrained, as this default mating condition is clearly defined by the simulator requirements. This figure is only used as an example to show mathematically how the datum reference frame is created by the datum features in an established order of precedence. If it is necessary to override the default mating conditions then a customized datum reference frame can be used. See pages 9.18/9.20

Workshop Exercise 6.4 - Datum Reference Frame Exercise

The table below has a variety of feature control frames that create different datum reference frames. Next to each feature control frame, identify the degrees of freedom (x,y,z,u,v,w) constrained by each referenced datum feature.

Remember, a lower precedence datum feature can not constrain any degrees of freedom that a higher precedence datum feature has already constrained. See the isometric view to the right for the labeled degrees of freedom.

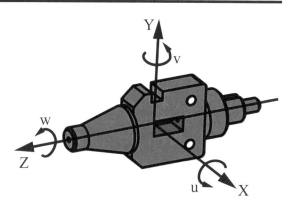

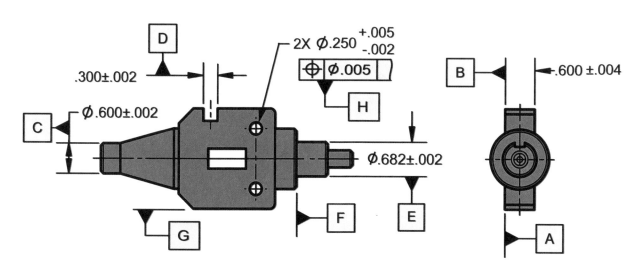

This datum reference frame	Constrained Degrees of Freedom (x,y,z,u,v,w)		
	Primary	Secondary	Tertiary
⊕ ⌀.005 A G D 1.			
⊕ ⌀.005 E F B 2.			
⊕ ⌀.005 F E B 3.			
⊕ ⌀.005 C-E D A 4.			
⊕ ⌀.005 A H 5.			

Unit 7

Datum Feature Modifiers

Datum Feature Modifiers - MMB, LMB and RMB
Calculating Boundaries for Datum Features at MMB
Workshop Exercise 7.1 - Calculating Datum Feature Boundaries
Development of a DRF at RMB
Development of a DRF at MMB
Comparison Between Datum Feature Modifiers RMB and MMB
Simultaneous Requirement Implied as Default
Separate Requirement Must be Specified
Datum Features Referenced Individually
Secondary Datum Feature at MMB - Paper Gage
Tertiary Datum Feature at MMB - Paper Gage
Workshop Exercise 7.2 - Calculating Allowable Datum Feature Shift

Datum Feature Modifiers - MMB, LMB and RMB

The drawing below shows a feature control frame referencing datum features of size. The process to establish a datum was explained in detail earlier in Unit 6. A datum feature simulator, which is the inverse of the feature, is mated with the datum feature. The datum axis, plane or point (or combination of) is then established from the datum feature simulator.

Since the datum feature has size (it can get larger and smaller), information is necessary on the size condition of the datum feature to which the datum feature reference applies. The modified condition of the datum feature (MMB, LMB, RMB) defines the size or condition of the datum feature simulator.

Datum feature modifiers applied to datum features of size

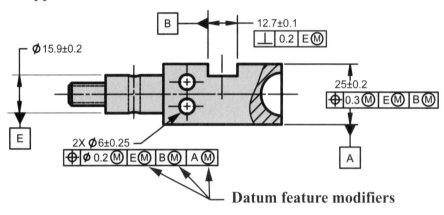

Datum feature modifiers

Datum feature modifiers define the size or condition of the datum feature simulator

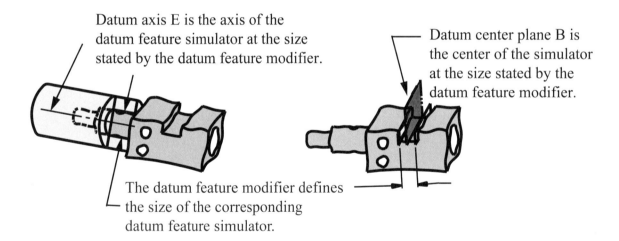

Datum axis E is the axis of the datum feature simulator at the size stated by the datum feature modifier.

The datum feature modifier defines the size of the corresponding datum feature simulator.

Datum center plane B is the center of the simulator at the size stated by the datum feature modifier.

Datum feature modifiers are Maximum Material Boundary (MMB), Least Material Boundary (LMB) and Regardless of Material Boundary (RMB). By default, all datum features apply at RMB. If it is desired that the datum feature apply at MMB or LMB, it must be stated in the feature control frame with the symbols circle M, or circle L, following the datum feature reference. The datum feature modifiers identify the engagement requirements of the simulators with the datum feature.

Specifying Datum Features at RMB. Where a datum feature is referenced at RMB in a feature control frame, the datum feature simulator geometry originates at the MMB and progresses proportionally through the tolerance zone to make maximum possible contact with the extremities of the datum feature or collection of features. If another fitting routine is required, it shall be stated on the drawing.

As a practical example, a machine element that is variable (such as a chuck, mandrel, vise, or centering device) is used to simulate a datum feature simulator of the feature and to establish the simulated datum.

Specifying Datum Features at MMB. Where MMB is applied to a datum feature referenced in a feature control frame, it establishes the datum feature simulator at the MMB boundary. The MMB boundary is determined by its collective effects of size and any applicable geometric tolerances relative to any higher precedence datums.

As a practical example, where a datum feature is applied on an MMB basis, machine and gaging elements in the processing equipment that remain constant (a fixed size gage pin) are used to simulate a theoretical datum feature simulator and to establish the simulated datum.

Specifying Datum Features at LMB. Where LMB is applied to a datum feature referenced in a feature control frame, it establishes the datum feature simulator at the LMB boundary. The LMB boundary is determined by its collective effects of size, and any applicable geometric tolerances relative to any higher precedence datums.

As a practical example, where a datum feature is applied on an LMB basis, machine and gaging elements in the processing equipment are very difficult to use because the LMB boundary is within the material. Often, CMM software is necessary to simulate the LMB boundary of a theoretical datum feature simulator that is used to establish the simulated datum.

In cases where the datum feature boundary needs clarification, or another boundary is desired, the size of the datum feature boundary may be included inside the feature control frame in brackets following the modifier. The term BASIC or abbreviation BSC may also be used in the feature control frame when it is necessary to establish the simulator at the basic location.

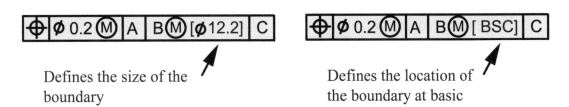

Defines the size of the
boundary

Defines the location of
the boundary at basic

Note: The terms MMB, LMB and RMB are new to the ASME Y14.5-2009 standard and replace the RFS, MMC and LMC datum feature modifiers used in previous editions of the standard. These terms were replaced because their meaning did not clearly represent the boundaries established by the collective effects of size, order of precedence and geometric tolerances.

Development of a Datum Reference Frame at RMB

Theoretical sequence of events in the development of a DRF, RMB on the alignment plate.

The initial establishment of a DRF on a part is done by the design engineer. The assembly is examined to determine the functional relationships between the parts. The design engineer will select the major mounting features of the part and establish them as the datum features. They are selected in an order of precedence, by placing them in a feature control frame, to establish a datum reference frame. The datum reference frame will arrest the six degrees of freedom on the part.

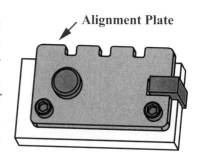

Alignment Plate

Alignment Plate Assembly

There are leading features, and following features on a part. The leaders are selected as the datum features to establish the DRF. The plane surface, the pin and the key, locate the alignment plate in the assembly. The features that follow are the two clearance holes and the grooved top surface.

Notice how the part mates in the assembly.

First, and most importantly, the part is bolted down tight to the back face. This planar face establishes the initial orientation of the part. This back face is identified as the primary datum feature A in the feature control frame. It constrains one translation and two degrees of rotation.

Second, the part mates to the assembly by the large hole fitting over the pin. The bolts do not locate the part, but keep the part clamped down to the face. Location up and down and fore and aft is achieved by the large hole fitting over the pin. The hole is a four way locator. The hole is identified on the drawing as datum feature B at RMB. It constrains two translations.

Third, the part mates to the assembly by the slot fitting over the key. The slot eliminates the rotation of the part. The slot is identified on the drawing as datum feature C at RMB. It constrains the final rotation. The feature control frames for the two holes and the slotted surface identify the DRF established in order, from datum features A, B and C.

Datum feature selection represents mounting conditions in the assembly

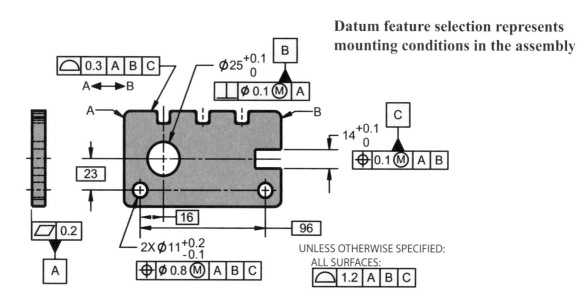

7.4

Mating perfect geometry at RMB with the imperfect geometry of the part.

In order to establish a DRF, the datum feature simulators are set up in accordance with the datum feature simulator requirements shown earlier in the datum section, page 6.13.

1. The simulator for primary datum feature A is a planar surface and establishes a datum plane that contacts the high points of the feature. This plane sets the orientation.

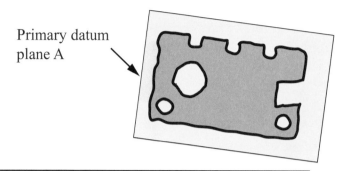

Primary datum plane A

2. The simulator for secondary datum feature B at RMB is an inscribed cylinder expanded to make maximum possible contact with the feature. Since B is a secondary datum, the simulator is oriented 90° basic to datum plane A, according to the simulator requirements. An axis is established from the center of the simulator. An origin point is established where this axis intersects datum plane A.

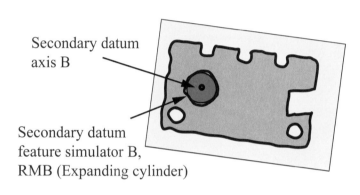

Secondary datum axis B

Secondary datum feature simulator B, RMB (Expanding cylinder)

3. The simulator for datum feature C at RMB is an inscribed width expanded about basic, to make maximum possible contact with the highest points on the surface. Since C is a tertiary datum feature, the block is oriented 90° basic to datum plane A and basically located and oriented to datum axis B. A center plane is established by the block that intersects the axis B and origin point.

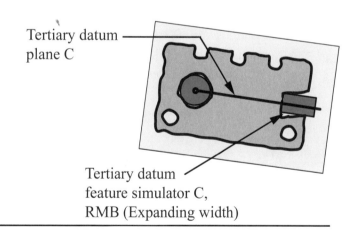

Tertiary datum plane C

Tertiary datum feature simulator C, RMB (Expanding width)

4. A datum reference frame is established from which all dimensions originate. This procedure will constrain the part six degrees of freedom and make it mathematically defined.

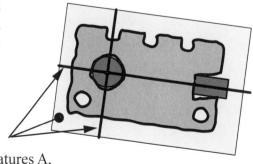

Datum reference frame established by datum features A, B at RMB and C at RMB

7.5

The theoretical procedure of establishing a DRF on a part begins with datum feature simulators, (which are the inverse of the datum feature) engaging with the datum features, shown in their order of precedence in the rear compartment of the feature control frame. The simulators establish datums which are a combination of planes, axes and/or points. The datums in their specified order of precedence establish a DRF. This DRF mated to the part will constrain the six degrees of freedom and make it clearly defined. All other features on the part are located and oriented to this DRF.

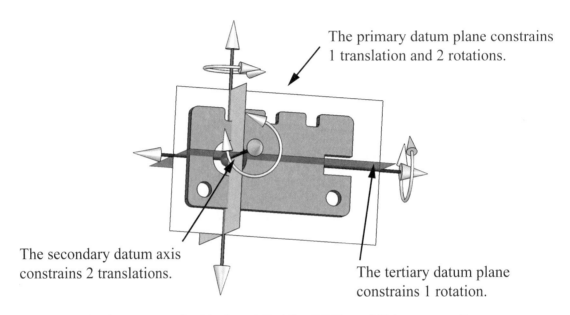

The primary datum plane constrains 1 translation and 2 rotations.

The secondary datum axis constrains 2 translations.

The tertiary datum plane constrains 1 rotation.

The alignment plate is shown mated with the DRF. The DRF establishes a coordinate system on the part that constrains the six degrees of freedom. Features on the part are related to this DRF by referencing the datum features in the stated order of precedence.

Development of a DRF at MMB

The drawing below is an example of the alignment plate shown earlier in this text. In the preceding example of the alignment plate, a datum reference frame was established with the datum features modified at RMB. In this example, the DRF is established in an identical manner except the datum features are modifed at Maximum Material Boundary (MMB). The object is to allow the reader to compare the two examples to help understand the difference between the datum feature modifier conditions. The DRF is established from datum feature simulators in accordance with the datum feature simulator requirements shown earlier in the datum section, page 6.13.

Alignment Plate with datum features modified at MMB

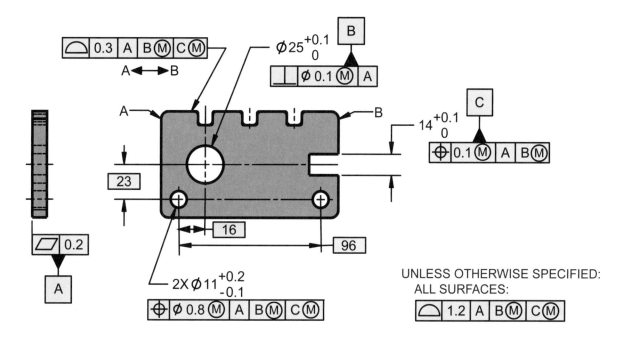

The datum reference frame establishment at RMB and MMB is identical except for the size of the simulators. RMB datum feature modifiers require the datum feature simulators to collapse/expand, making maximum possible contact to center the datum features according to their order of precedence. MMB datum feature modifiers require the datum feature simulators to be fixed in size at their MMB. The MMB for datum feature B is 24.9 (25 MMC - 0.1 perp) considering the size and perpendicularity relative to datum feature A. The MMB for datum feature C is 13.9 (14 - 0.1), considering the size and position to datum feature A and B.

In the engagement process with the datum feature simulators, if datum features are produced at their MMB, they will be centered by their respective datum feature simulators. However, if the datum features depart from their MMB, they will "rattle around" or be free to shift or displace within the confines of the datum feature simulators at MMB. The amount of shift is dependent on the amount of departure.

If the design engineer determines that this shifting and adjustment is acceptable in the assembly, then the MMB modifier is applied to the datum features. If the design engineer determines that the part will not be adjustable in the assembly, as in a press or interference fit, the design engineer will leave it implied to apply at RMB for the datum features.

The figures below show the DRF established with datum feature simulators fixed at the MMB. **The DRF is established from the datum feature simulators and not the datum features.** All dimensions originate from the DRF. The 4 figures below illustrate the principle of datum feature shift. The datum feature simulator for the hole is set at a MMB diameter of 24.9 and the datum feature simulator for the slot is set at a MMB width of 13.9. As the datum feature departs from its MMB, the datum feature is allowed to shift or displace within the confines of the MMB datum feature simulator. The amount the datum feature displaces is dependent on the datum features departure from MMB.

In all the figures below, the datum features have departed from their MMB and are displaced or shifted within the confines of the datum feature simulator. As you can see from the figures, the datum feature displacement can not be just added to the position tolerance for the holes or profile tolerance. If the part were checked with a functional gage, the datum feature displacement would be factored in automatically by adjusting the part and shifting the datum features in the simulators to meet the requirements for the remaining toleranced features. Factoring in the datum feature shift in an open set up inspection or CMM is more difficult. Paper gaging, holding fixtures or more sophisticated CMM software will also assist in the evaluation.

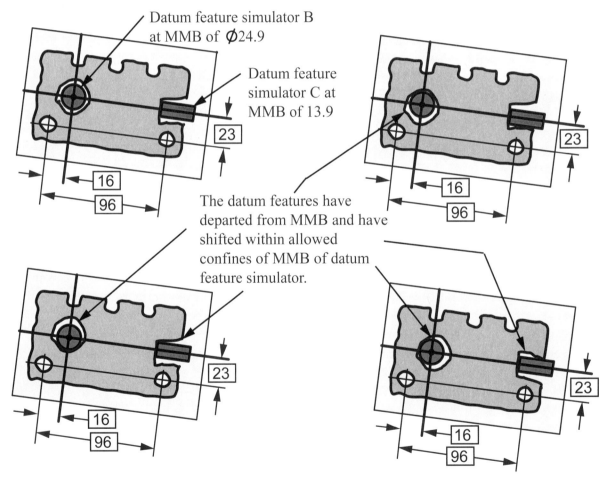

The 4 figures above represent 4 of many possible datum feature displacement possibilities. The datum features have departed from MMB and are allowed to shift or displace within the confines of the maximum material boundary defined by the datum feature simulator. The dimensions originate from the datum reference frame established from the simulators at MMB.

Comparison Between Datum Feature Modifiers RMB and MMB

The application of the datum feature modifiers MMB, LMB and RMB will have an effect on how the DRF is established relative to the datum features. The difference between RMB and MMB applied to the datum feature are explained on the alignment plate graphics below.

RMB applied/implied to a datum feature requires the datum feature simulator to expand, collapse or progress until it makes maximum contact with the datum feature. The DRF is established from the datum feature simulators in accordance with the datum feature simulator requirements defined on page 6.13.

Implied RMB

⊕ | Ø 0.8 Ⓜ | A | B | C

Datum reference frame established by datum feature simulators at RMB.

MMB applied to a datum feature requires the datum feature simulator to be fixed in size at the MMB boundary. As the datum features depart from MMB they are allowed to shift or displace within the confines of the maximum material boundary defined by the datum feature simulator. The dimensions originate from the datum reference frame established from the datum simulators, not the datum features.

Specified MMB

⊕ | Ø 0.8 Ⓜ | A | B Ⓜ | C Ⓜ

Datum feature simulator B fixed in size at MMB of Ø24.9

Datum feature simulator C fixed in size at MMB of 13.9

23

Datum reference frame established by datum feature simulators fixed in size at MMB.

16

96

Verification of MMB can be easily accomplished with the use of a functional gage but is often more challenging using a CMM. Conversely, the verification of RMB can be easily accomplished with the use of a CMM, but often more challenging using a functional gage. The datum feature modifier MMB is typically used in clearance conditions or when the part can be adjusted in the assembly to an optimum location. If adjustment of the part in the assembly is not possible or undesirable, and where centering is important, the RMB datum feature modifier should be applied.

Calculating Boundaries for Datum Features at MMB

An analysis of geometric tolerances applied to a datum feature is necessary to determine the size of a datum feature's boundary and size of its corresponding simulator. A feature or pattern of features of size serving as a datum feature may have several MMB. These include the MMC of a datum feature and the collective effects of geometric tolerances, while respecting datum feature precedence. Since a feature can have several MMB, it is important to calculate the correct MMB for each datum feature reference. The guidelines below will insure that you calculate the correct MMB.

For an internal datum feature of size, it is the largest MMB that the datum feature(s) of size will contain while respecting the datum feature precedence. (MMC minus relationship to datums)
For an external feature of size, it is the smallest MMB that will contain the datum feature(s) of size while respecting the datum feature precedence. (MMC plus relationship to datums)

The 9mm holes on the drawing below are shown with 4 different feature control frames, all with different datum references. The correct calculated MMB for the datum feature is shown to aid the user in calculating the correct MMB for each DRF.

Calculating Maximum Material Boundary (MMB)

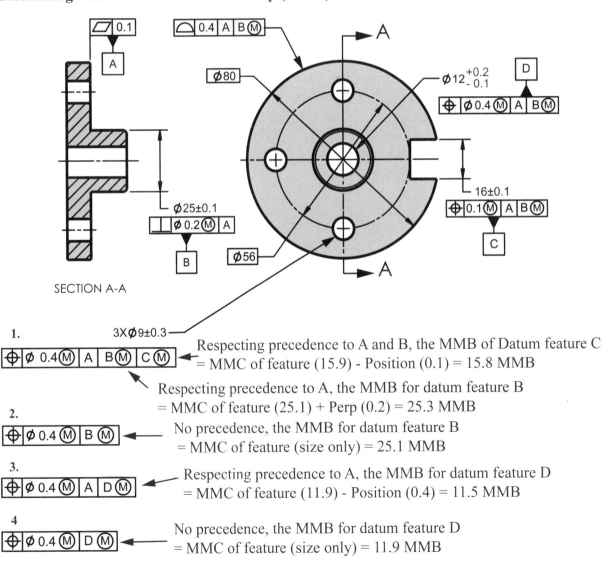

1.
Respecting precedence to A and B, the MMB of Datum feature C
= MMC of feature (15.9) - Position (0.1) = 15.8 MMB

Respecting precedence to A, the MMB for datum feature B
= MMC of feature (25.1) + Perp (0.2) = 25.3 MMB

2.
No precedence, the MMB for datum feature B
= MMC of feature (size only) = 25.1 MMB

3.
Respecting precedence to A, the MMB for datum feature D
= MMC of feature (11.9) - Position (0.4) = 11.5 MMB

4
No precedence, the MMB for datum feature D
= MMC of feature (size only) = 11.9 MMB

On the drawing below, there are four feature control frames for the 9mm holes. Each feature contol frame has a different datum feature reference. Using the information on the drawing, calculate the MMB boundary for the secondary and tertiary datum features, and enter it in the spaces below. Show your work.

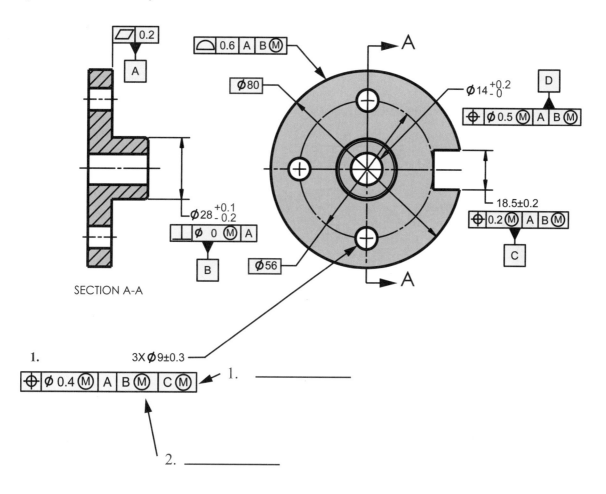

SECTION A-A

1.   3X Ø9±0.3

1. _____

2. _____

2.

3. _____

3.

4. _____

4

⊕ | Ø 0.4 Ⓜ | D Ⓜ

5. _____

Simultaneous Requirement is Implied as Default

In the part below, all the features are related with basic dimensions using the same datum feature reference, in the same order of precedence, with the same datum feature modifiers. This constitutes a single pattern and all the features must rotate together as a group simultaneously.

The groups of features must also shift or displace simultaneously as datum feature B departs from MMB. In this case, since all features are considered a single pattern, another datum to control the rotation or shift of the features to each other is not necessary. This implied simultaneous requirement applies to the two position specifications, as well as the profile specification.

Simultaneous requirement is implied by default

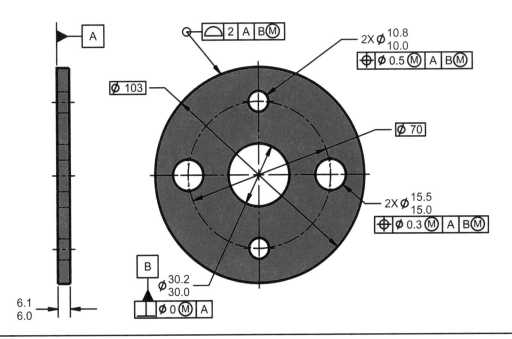

This sample functional gage is shown to illustrate the concept of the implied simultaneous requirement. It is designed to check all the features at once. This will not let the two groups of holes and the profile specification to rotate or shift relative to one another, more than the allowable position and profile tolerances.

In a rotational surface plate set-up or CMM check all the features must be balanced to each other and checked as if they were one single pattern.

Hole location and profile specifications are implied simultaneous and may not rotate or shift to each other.

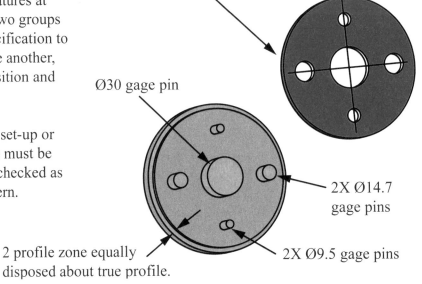

Ø30 gage pin

2X Ø14.7 gage pins

2 profile zone equally disposed about true profile.

2X Ø9.5 gage pins

7.12

Separate Requirement Must be Specified

In the part below, all the features are related with basic dimensions using the same datum features in the same order of precedence with the same datum feature modifiers. By default, this constitutes a single pattern and all the features must rotate and shift together as a group simultaneously.

In order to override this simultaneous requirement, the notation SEP REQT (Separate Requirement) is placed under the feature control frames. This constitutes three separate patterns and all the features may rotate, and/or shift independently to each other as datum feature B departs from MMB. If orientation or datum feature shift is allowed between the features, it must be specified.

Separate requirement must be stated

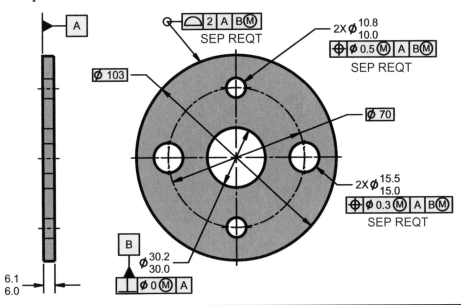

These sample functional gages are shown to illustrate the concept of the specified separate requirement. There are three separate gages to check each individual requirement. This allows the two groups of holes and the profile specification to rotate or shift to one another as they are now separate requirements.

In a rotational surface plate set-up or CMM check, all the features may be checked individually for conformance without regard to the other features labeled SEP REQT.

Hole location and profile specifications are not related and may rotate or shift to each other.

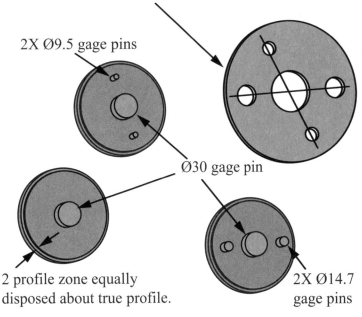

2X Ø9.5 gage pins

Ø30 gage pin

2 profile zone equally disposed about true profile.

2X Ø14.7 gage pins

Secondary Datum Feature at MMB - Paper Gage Calculations

This problem was shown earlier on page 4.13 with the datum feature at RMB. This example is intended to illustrate the calculations necessary to factor in tolerance gained from MMB datum feature modifiers. The produced part has been set up in inspection by "zeroing" on datum features A, B and C. Inspection data has been collected from the workpiece and the results are shown in the table below. It appears from the data collected that the location of the 3 holes are out of position. Notice however that datum features B and C are modified at MMB. Additional tolerance may be available from the datum features.

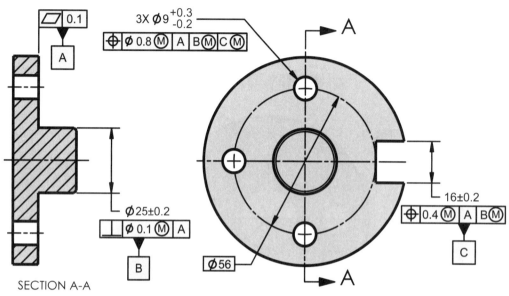

SECTION A-A

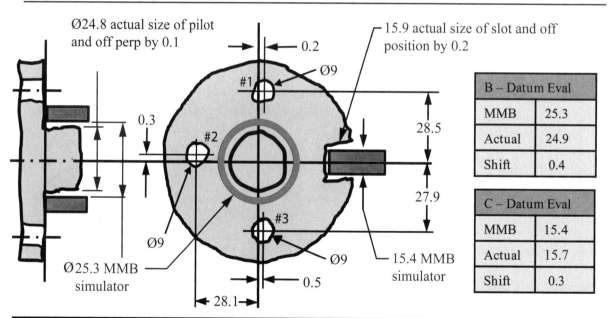

Ø24.8 actual size of pilot and off perp by 0.1

15.9 actual size of slot and off position by 0.2

Ø25.3 MMB simulator

15.4 MMB simulator

B – Datum Eval	
MMB	25.3
Actual	24.9
Shift	0.4

C – Datum Eval	
MMB	15.4
Actual	15.7
Shift	0.3

No.	MMC Size	Act Size	Allow Pos	"X" Dev	"Y" Dev	Act Pos	Acc Rej	Datum Feature Evaluation	Acc Rej
1	8.8	9	1	+0.2	+0.5	1.077	R	Datum feature B - Ø0.4 departure from MMB allows Ø0.4 shift of the datum feature. Paper gage evaluation accepts location of 3 holes.	A
2	8.8	9	1	-0.1	+0.3	0.632	A		A
3	8.8	9	1	+0.5	+0.1	1.020	R		A

7.14

Both datum features B and C have the MMB modifier. Datum feature B, the pilot, departed from the MMB of 25.3. Additional tolerance is available from the pilot since its actual related mating size was produced at 24.9. The inspection process is allowed to shift the axis of datum feature B back and forth within the MMB of 25.3 to obtain conformance on the position tolerance for the 3 holes. This datum feature departure tolerance CAN NOT be just added to the position tolerance for the holes. It must be plotted and factored in as shown below.

This bonus tolerance from the pilot is accommodated automatically if verified with a functional hard gage. A good CMM operator can also manipulate the software to factor in this additional tolerance by figuring it as a pattern of 4 holes, rather than just 3 holes. The paper gage concept is another manner in which the effect of the MMB modifier on datum feature B can be evaluated.

The 3 holes are plotted on the paper gage coordinate using the data shown on the inspection report. Squares are in increments of 0.1 mm and circles are in 0.2 mm. The locations are shown labeled with numbers corresponding to their numbering on the produced part. Looking at the paper gage polar graph before shifting, we see two of the holes are outside their allowed position tolerance of 1 mm.

Looking at the geometry of the part, we can see that since the actual related mating size of the produced pilot is 0.4 smaller than the MMB. The 3 holes could shift about the datum B axis by a dia of 0.4. Looking at the graph, after shifting the datum feature off center by 0.4 dia (0.2 radius), the 3 holes are well within the 1 mm position tolerance. To prove the results obtained by the paper gage, the part could be reset in the inspection procedure and the origin of the pilot could be offset the center by 0.4 dia. (0.2 R).

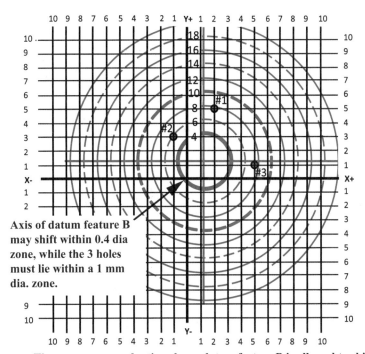

Datum feature C also departed from MMB. This datum feature shift was not used, as the datum feature B shift was enough to accept the part. See next page for an evaluation of a tertiary datum feature modified at MMB.

The paper gage calculations to the right are an approximate evaluation of the hole pattern. Other factors, such as form, orientation and number of datum points, will affect the outcome. The example is intended to illustrate the general concepts. Data may also be evaluated by appropriate CMM software or on a CAD system.

Axis of datum feature B may shift within 0.4 dia zone, while the 3 holes must lie within a 1 mm dia. zone.

The paper gage evaluation shows datum feature B is allowed to shift from the axis of the datum feature B simulator by 0.4 dia. and allows the acceptance of the location of 3 holes because of the MMB modifier.

This example is intended to illustrate the calculations necessary to factor in tolerance gained from datum feature modifiers. The produced part has been set up in inspection by "zeroing" on datum features A, B and C. Inspection data has been collected from the workpiece and the results are shown in the table below. It appears from the data that the location of the 3 holes are out of position. Notice however that datum features B and C are modified at MMB. Additional tolerance may be available from the datum features.

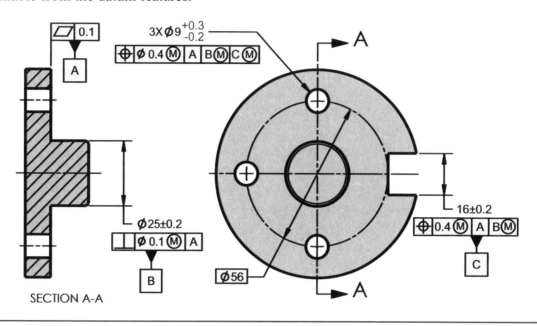

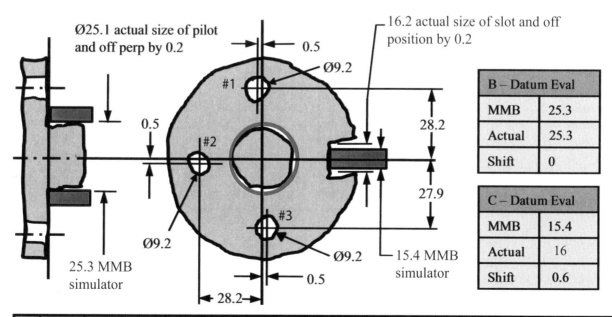

B – Datum Eval	
MMB	25.3
Actual	25.3
Shift	0

C – Datum Eval	
MMB	15.4
Actual	16
Shift	0.6

No.	MMC Size	Act Size	Allow Pos	"X" Dev	"Y" Dev	Act Pos	Acc Rej	Datum Feature Evaluation	Acc Rej
1	8.8	9.2	0.8	-0.5	+0.2	1.077	R	Datum feature C- 0.6 departure from MMB allows 0.6 rotation at slot. Paper gage evaluation accepts location of 3 holes.	A
2	8.8	9.2	0.8	-0.2	-0.5	1.077	R		A
3	8.8	9.2	0.8	+0.5	+0.1	1.020	R		A

Both datum features B and C have the MMB modifier. Since datum feature B did not depart from its MMB, there is no additional tolerance available from this datum feature.

However, datum feature C, the slot, did depart from the MMB of 15.4. Additional tolerance is available from the slot since its actual related mating envelope was produced at 16. The inspection process may rotate the part about the axis of datum B back and forth while still containing the produced slot size of 16 within the MMB of 15.4 to obtain conformance on the position tolerance for the 3 holes. The datum feature shift tolerance CAN NOT be just added to the position tolerance for the 3 holes, it must be plotted and factored in as shown below.

This departure tolerance from the slot is accommodated automatically if verified with a functional hard gage. It is very difficult however, for a open set up or CMM software to factor in this bonus tolerance. The paper gage concept is a manner in which the effect of the MMB modifier on datum feature C can be evaluated.

The 3 holes are plotted on the paper gage coordinate using the data shown on the inspection report. Squares are in increments of 0.1 mm and circles are in 0.2 mm. The hole locations are shown labeled with numbers corresponding to their numbering on the produced part. Looking at the paper gage polar graph before shifting, we see all the holes are outside their allowed position tolerance of 1 mm.

Looking at the geometry of the part, we see that since the actual related mating envelope of the produced slot is 0.6 larger than the MMB, functionally, this will allow the 3 holes to rotate back and forth about the datum B axis approximately 0.6, considering the distance of the slot

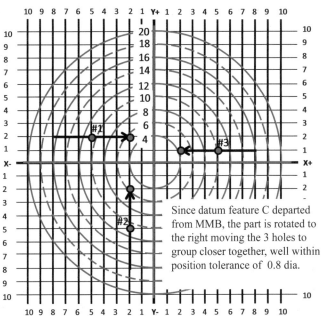

Since datum feature C departed from MMB, the part is rotated to the right moving the 3 holes to group closer together, well within position tolerance of 0.8 dia.

The paper gage evaluation shows 3 holes can be rotated 0.6 (0.3 each way) to an acceptable position tolerance because datum feature C departed from MMB.

from the center hole is approximately the same as the distance of the 3 holes from the center hole. (If the distance were not the same, then a ratio needs to also be factored in the calculations.)

If the part was rotated to the right, the location of the holes would move on the coordinate system accordingly. Notice that each hole on the graph moved in a direction as if the part were rotated clockwise. Hole #1 moved to the right by 0.3 (1/2 datum feature departure), hole #2 moved up, hole #3 moved to the left. After the rotation shown by the arrows on the graph, the 3 holes are well within the 1 mm position tolerance. To prove the results obtained by the paper gage, the part could be reset in the inspection procedure with the new datum feature alignment.

The paper gage calculations above are an approximate evaluation of the hole pattern. Other factors, such as orientation, form and number of datum points will affect the outcome. The example is intended to illustrate the general concepts. Data may also be evaluated by appropriate CMM software or on a CAD system.

Shown below are two separate parts produced from the drawing shown earlier on page 7.16. Evaluate the departure from MMB for the datum features on these two produced parts. Calculate the actual MMB and determine the amount of datum feature shift that is allowed for each datum feature. Record your answers for each datum feature in the tables below to the right.

Part 1

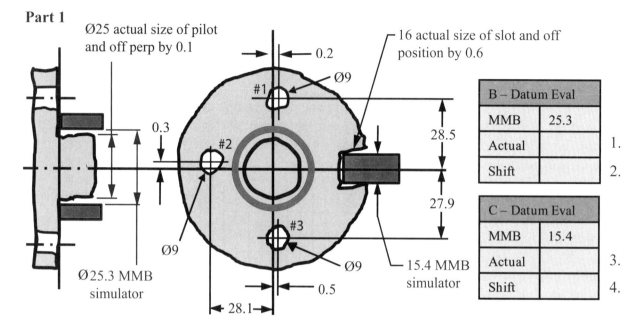

B – Datum Eval	
MMB	25.3
Actual	
Shift	

1.
2.

C – Datum Eval	
MMB	15.4
Actual	
Shift	

3.
4.

Part 2

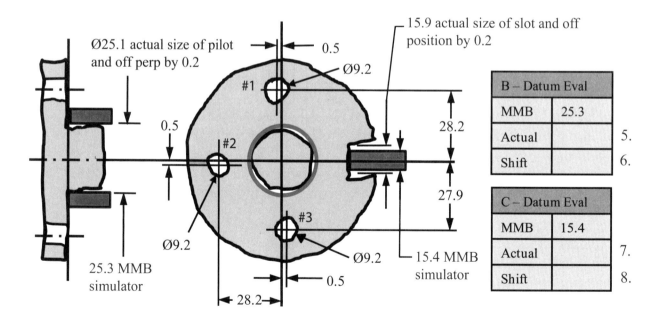

B – Datum Eval	
MMB	25.3
Actual	
Shift	

5.
6.

C – Datum Eval	
MMB	15.4
Actual	
Shift	

7.
8.

9. Features related with basic dimensions using the same datum feature reference, in the same order of precedence, with the same datum feature modifiers, constitute a simultaneous requirement for position and profile tolerances.   True   or   False

Unit 8

The Datum Reference Frame II
Targets and Irregular Surfaces

Inclined Datum Feature
Contoured Datum Feature
Datum Centerplanes
Datum Targets - Definitions
Datum Targets - Turbine Blade
Datum Targets - Hood Panel Outer
Datum Targets - Pillow Block
Datum Targets - Casting to Machining Calculations
Datum Targets - Movable Datum Target 1
Datum Targets - Movable Datum Target 2
Workshop Exercise 8.1
Workshop Exercise 8.2

Inclined Datum Features

The datum reference frame is always three mutually perpendicular planes establishing an orthogonal coordinate system. However, the datum features may not necessarily be mutually perpendicular to each other.

In some cases, it may be necessary to establish a datum reference frame from datum features that are not mutually perpendicular to each other. The wear plate assembly is a good, simple example. Notice the wear plate mounts on the back surface, the bottom surface and the inclined 45° surface. The three holes do not locate the part, they just clamp the plate in position. Based on functional requirements, a datum reference frame is developed using the non-orthorgonal surfaces.

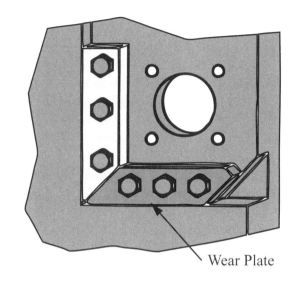

Wear Plate

Wear Plate Assembly

On the drawing below, all the angles between the datum surfaces are indicated 45° or implied 90° basic. The primary datum feature A is qualified with a flatness specification. The secondary datum feature B is qualified back to the primary datum with an angularity tolerance. The tertiary datum feature C is qualified back to the primary and secondary datum with an angularity tolerance.

Note: A perpendicularity tolerance could have also been used to qualify the secondary and tertiary datum features. However, the ASME Y14.5-2009 standard allows the use of angularity to be used for all angles to simply control the orientation.

Wear plate drawing with an inclined datum feature

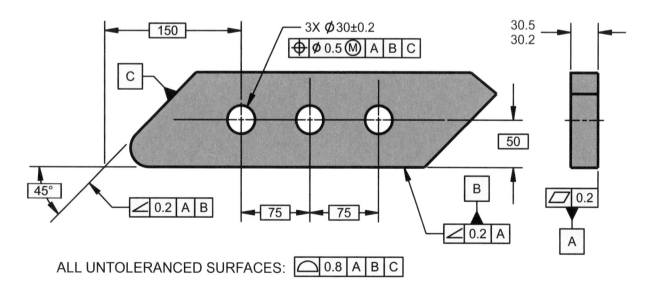

The planes in the datum reference frame are always 90° to each other. The datum features may be at any angle, even compound angles, as long as all angles are basically oriented to each other. The datum simulator is oriented relative to the DRF at the specified angle.

3D illustration of DRF established from inclined datum feature

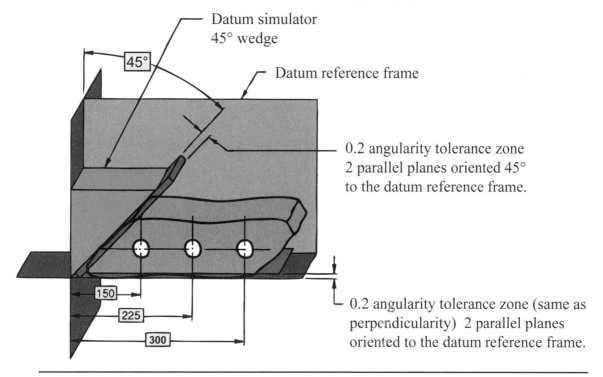

Datum simulator 45° wedge

45°

Datum reference frame

0.2 angularity tolerance zone
2 parallel planes oriented 45°
to the datum reference frame.

150

225

300

0.2 angularity tolerance zone (same as perpendicularity) 2 parallel planes oriented to the datum reference frame.

Since the datum features are 3 planar surfaces, the part will make contact with the primary datum a minimum of 3 points, the secondary at 2 points contact and the tertiary at 1 point of contact. All measurements originate from the datum reference frame and not the part. The part may be verified in a functional gage, coordinate measuring machine (CMM) or rotated on a sine plate in an open set-up.

2D illustration of DRF established from inclined datum feature

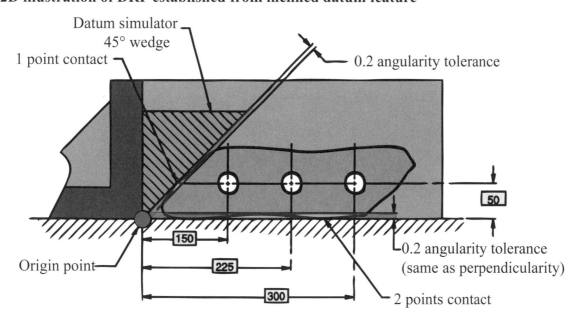

Datum simulator
45° wedge

1 point contact

0.2 angularity tolerance

50

150

225

300

Origin point

0.2 angularity tolerance
(same as perpendicularity)

2 points contact

Access Door Application

The access door mounts on a partial contoured surface as shown. The door is fastened down with 12 screws. The contoured surface geometry, the hole locations and other product definition are defined in a CAD file.

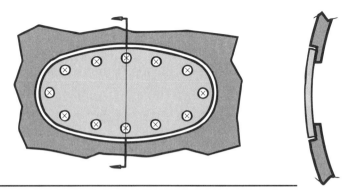

The engineering drawing below reflects the functional requirements of the access door assembly. In this case, rather than using datum targets, the partial area of the contoured surface was selected as the primary datum feature. The datum feature A simulator is set at the basic math data defined in the CAD file. Datum feature A is controlled with a unilateral .006 inside profile tolerance. This allows the surface to vary inside material from the basic math data by .006. The 12 holes are positioned to each other and datum feature A and are used as the secondary datum feature B. Datum feature B is modified at MMB, which establishes the 12 datum simulator pins at .170 dia. (.177 minus position tolerance .007). The all around edge of the access door, and any other features, are profiled to this DRF.

This on the Drawing

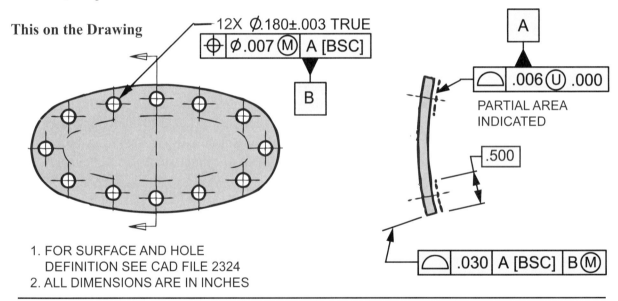

1. FOR SURFACE AND HOLE
   DEFINITION SEE CAD FILE 2324
2. ALL DIMENSIONS ARE IN INCHES

Sample check/holding fixture for access door

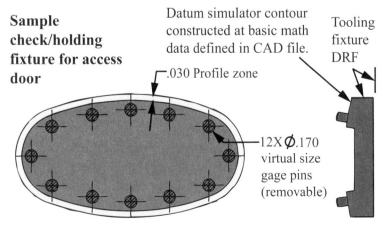

Datum simulator contour constructed at basic math data defined in CAD file.

.030 Profile zone

12X⌀.170 virtual size gage pins (removable)

Tooling fixture DRF

The access door contour geometry on the check fixture was constructed from the basic CAD data using a DRF established from the bottom surface and sides of the fixture. This check/holding fixture DRF can be used in a open set up or by a CMM to establish the part DRF for the contoured surface and holes of the access door.

Datum Centerplanes

The part below has a datum reference frame established from its mating features, the back face and the two outside widths. These two outside widths create datum center planes B and C. The profile and position tolerance zones are basically located from these center planes. Datum center planes keep the tolerance zones symmetrical about the outside shape. This reference frame is also consistent with the way the part fits in its mating pocket.

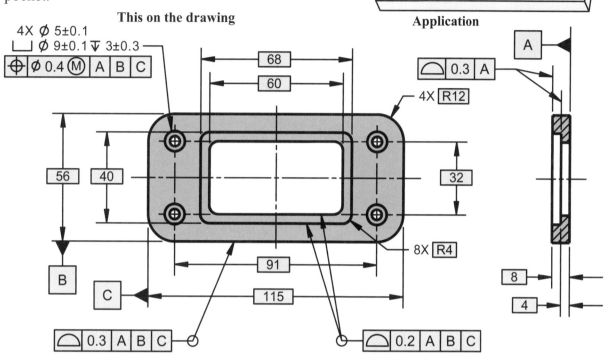

This on the drawing   **Application**

Datum center planes with position and profile tolerance zones

Means this

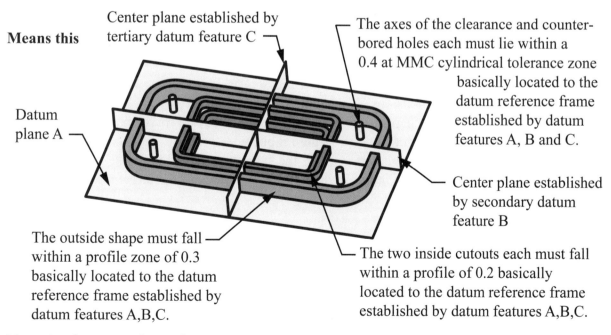

Center plane established by tertiary datum feature C

Datum plane A

The outside shape must fall within a profile zone of 0.3 basically located to the datum reference frame established by datum features A,B,C.

The axes of the clearance and counter-bored holes each must lie within a 0.4 at MMC cylindrical tolerance zone basically located to the datum reference frame established by datum features A, B and C.

Center plane established by secondary datum feature B

The two inside cutouts each must fall within a profile of 0.2 basically located to the datum reference frame established by datum features A,B,C.

Note: A reference to datum feature C was added to the outside profile to eliminate stack up between the outside shape and the other features because of different datum reference frames.

Datum Target Definition

In establishing a datum reference frame, there are applications where the designer would prefer to select specific points of contact on a surface rather than use an entire surface. This may occur because of manufacturing concerns or surface irregularities that make the use of the entire surface impractical. Examples of these type of applications might be castings, forgings, sheet metal, plastic parts and weldments.

In the past, these areas of contact were called set up points, tooling points, fixture points, principle locating points. All of these names have been discarded and we now call these contact areas, datum targets. Datum targets can be points, lines or areas of contact.

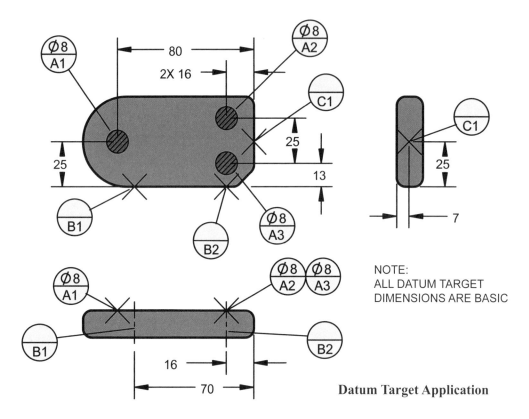

NOTE:
ALL DATUM TARGET
DIMENSIONS ARE BASIC

Datum Target Application

Area Contact: If the target is an area contact, the area is indicated by section lines inside a phantom outline with size dimensions. If the definition of the target area is clear, the section lines may be omitted. If the target shape is a circle or square, the size and shape may be designated in the upper half of the datum target symbol. Where it becomes impracticable to define the size of the target in the upper half of the target symbol, the size may defined in a note or with a leader directed to the upper half of the symbol. In the bottom or side view, the target area is designated as an "X". See datum targets "A".

Line Contact: If the target is a line contact, it is indicated by an "X" on the edge in the plan view. In the bottom view, it is indicated with a phantom line. See datum targets "B". Where the length of the line contact is important, the length must be dimensioned.

Point Contact: If the target is a point contact, the point is indicated by an "X" in the side view. It is also indicated with an "X" in subsequent views. See datum target "C".

The datum target symbol is placed on the outside of the part outline. If the target area is on the near visible side of the part, a solid leader line is used. **If the target area is on the far (hidden) side of the part, the leader line is dashed.**

The location and size of the datum targets are usually defined with basic dimensions. If targets are defined with basic dimensions, tooling and gaging tolerances are assumed to apply. If necessary, the targets can also be defined with toleranced dimensions.

Datum targets can be used alone to establish a DRF or in combination with datum features. The datum target is not a spot on the part, it defines the datum simulator. Targets can be any size or shape as long as they are clearly defined on the drawing.

Datum targets are labeled with letters of the alphabet. In some company practices, datum features are labeled A, B and C on the machining drawings and X, Y and Z on casting drawings. This is done to differentiate the DRF labeling from the machining to the casting.

Datum Targets at MMB, LMB and RMB

Where datum targets establish a center point, axis or centerplane, the datum feature modifiers MMB, LMB and RMB may be used. When MMB or LMB are applied to the datum features, the datum feature modifiers are fixed at their designated size or location. If datum feature modifiers are applied on a RMB basis, the datum feature simulator expands/ collapses or progresses normal to the true profile. A movable datum target symbol may be used for clarification.

Where the datum feature simulator is required to move, and the movement is not normal to the true profile, the movable datum target symbol should be used with the movement direction clearly defined. See movable datum target later in this unit.

Datum Targets - Turbine Blade

This is a engineering drawing of a turbine blade with datum targets applied. Datum targets have been applied to clearly establish a DRF on the part. Datum targets X1 and X2 define line contacts. Datum target X3 defines a point contact. Datum targets Y1 and Y2 define line contacts. Datum target Z is a point contact. The product definition of the blade can be found in the math data of the CAD model.

Drawing of turbine blade casting with datum targets

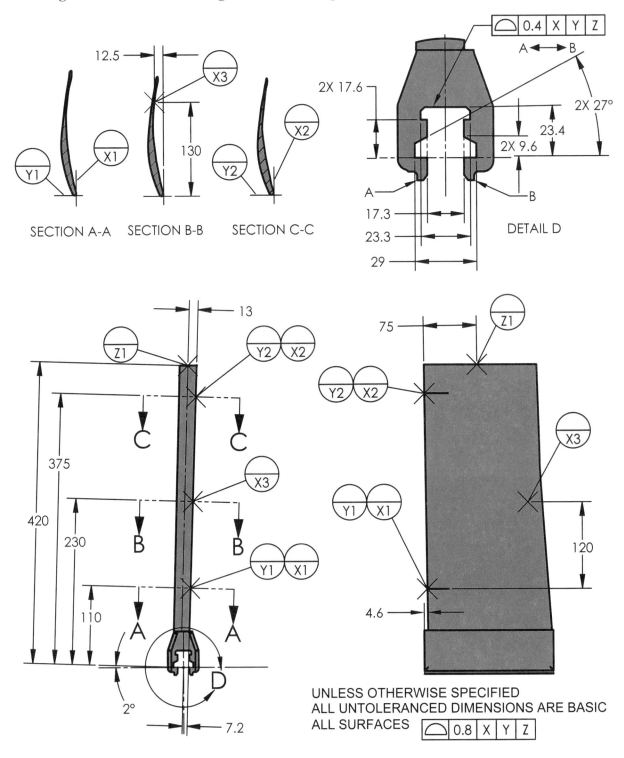

SECTION A-A SECTION B-B SECTION C-C

DETAIL D

UNLESS OTHERWISE SPECIFIED
ALL UNTOLERANCED DIMENSIONS ARE BASIC
ALL SURFACES

Datum targets are useful for establishing a DRF on parts with irregular surfaces. The turbine blade shown is a good example of an application with datum targets.

The turbine blade is produced with inherent variations due to the casting process. A datum reference frame must be established on the part in order to define design requirements. Datum target lines and points are established on the blade surface to create the datum reference frame.

The targets in this application were selected based on the holding fixture for machining the casting. The check fixture below simulates the datum targets and is a replica of the maching fixture. This DRF is necessary for product definition, verification and manufacturing.

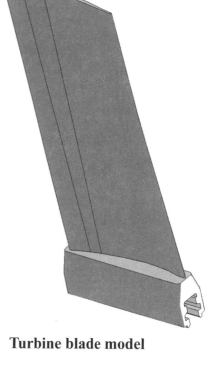

Turbine blade model

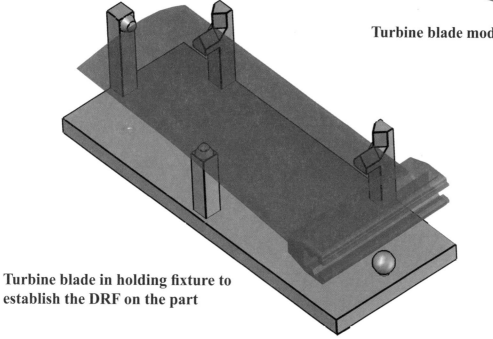

Turbine blade in holding fixture to establish the DRF on the part

The figure above represents an "L" trap fixture to contain the turbine blade, as defined by the datum targets specified on the drawing. The size, shape and location of the datum target locators are practically represented in this fixture. The part may be placed in this fixture to simulate the DRF and establish it on the part. Measurements can be made from the planar face and tooling ball on the fixture because of the known relationship between the locators and the tooling ball. In some cases, it may be necessary to define a restraint requirement.

The datum reference frame for this outer hood panel was applied using datum targets. These datum targets were selected based on how the part loads in the assembly fixture with the hood panel inner reinforcement. The targets are located relative to the body grid system. Since more than 3 targets were selected as the primary datum, a restraint requirement was added. See the profile drawing for the tolerances. The section views show the part mounting in the check/assembly fixture.

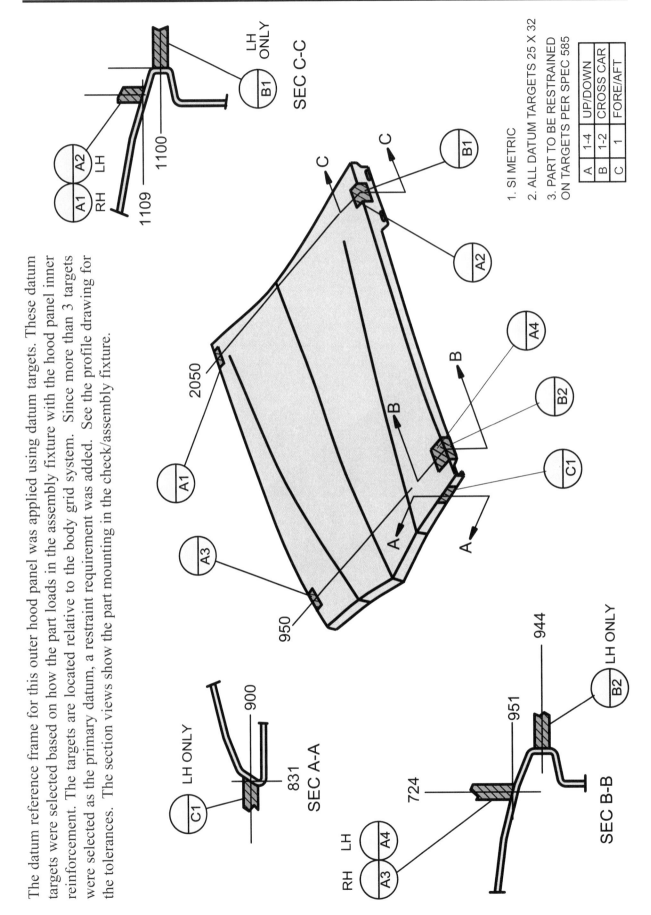

1. SI METRIC
2. ALL DATUM TARGETS 25 X 32
3. PART TO BE RESTRAINED ON TARGETS PER SPEC 585

A	1-4	UP/DOWN
B	1-2	CROSS CAR
C	1	FORE/AFT

SEC C-C

SEC A-A

SEC B-B

Profile Application - Hood Panel Outer

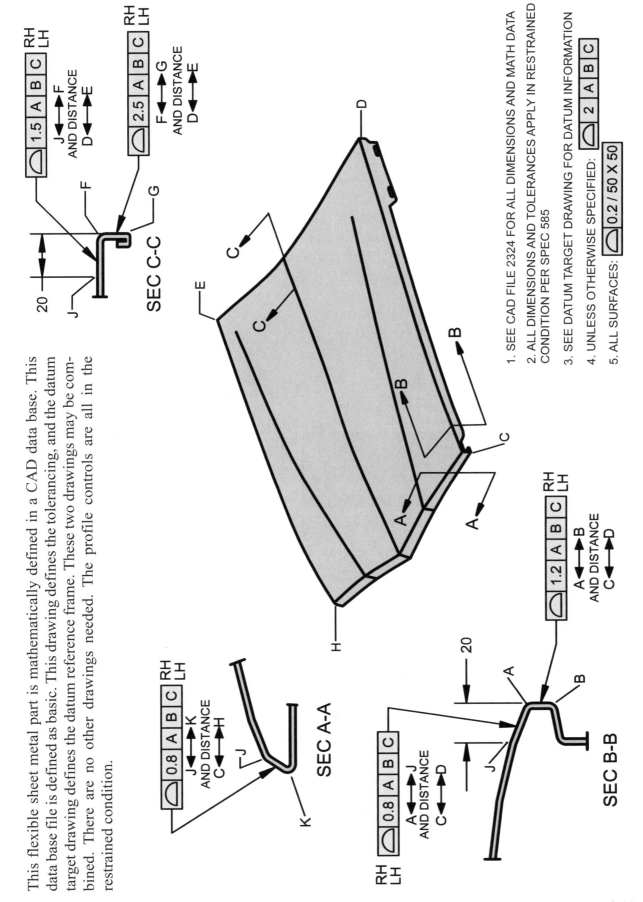

This flexible sheet metal part is mathematically defined in a CAD data base. This data base file is defined as basic. This drawing defines the tolerancing, and the datum target drawing defines the datum reference frame. These two drawings may be combined. There are no other drawings needed. The profile controls are all in the restrained condition.

SEC C-C

SEC A-A

SEC B-B

1. SEE CAD FILE 2324 FOR ALL DIMENSIONS AND MATH DATA
2. ALL DIMENSIONS AND TOLERANCES APPLY IN RESTRAINED CONDITION PER SPEC 585
3. SEE DATUM TARGET DRAWING FOR DATUM INFORMATION
4. UNLESS OTHERWISE SPECIFIED:
5. ALL SURFACES:

Datum Targets- Pillow Block

The machined pillow block shown in the assembly mounts on the bottom face, the side face and is centered in the assembly by the large hole. The pattern of three holes bolt down and fix the part in the assembly.

The engineering drawing below shows the datum features selected based on function and mating conditions. The bottom face is datum feature A, the side face is datum feature B and the 13.2/13.4 mm hole is datum feature C. These features, in order, establish the DRF. The datum features are qualified with a flatness, perpendicularity and position tolerance respectively. The datum reference frame symbol is shown on the drawing and represents the DRF for the machining operations. All remaining machined features are located to this DRF. The pillow block will be made from a casting, which is shown on the next pages.

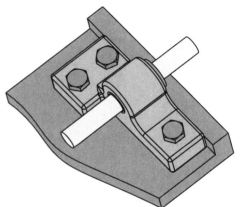

Pillow block assembly

Machining drawing

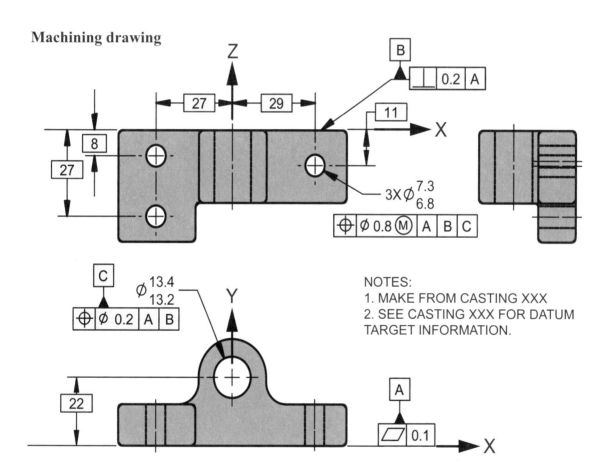

NOTES:
1. MAKE FROM CASTING XXX
2. SEE CASTING XXX FOR DATUM TARGET INFORMATION.

8.12

The machined part has an A, B, C datum reference frame established, and all the machined features on the part are located to this DRF. The cast part has an X, Y, Z datum reference frame and all the cast features are located to this DRF.

In order to make stock removal and resulting wall thickness calculations, the machine DRF must be located to the cast DRF. The bottom plane surface on the machine part is the primary datum feature, and it has a profile tolerance applied to locate it to the cast X, Y, Z DRF. The rear plane surface on the machine part is the secondary datum feature, and it also has a profile tolerance applied to locate it to the cast DRF. The 13.2/13.4 mm hole is the tertiary datum feature on the machined part and it is positioned to the cast DRF.

These profile and position tolerances (indicated by note 2) locate the machine DRF to the cast DRF. This is what we formerly called the "first machine cut."

Machining drawing showing location from casting

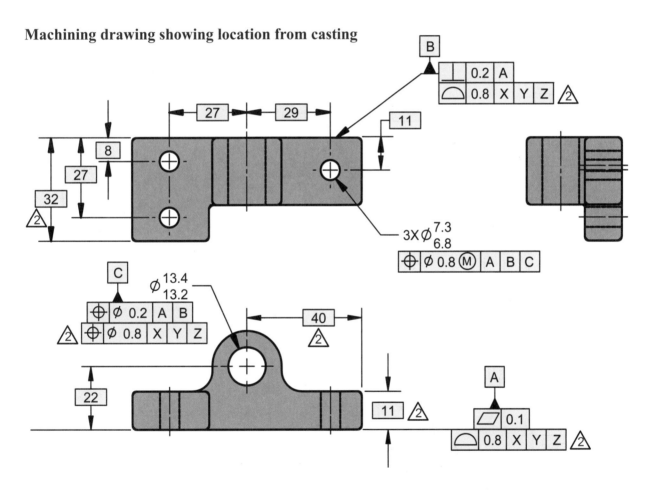

NOTES:
1.   MAKE FROM XXX

2. DESIGNATES THE LOCATION OF THE MACHINING TO CASTING DRF.

8.13

The casting has been defined on the drawing below. Manufacturing, in consultation with design, determines the manufacturing process and fixturing necessary to machine the pillow block. Manufacturing has decided to clamp the casting in a machining fixture and perform all machine operations in one set-up.

The datum targets on the drawing correspond to the datum targets on the fixture.

The "X" datum targets are located and have their centers removed to accommodate the drilling operation of the 3 holes. The datum reference frame symbol is shown on the drawing and represents the DRF established by the datum targets.

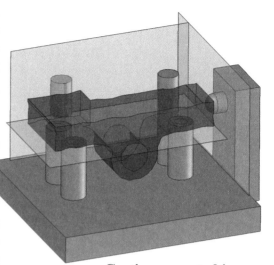

Casting mounted in machining fixture

Casting drawing

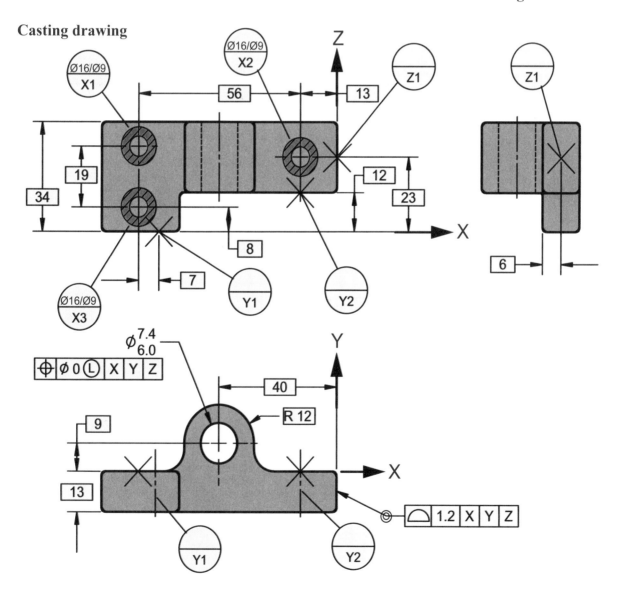

Calculations between datum targets on casting [X, Y, Z –DRF] and the machined surface [A, B, C- DRF]

The geometry on the pillow block casting is defined with a 13 mm basic dimension originating from the DRF established by the datum targets X, Y and Z. A profile tolerance of 1.2 mm is applied to the casting surface. The location of the machined surface is defined on the machine drawing with an 11 mm basic dimension originating from the DRF established by the datum targets X, Y and Z. A profile tolerance of 0.8 mm provides the location of this machine surface.

This machined surface is ultimately used to establish the A, B, C -DRF on the machined part.

The part is mounted to the datum target locations in the machine fixture and the machining operation is given an 11 mm basic dimension and allowed a 0.8 mm profile tolerance to locate the datum A surface. This arrangement allows a minimum of 1 mm clean-up on the casting.

Cast basic dimension	13	**Expanded view showing casting mounted**
- ½ cast profile tol	-0.6	**on datum targets, with calculations to**
- 1 cast stock removal	-1	**ensure adequate machine stock**
- ½ machine location tol	-0.4	
Machine basic dimension	11	

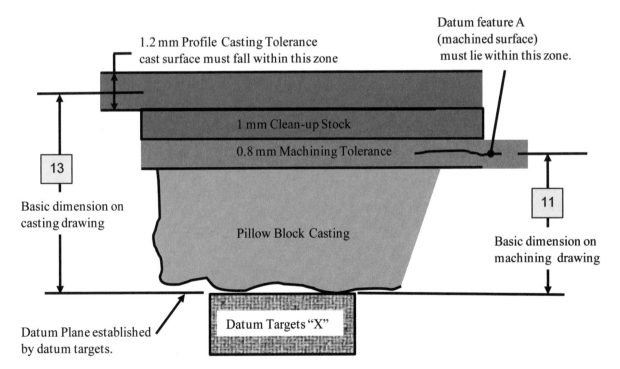

1.2 mm Profile Casting Tolerance
cast surface must fall within this zone

Datum feature A
(machined surface)
must lie within this zone.

1 mm Clean-up Stock

0.8 mm Machining Tolerance

13

Basic dimension on
casting drawing

Pillow Block Casting

11

Basic dimension on
machining drawing

Datum Targets "X"

Datum Plane established
by datum targets.

This is an expanded view of the part mounted on the datum targets which establish the X,Y, Z DRF. The casting is defined and the machining operations are located from this DRF. The datum targets represent the size and location of the datum locators on both the machining fixture and the casting inspection fixture. The locators on the machining fixture and the inspection fixture (for the casting) should be identical.

Movable Datum Targets 1

The datum feature simulator requirements on page 6.13 state that all simulators are fixed at basic. In some cases, it may be necessary to designate that a datum simulator should move to engage the datum feature. It this case, a movable datum target may be used. The movable datum target symbol, shown below on C1 and C2 on the left view, is differentiated from the datum target symbol with a triangle attached to the side. (Also, sometimes called a bird beak).

The movable datum target designates that the datum simulator moves to engage the part. In cases where the movement direction is not clear, it should be defined on the drawing. The movable datum target symbol is new for the ASME Y14.5-2009 standard. It was brought over from the ASME Y14.8 casting standard.

In this example, datum targets C1 and C2 in the front view are point contact and are required to move to engage the datum feature. If it were necessary to have the entire surface referenced as a datum feature, the translation symbol could be applied following datum feature C in the feature control frame. See unit 9 for more information on the datum feature translation symbol.

This on the drawing

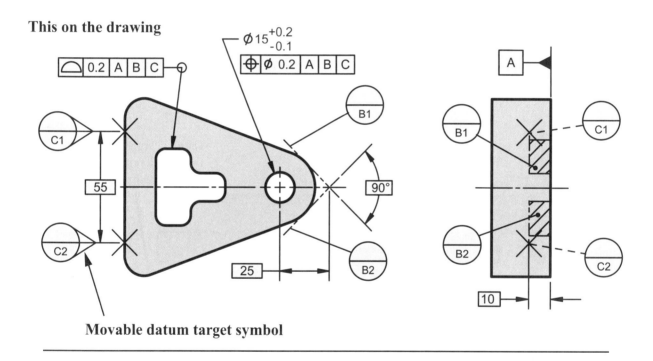

Movable datum target symbol

Defines this holding/checking fixture

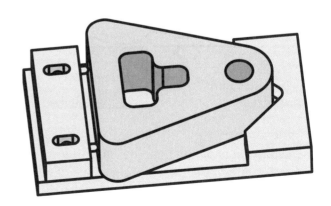

Holding/checking fixture with movable rail and 2 point locators to make contact with datum targets C1 and C2 and establish a DRF on the part.

The datum feature simulator requirements, on page 6.13, state that all simulators are fixed at basic. In some cases, it may be necessary to designate that a datum simulator should move to engage the datum feature. It this case, a movable datum target may be used. The movable datum target symbol is differentiated from the datum target symbol with a triangle attached to the side. The movable datum target designates that the datum feature simulator moves to engage the part.

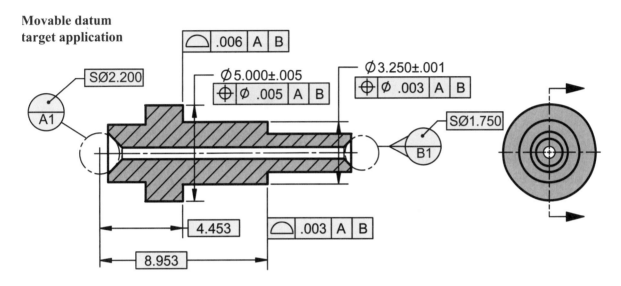

Movable datum target application

The drawing shows a part with datum targets applied to establish a datum reference frame. Datum target A is defined as a spherical diameter of 2.200. Datum target B is defined as a spherical diameter of 1.750 and is movable. The direction of movement is clear, so no dimensions showing movement are necessary.

Datum reference frame established by datum target A and movable datum target B

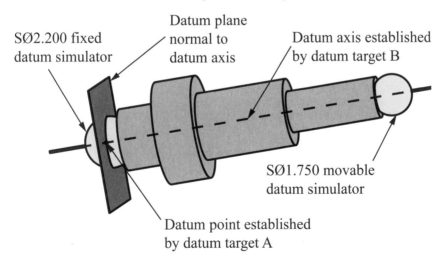

The graphic illustrates the datum reference frame established by datum targets A and B. Datum target A is a spherical simulator 2.200 in diameter and establishes a datum point and origin. Datum target B is a spherical simulator 1.750 in diameter, moves toward datum A and establishes a point. The two datum points in combination establish a DRF. All measurements originate from this DRF.

1. True or False? Datum planes in a datum reference frame are always 90 degrees apart, but datum features may be at angles other than 90 degrees to each other.

_____

2. Sketch a datum target symbol X1 that creates a square target of .250 on the feature below.

3. What is the minimum amount of datum target points required to establish a primary datum plane?

_____

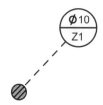

4. What does it mean when the leader line to a datum target is dashed (shown as a hidden line) as sketched above?

_____

5. Sketch a movable datum target symbol.

6. Match the sketches below with the following types of datum targets.

datum target point   datum target area   datum target line

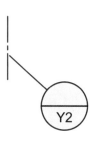      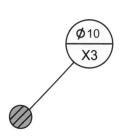

Workshop Exercise 8.2

Study the fixture and setup for the part below. Apply datum targets to the drawing views to simulate the fixture setup shown. Make sure to dimension the targets to each other by simply estimating the distances. Assume all round pads are 10 mm in diameter. Then apply an all over profile in reference to the datum features within 1.5 mm.

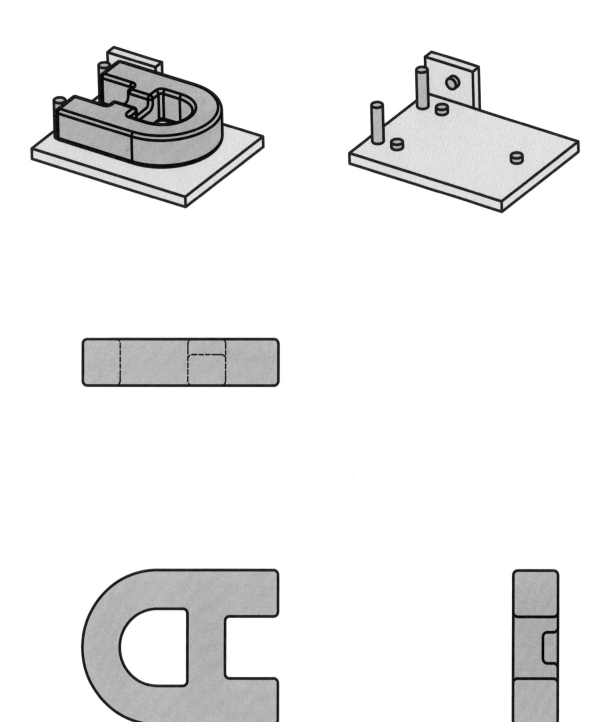

Unit 9

The Datum Reference Frame III
Advanced Concepts

Two Holes as a Datum Feature
Hole and Slot as Datum Features
Pattern of Holes as a Datum Feature
   Paper Gage Evaluation
Coaxial Holes as a Datum Feature
Rotational Control - Opposed Surface
Rotational Control - Opposed Contoured Surface
Rotational Control - Unopposed Planar Surface
Rotational Control - Unopposed Planar Surface at MMB and RMB
Rotational Control - Unopposed Planar Surface at Basic
Irregular Datum Feature / Workshop Exercise 9.1
Rotational Control - Slot at RMB w/ and w/o Translation Modifier
Translation Modifier
Customized Datum Reference Frame - Introduction
Customized Datum Reference Frame - Square Hole and Slot
Customized Datum Reference Frame - 4 Datum Feature References
Multiple Datum Reference Frames
Establishing a Relationship Between Two DRF
Workshop Exercise 9.2 - Customized Datum Reference Frame

Two Holes as a Datum Feature

Below is an assembly of a base and sheet metal cover plate. It is important that the three 4 mm holes in the base are lined up with the three holes in the cover plate when assembled. Two alignment pins and holes are used to align the parts together. There are two examples of how to tolerance the mating sheet metal plate below. Tolerancing Example II gives the most location tolerance between the alignment holes, and the datums more accurately represent the mating condition. Detailed explanations are shown on the following page.

Mating Part

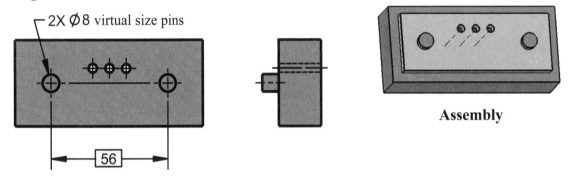

Assembly

Example I **One hole is B, the other is C**

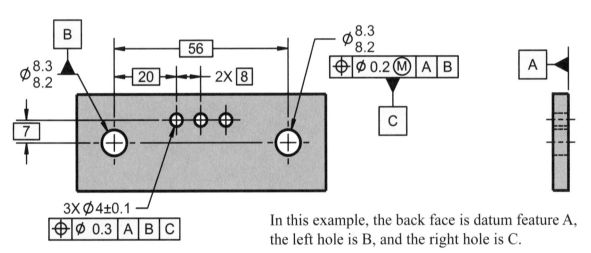

In this example, the back face is datum feature A, the left hole is B, and the right hole is C.

Example II **Both holes as datum feature B provides most location tolerance**

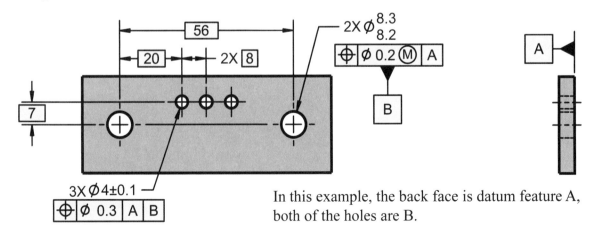

In this example, the back face is datum feature A, both of the holes are B.

9.2

Study the tolerancing examples on the previous page. The size and distance between two holes are measured on the manufactured part. What is the difference in the two tolerancing examples?

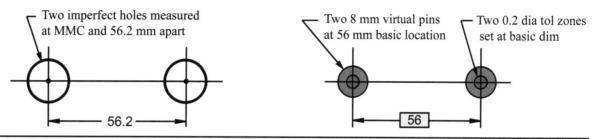

Two imperfect holes measured at MMC and 56.2 mm apart

56.2

Two 8 mm virtual pins at 56 mm basic location

Two 0.2 dia tol zones set at basic dim

56

Example I   One hole is B, the other is C

Datum B is selected as the left hole. This sets the origin of the reference frame in the center of the left hole, regardless of material boundary. The right hole is called off as datum C and only stops rotation of the datum reference frame. The position tolerance is calculated for datum feature C, making the virtual size of the hole the same as the mating virtual pin. A perpendicularity tolerance was not given to datum feature B because the sheet metal plate is so thin. A perpendicularity of zero at MMC could also have been added. A position tolerance of zero at MMC could also be added to the width of the slot on the next page. These controls were eliminated for simplicity.

This tolerancing example is selecting one of the holes as the origin and locating the other hole to this origin. When the distance varies between the holes, all the error is placed on one hole. This example allows less location tolerance between the holes, only 0.2 at MMC for the right hole.

The datum planes are aligned with the axis of datum feature B. No position tolerance is given to this hole.

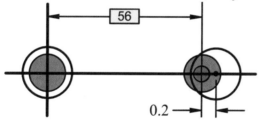

56

0.2

All the distance error between the holes is shown on this one hole. The measured part would fail inspection.

Example II   Both holes as datum feature B   (gives most location tolerance between holes)

Datum B is selected as both holes. This sets the origin of the reference frame as the average of the holes. The two holes will locate and stop rotation of the part together. A position tolerance is calculated for the datum features, making the virtual sizes of the holes equal to the mating pins.

This tolerancing example is selecting the pattern of holes as the datum and locating the holes to each other. When the distance varies between the two holes, the error is balanced between the two holes. This example allows more location tolerance between the holes, 0.2 at MMC for each.

If the part is checked using this scheme, it will also give a more accurate representation of where the three 4 mm holes will be when assembled.

The datum planes are set as the average between the two holes. This more accurately represents how the part will probably mount in the assembly.

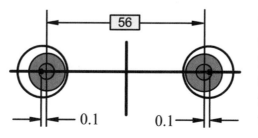

56

0.1   0.1

The distance error between the holes is balanced amid the two holes and a position tolerance is given to each of them. The measured part would pass inspection.

9.3

Hole and Slot as Datum Features

Below is an assembly of a base and sheet metal cover plate as an alternative design to the assemblies on the previous pages. It is important that the three 4 mm holes in the base are lined up with the three holes in the cover plate when assembled. Two alignment pins on the base and a hole and slot on the cover plate are used to align the parts together. The drawing below shows the tolerancing example for a hole and slot application.

A hole and slot design is the best way to precisely align parts together. Because of the slotted geometry of the second hole, it makes the location between the alignment holes unimportant and also allows smaller hole sizes for a more precise fit.

Mating Part

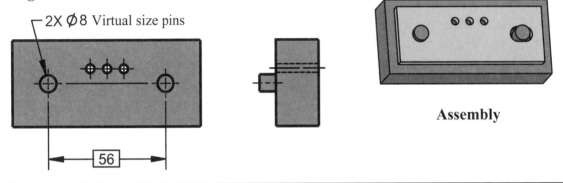

Assembly

Example III In this example, the back face is datum feature A, the left hole is B, and the width of the slotted hole is C.

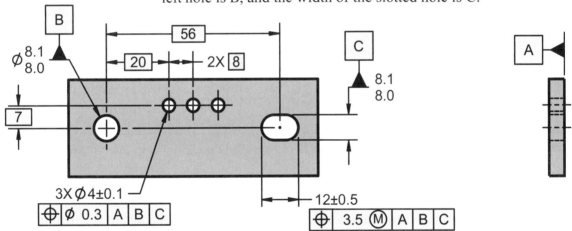

In this example, datum feature B is selected as the left hole (stopping movement in two directions), and datum feature C is the width of the slot (stopping rotation). The slotted geometry on the right hole allows for a large position tolerance on the length of the slot. It also allows hole sizes closer to the mating pin size for less "slop" in the assembly.

Compare this drawing to the similar assemblies on the previous pages to see why a hole and slot is the best way to locate two parts together. Manufacturing and quality receive larger position tolerance and design ends up with a better fitting assembly.

Note: Hole and slot designs work well on sheet metal and plastic, but slots can be difficult to produce on machined parts. Another option to achieve the same result is using two round holes with a pin and diamond pin. The pin locates in two directions and the diamond pin stops rotation.

Pattern of Holes as a Datum Feature

When function dictates, multiple features of size, such as pattern of holes, may be used as a datum feature to establish a DRF. On the part below, the back face is identified as datum feature A, establishes the primary datum and constrains one translational and two rotational degrees of freedom. The 4 holes are positioned to each other with respect to datum feature A and are identified as secondary datum feature B. The axes of the 4 holes must fall within the position tolerance zones. The tolerance zone cylinders are basically located to each other, and in effect, establish the second and third planes of the DRF. This datum feature constrains two translational and one rotational degree of freedom.

Application

The light switch cover mounts on the four holes and must clear the switch. The outside edges of the cover are not important.

This on the drawing

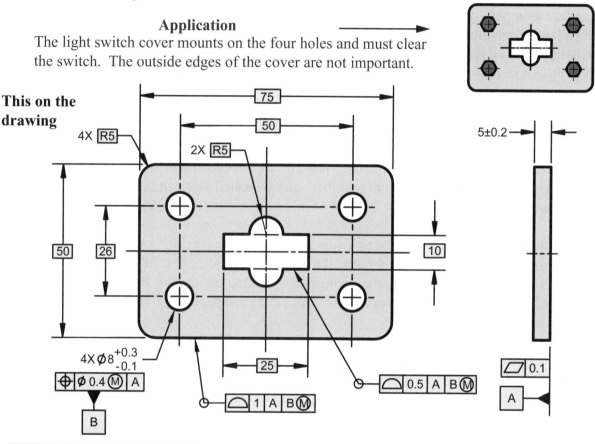

Means this

Four Ø0.4 position tolerance cylinders at MMC within which the axes of the four holes must lie. The tolerance zone cylinders are basically located to each other, and in effect, establish the second and third planes of the DRF.

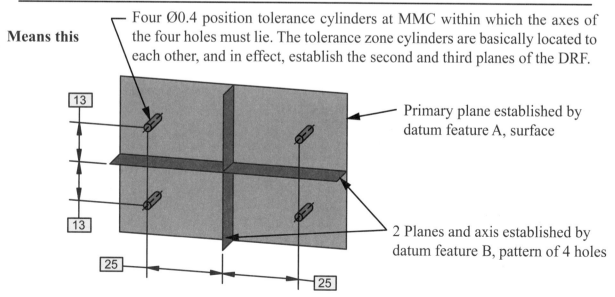

Primary plane established by datum feature A, surface

2 Planes and axis established by datum feature B, pattern of 4 holes

9.5

A pattern of holes used to establish a datum is a common application. Many parts mount on a group of holes. There is no need to identify two holes of the pattern as datum features. The four holes restrict the movement of the part in the X and Y direction and rotationally. Function dictates that the entire "best fit" pattern establishes the DRF.

A pattern of holes will provide function, as well as obtaining maximum manufacturing tolerance, since a pattern of holes is "best fit" during the manufacturing process. Holes are usually processed in one operation. Good parts may show bad if inspection aligns on only two holes. In our part, the edges are unimportant to function. The part mounts on the holes and the surfaces are located from the pattern. Basic dimensions on the drawing, locating other features, may originate from any of the holes, since all features are connected with basic dimensions. The basic dimensions originate from the axis of the tolerance cylinders, not the individual actual produced hole.

To establish a DRF from a pattern of holes on a CMM or open set-up inspection, the pattern of holes must be "averaged, balanced or best fit" to insure conformance to the position requirement. If the position requirement is met, then a DRF can be established on the average of the pattern; without reorienting the part, the remaining features are checked relative to this set-up. Once the DRF origin has been established, it may be basically relocated since all the features are related to each other with basic dimensions.

If inspection chooses only two holes to set-up the part and the part checks good, it is good. If the part checks bad while set up to the 2 holes, it may be "balanced or best fit" similar to a functional gage to insure acceptance. This balancing method of "eking out the last bit of tolerance" may be accomplished by CMM's with appropriate software or by the paper gage method.

A pattern of holes establishing a datum can be referenced at MMB, LMB or RMB. In our example, they are referenced at MMB. The figure below represents a functional gage used to verify features related to the pattern of holes at MMB. The four MMB size pins verify the position tolerance, as well as establish the DRF from which the other features are located.

Functional gage to check location of pattern of holes and related profile specifications

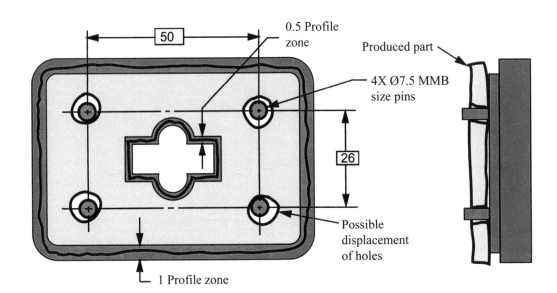

Pattern of Holes as a Datum Feature - Paper Gage Evaluation

This is a simple illustration on how to use the paper gage concept to evaluate a pattern of holes and then use this pattern as a datum feature. The product drawing and the associated discussion of the produced part is shown on the previous pages. All the holes are shown produced at Ø8.3 to keep the example simple.

Notice on the example below, the 4 holes were balanced or aligned in the initial inspection set-up by establishing a center between the top two holes and the bottom two holes. Then, hole #1 was used as the origin to collect data. The data is recorded in the chart below. In this initial evaluation, hole #4 is shown out of location with a position of 0.89. But, when the holes are plotted and evaluated with the paper gage, it can be seen that all the holes as a group are within a 0.6 tolerance zone. See Figure 1 on the next page. After evaluation with the paper gage, the 4 holes are all within their acceptable limit of location.

A pattern of holes as a datum feature can be modified at MMB, LMB or RMB. If a pattern of holes serving as a datum feature are modified at RMB, the datum feature simulators are fixed in relation to each other. The 4 datum feature simulators would expand simultaneously from their MMB to make maximum contact with the datum features to establish the DRF.

In our example, the pattern of holes serving as a datum feature were modified at MMB. This may result in additional datum feature shift tolerance. According to our calculations in the chart, the holes are allowed to lie within a position ring of 0.8 and after evaluation, they are within a position ring of 0.6. When referenced as a datum feature at MMB, the pattern of holes are allowed a displacement or shift equal to the difference between the MMB calculated from the drawing and their actual MMB.

Produced part

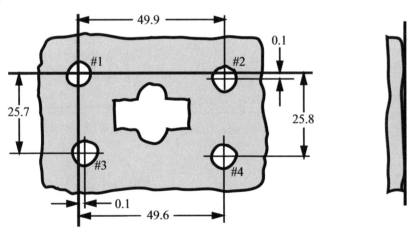

Hole No.	MMC Size	Actual Size	Allowed Position	"X" Dev	"Y" Dev	Actual Position	Acc Rej	Paper Gage Evaluation	Acc Rej
1	7.9	8.3	0.8	0	0	0	A	The pattern of holes fall within 0.6 tol zone to each other after paper gage evaluation.	A
2	7.9	8.3	0.8	-0.1	-0.1	0.28	A		A
3	7.9	8.3	0.8	+0.1	+0.3	0.63	A		A
4	7.9	8.3	0.8	-0.4	+0.2	0.89	R		A

The paper gage evaluation shows the four holes can all be contained within a 0.6 tolerance zone. This is the hole location to each other. According to the tolerance in the feature control frame plus the departure from MMC, the holes are actually allowed up to 0.8 position tolerance and are well within their acceptable limits of location. The origin of the DRF established by the pattern of holes is the center of the 0.6 tolerance ring. The center of the 0.6 ring is approximately -0.13 in the x direction and +0.16 in the y direction from the #1 hole. To establish the DRF from the pattern of 4 holes and make measurements to other features, reset the origin of measurement to this point from any of the actual hole locations.

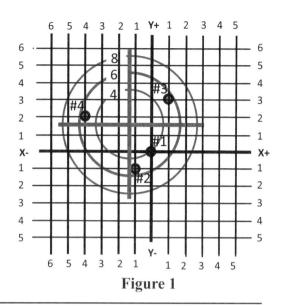

Figure 1

When datum feature B, the pattern of holes, is referenced at MMB, it allows a possible datum feature shift in an amount equal to the difference between the MMB and the datum feature(s) actual MMB. The darkened area in figure 2 is a zone created by striking a 0.8 dia position tolerance zone from the actual location of each hole. The area circumscribed within the arcs is the boundary within which the origin of the DRF may displace. The paper gage mathematically reflects the mechanics of a part fitting in a functional gage. Notice hole #1 falls well within the 0.6 zone and does not contribute to the location of the part.

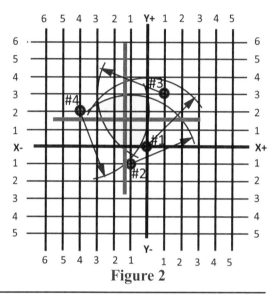

Figure 2

The graphics below represent the 4 datum simulators and the 4 holes all shifted by their basic locations and superimposed upon one another.

Datum feature B modified at MMB

4 virtual pins

If the MMB datum feature modifier is applied to datum feature B, it requires the 4 datum feature simulators to be fixed at basic and be fixed in size at the MMB of 7.5. The datum features can displace outside this boundary. Clearance between the simulators and datum features will allow datum feature shift. Accordingly, it will also allow the part to move or rattle in its mating condition.

Datum feature B modified at RMB

4 expanding pins

If the RMB datum feature modifier is implied for datum feature B, it requires the 4 datum feature simulators to be fixed at basic and expand or progress simultaneously from the MMB to make maximum contact with the datum features. Datum feature shift is nonexistent.

Coaxial Holes as a Datum Feature

Application

The seat latch bracket mounts on a single pin through the two coaxial holes and fits inside the frame. The three square slots engage in the seat rack. The adjustment lever attaches to the hole in the seat latch bracket and provides engagement and disengagement for various locked seat positions.

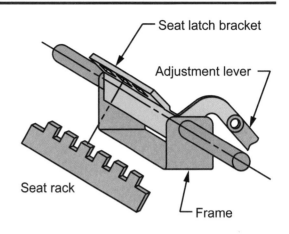

This on the drawing

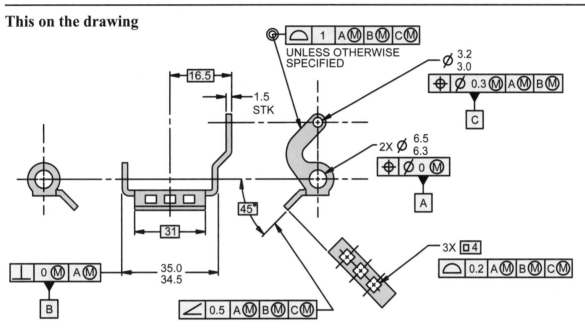

Means this

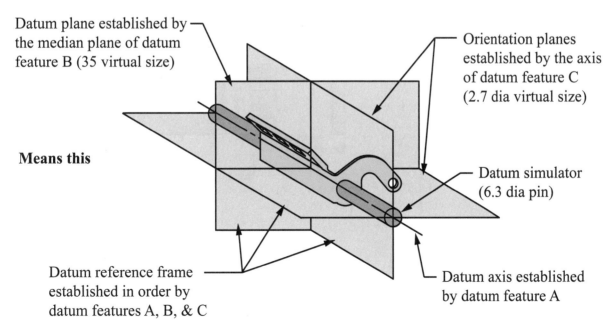

Datum plane established by the median plane of datum feature B (35 virtual size)

Orientation planes established by the axis of datum feature C (2.7 dia virtual size)

Datum simulator (6.3 dia pin)

Datum reference frame established in order by datum features A, B, & C

Datum axis established by datum feature A

On both drawings below, the rear face is defined as datum feature A and the center hole as datum feature B at RMB. The periphery of the part is located with a 2 mm all around profile tolerance to this A, B DRF. The bottom surface is identified as datum feature C to stop rotation.

On the upper part, the four holes are positioned to the A, B, C **at MMB** DRF. On the lower part, the four holes are positioned to the A, B, C **at RMB** DRF. This example illustrates the difference between a tertiary datum feature modified at MMB and another modified at RMB.

On both parts, the simulator for datum feature A is a plane and the simulator for datum feature B, at RMB, is an expanding cylinder.

Effect of tertiary datum plane surface modified at MMB

The simulator for datum feature C is a **plane fixed at 66 MMB** (65 from the center of datum B plus ½ profile tolerance). Datum feature C may rotate and is free to move within the confines established by the MMB boundary. This constrains the rotation of the part.

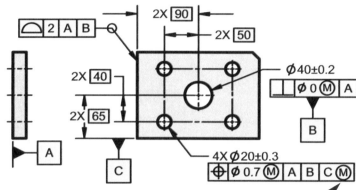

Datum feature C, modified at MMB, establishes a fixed simulator at 66 MMB to orient the DRF to the part.

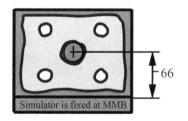

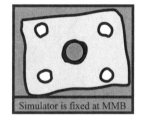

Effect of tertiary datum plane surface modified at RMB

The simulator for datum feature C is a **plane that moves or progresses** from MMB to make maximum contact with the datum feature. This constrains the rotation of the part.

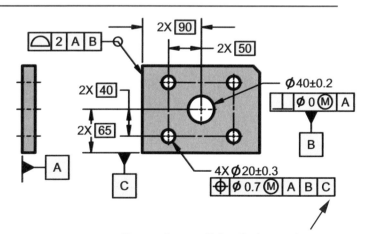

Datum feature C, implied to apply at RMB, establishes a simulator that moves/progresses normal to make maximum contact with the datum feature to orient the DRF to the part.

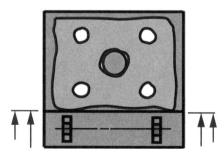

RMB simulator moves normal to make maximum contact, and level the datum feature.

On both parts below, the rear face is defined as datum feature A, the center hole is datum feature B at RMB. The Ø50 surface is located with a 0.2 mm profile tolerance to the A, B DRF. The Ø50 surface is identified as datum feature C to stop rotation.

On the upper part, the remaining contour is located to the A, B, C **at MMB** DRF. On the lower part, the remaining contour is located to the A, B, C **at RMB** DRF. This example illustrates the difference between a tertiary datum feature modified at MMB and another modified at RMB.

Effect of MMB modifier on contoured tertiary datum feature.

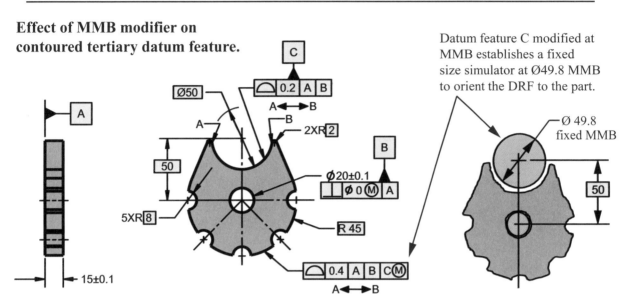

Datum feature C modified at MMB establishes a fixed size simulator at Ø49.8 MMB to orient the DRF to the part.

Datum feature C at MMB establishes the orientation planes of the DRF by a cylindrical simulator fixed at the MMB to constrain rotation of the part. If the datum feature departs from MMB it may rotate within the boundary confines of the datum feature simulator at MMB to establish the DRF.

Effect of RMB modifier on contoured tertiary datum feature.

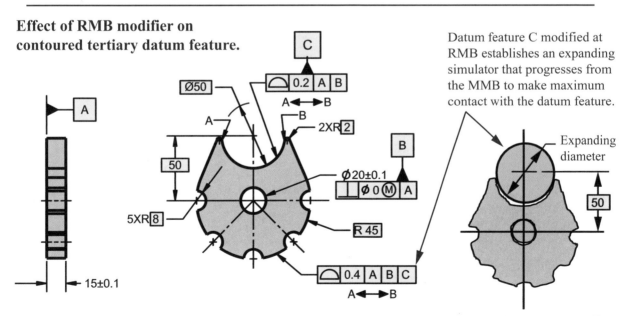

Datum feature C modified at RMB establishes an expanding simulator that progresses from the MMB to make maximum contact with the datum feature.

Datum feature C at RMB establishes the orientation planes of the DRF by an expanding cylindrical simulator that makes maximum contact with the datum feature to constrain rotation of the part.

This assembly contains a rocker that mounts through the large hole with a press fit on the shaft. The rotation of the rocker is based on the engagement of the bottom plane surface with the rest plate. Based on these mating requirements, the A, B DRF is established and shown on the rocker drawing below. The large hole is defined as datum feature A, at RMB. The bottom plane surface is defined as datum feature B to stop rotation.

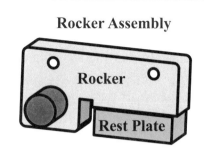

Rocker Assembly

Modifying datum feature modifiers at MMB, RMB, and Basic

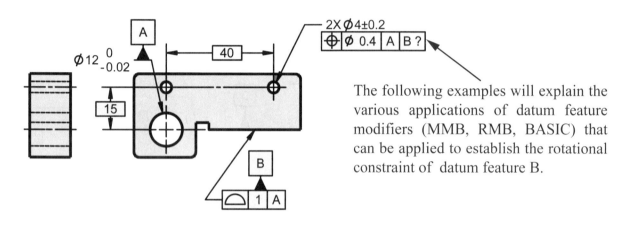

The following examples will explain the various applications of datum feature modifiers (MMB, RMB, BASIC) that can be applied to establish the rotational constraint of datum feature B.

The large hole is identified as the primary datum feature A, implied at RMB. This establishes an expanding cylindrical datum feature simulator inside datum feature A. The axis of this cylinder is datum axis A.

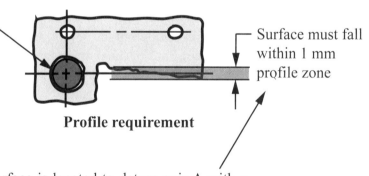

Profile requirement

Surface must fall within 1 mm profile zone

The bottom plane surface is located to datum axis A with a profile tolerance. The surface may lie anywhere within the profile zone. The bottom surface is then referenced as secondary datum feature B and will serve to orient the DRF.

The engagement of datum feature B with its simulator and the ultimate establishment of the DRF is based on the modifier in the feature control frame. The modifier defines the boundary or engaging requirements of the datum feature simulator to datum feature B.

Datum feature B can be modified at MMB, LMB, RMB or set at basic. The illustrations on the next pages define the applications of a few these datum feature modifiers.

Plane surface, datum feature B referenced at MMB

This on the drawing

Means this

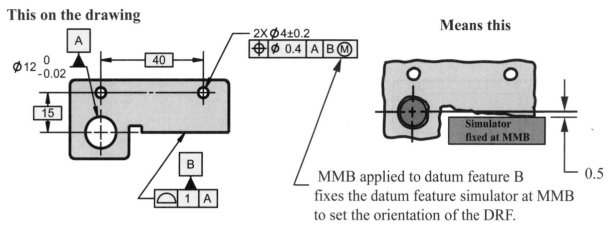

MMB applied to datum feature B fixes the datum feature simulator at MMB to set the orientation of the DRF.

Datum feature B referenced at MMB establishes a datum feature simulator fixed at the MMB of the datum feature, 0.5 below basic (1/2 the profile tol). If the surface is produced high or low, within the profile tolerance, it must always rotate to a fixed stop at the MMB to establish the DRF. Since the datum feature can not be contained by, or contain, an actual mating envelope, it may not pull away, and must always remain in contact with the simulator.

Datum feature B modified at MMB might be selected if the corresponding mating feature to datum feature B were fixed in the assembly at MMB.

Plane surface, datum feature B referenced at RMB

This on the drawing

Means this

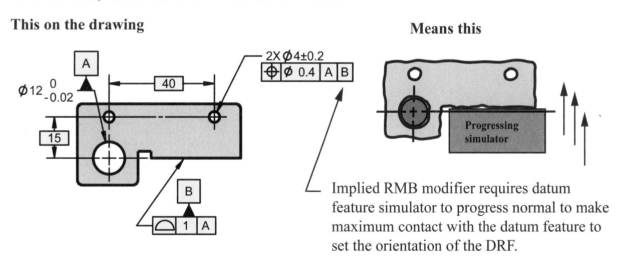

Implied RMB modifier requires datum feature simulator to progress normal to make maximum contact with the datum feature to set the orientation of the DRF.

Datum feature B implied at RMB requires the simulator to grow/expand/progress from the MMB, within the allowable profile tolerance, to make maximum contact with the datum feature. This will result in the datum feature rotating level, or parallel, to make maximum contact with the simulator. The DRF is oriented (parallel) to the datum feature simulator. The datum feature may only be rotated within the boundaries of the profile tolerance.

Datum feature B modified at RMB might be selected if the corresponding mating feature to datum feature B were adjustable to make it level or parallel in the assembly.

Plane surface, datum feature B referenced at basic

This on the drawing

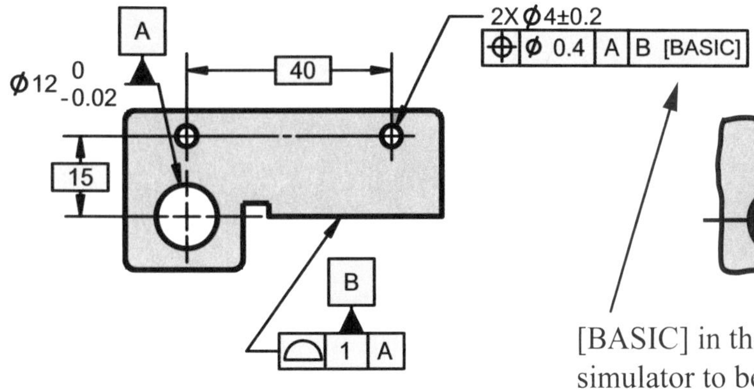

Means this

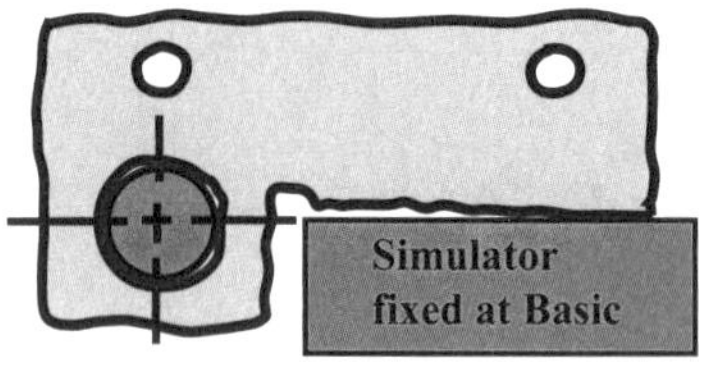

2X ⌀4±0.2

| ⊕ | ⌀ 0.4 | A | B [BASIC] |

⌀12 $^{0}_{-0.02}$

40

15

A

B

| ⌒ | 1 | A |

Simulator
fixed at Basic

[BASIC] in the feature control frame requires the simulator to be fixed at basic to set the orientation of the DRF.

Datum feature B referenced in brackets at [BASIC], or abbreviated [BSC], establishes the datum feature simulator at the basic location of zero from datum axis A. The feature must always stay in contact with the datum feature simulator. If the surface is manufactured high or low, within the profile tolerance, it must always rotate to a fixed stop at basic to establish the DRF.

Datum feature B, referenced at [BASIC], might be selected if the corresponding mating feature to datum feature B were fixed in the assembly at basic. This is the most common reference for unopposed surfaces controlling orientation of a DRF.

Compartment Lid Assembly

The compartment lid mounts on the pivot pin on the three radii in the tabs. The geometry of the radii and the three tabs are enough to contain the pin. The lid rotates on the pivot pin to open and close a compartment.

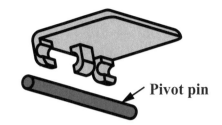

Pivot pin

The drawing of the compartment lid has a datum reference frame established based on how the part mounts in the assembly. The three radii are defined with basic dimensions and have a profile tolerance applied to allow variation from basic. The three radii are identified as datum feature A. Datum feature A is a good example of how an irregular feature of size can be toleranced and then used as a datum feature.

The inside width of the outer two tabs are identified as datum feature B. This datum feature is qualified back to datum A with an orientation (perpendicularity) of zero tolerance at MMC. The inside face of the lid is located to datum A and B with a profile tolerance and is identified as datum feature C. This datum feature establishes the orientation of the datum reference frame. All remaining features are located (profiled) to the A, B, C DRF.

Drawing for compartment lid based on mounting conditions

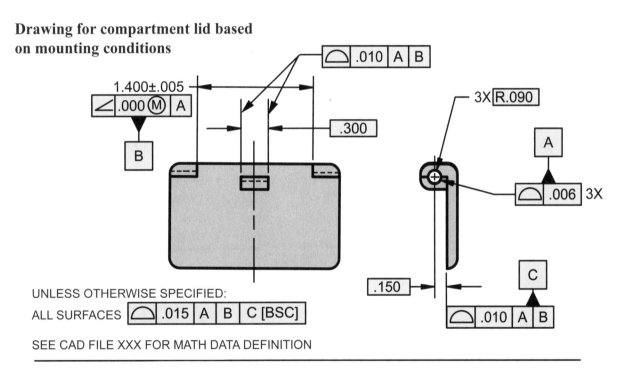

UNLESS OTHERWISE SPECIFIED:

ALL SURFACES ⌓ .015 | A | B | C [BSC]

SEE CAD FILE XXX FOR MATH DATA DEFINITION

Workshop Exercise 9.1 Circle the correct answer.

1. What is the min/max diameter of the pivot hole boundary considering the profile tolerance?
   .174/.186   **.177/.183**   **.168/.192**   **.084/.096**

2. What is the max/min of the center tab?
   .290/.310   **.300**   **Not defined**   **.295/.305**

3. How many translation/rotation degrees of freedom does primary datum feature A constrain?
   2 trans/1 rot   **0 trans/2 rot**   **2 trans/2 rot**   **1 tran/1 rot**

4. What is the maximum flatness allowed on datum feature C?
   .015   **.010**   **.005**   **Not defined**

Rotational Control - Slot at RMB with and without Translation Modifier

The two parts below are identical except for the establishment of the A, B, C, DRF on the position tolerance for the two holes. In both parts, the back face is the primary datum feature A and it establishes a plane, the center hole is the secondary datum feature B, at RMB, which establishes an axis. The tertiary datum feature C is a slot that establishes the orientation. The upper part has datum feature C modified at RMB. The lower part has datum feature C modified at RMB with a translation modifier applied. The illustrations below explain the differences in establishing the DRF with a tertiary datum feature applied with and without a translation modifier according to the datum feature simulator requirements defined earlier on page 6.13.

Tertiary datum feature implied at RMB

Without translation modifier, datum axis B and center plane of simulator C are coincident.

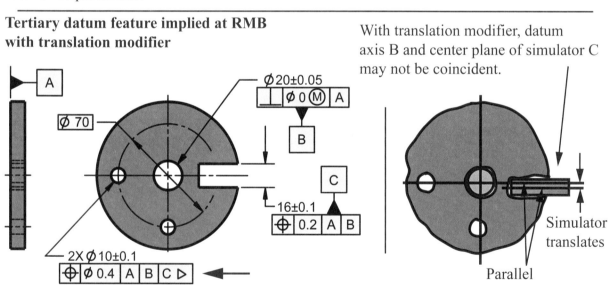

Datum feature C implied at RMB requires the simulator to expand into the slot while fixed at basic location. Since the simulator is fixed at basic and expands, it may engage a "cocked" slot to establish the DRF. This is the most common application and is usually used for key seats and location requirements.

Tertiary datum feature implied at RMB with translation modifier

With translation modifier, datum axis B and center plane of simulator C may not be coincident.

Simulator translates

Parallel

Datum feature C implied at RMB requires the simulator to expand into the slot. The addition of the translation modifier does not require the simulator to be fixed at basic. It allows the simulator to translate up and down from the basic location within the position tolerance using the slot orientation to establish the DRF. This is a less common application and is usually used when the slot width is longer in length and is used for orientation.

Translation Modifier

The datum feature simulator requirements state that during datum reference frame construction, all simulators are fixed at basic unless otherwise specified. There are some cases at RMB, where it is desired to allow a simulator to translate. This can be accomplished by referencing the datum feature with a translation modifier. Translation modifiers are usually not necessary for datum features modified at MMB or LMB because the boundary for the datum feature includes the location error.

In the example below, datum feature A is the back surface, datum feature B is the center hole at RMB, and datum feature C is the width of the bar at RMB. Two comparison feature control frames are shown for the two .313 holes. The explanation for each feature control frame is shown below.

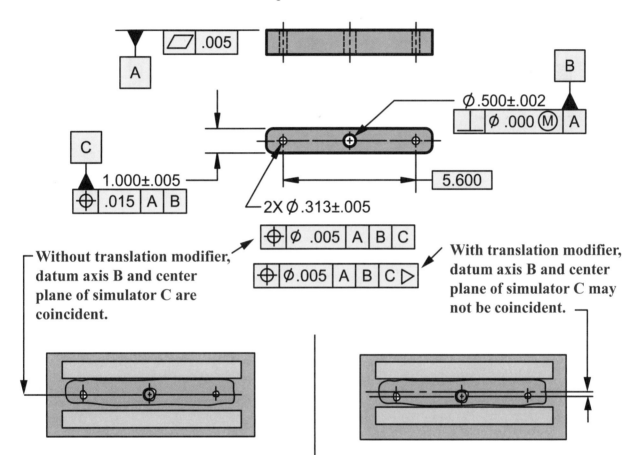

Without translation modifier, datum axis B and center plane of simulator C are coincident.

With translation modifier, datum axis B and center plane of simulator C may not be coincident.

In the upper feature control frame, datum feature C is modified at RMB. This requires the simulator for datum feature C to collapse about the feature while remaining perpendicular to datum plane A and centered about datum axis B.	In the lower feature control frame, the translation modifier follows the tertiary datum feature C at RMB. This translation modifier allows the datum feature simulator to translate or move and not be fixed.
If datum feature C is off center to datum axis B, the simulator will only contact one side of datum feature C. The median plane of the simulator establishes the orientation of the DRF.	Datum feature C simulator will collapse while remaining perpendicular to datum feature A and be able translate about datum axis B in an amount equal to the .015 position tolerance. The median plane of the simulator establishes the orientation of the DRF.

Datum features referenced in a feature control frame establish a datum reference frame based on an order of precedence. These datum features constrain the six degrees of freedom on a part. See datum feature simulator requirements on page 6.13.

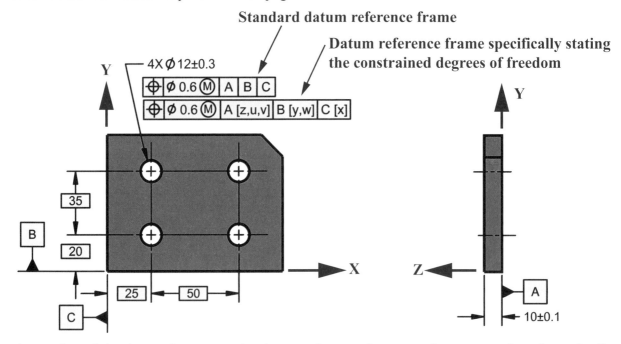

The mating of the datum features to the datum reference frame can be expressed mathematically as shown by the lower datum reference frame for the 4 holes. The lower datum reference frame is an example to explicitly state the degrees of freedom constrained by each datum feature shown in the upper feature control frame.

It is usually not necessary to state the degrees of freedom constrained, as this default mating condition is clearly defined by the simulator requirements. However, in some unusual circumstances, it may be necessary to override the standard reference frame requirements and impose other mating constraints. This overriding of the standard datum reference frame can be accomplished with a customized datum reference frame. When using a customized DRF, it is mandatory to label the x, y and z axes as shown above.

Customized datum reference frames are not common and should only be used lightly and by an experienced geometric tolerancing user when it is necessary to override the constraints dictated by a standard datum reference frame.

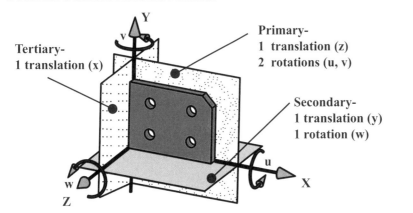

Translation Controls
X= Translation in "X" direction
Y= Translation in "Y" direction
Z= Translation in "Z" direction

Rotation Controls
u= Rotation about "X" axis
v= Rotation about "Y" axis
w= Rotation about "Z" axis

Customized Datum Reference Frame - Square Hole and Slot

The drawing below illustrates an example of a customized DRF, at RMB, being applied to a leveling bar. The back face on the bar is identified as datum feature A and constrains translation z and orientation u and v. The square hole is identified as datum feature B. Because of its physical geometry, and being a secondary datum feature, the square hole will constrain translations x and y and orientation w. In a standard DRF, all degrees of freedom would be constrained.

In this particular, special requirement, it was necessary to use the slot at the end for orientation rather than the shorter, non-functional square hole. The feature control frame for the two holes is a customized datum reference frame that specifically states the constrained degrees of freedom for each datum feature.

A customized datum reference frame specifically states the degrees of freedom that are constrained by each datum feature.

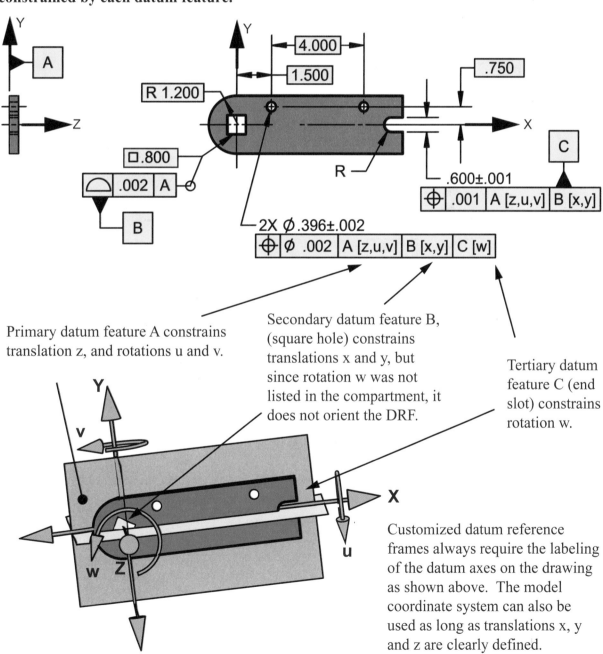

Primary datum feature A constrains translation z, and rotations u and v.

Secondary datum feature B, (square hole) constrains translations x and y, but since rotation w was not listed in the compartment, it does not orient the DRF.

Tertiary datum feature C (end slot) constrains rotation w.

Customized datum reference frames always require the labeling of the datum axes on the drawing as shown above. The model coordinate system can also be used as long as translations x, y and z are clearly defined.

This Sight Holder is an example of a more complex application of a customized DRF. The part mounts on the bottom surface, datum feature A and the width, datum feature B. The mating part allows adjustment of the Sight Holder up, down, and side-to-side, but there is no rotational adjustment.

The desire is to have the bottom mounting face and the width set orientation, but not location. Location for the two .187 holes is to originate from the large hole, datum feature C, and the datum feature D face.

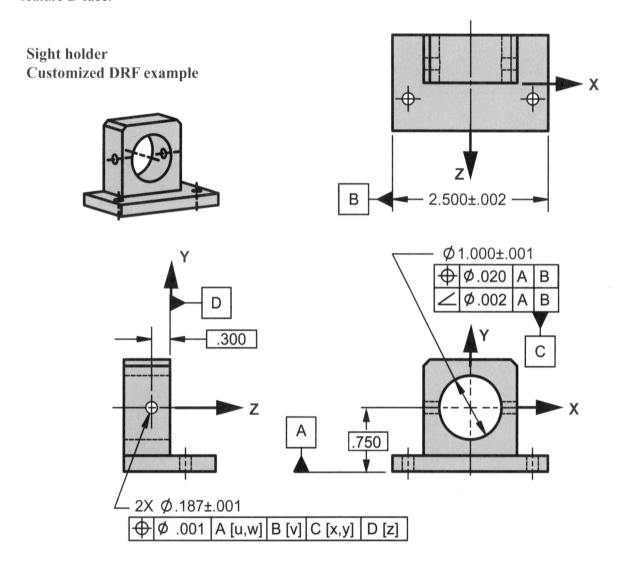

Sight holder
Customized DRF example

The customized datum reference frame for the two .187 holes specifically states the requirement for each datum feature. Datum feature A constrains rotation u and w, but not translation y. Datum feature B constrains rotation v, but not translation x. All rotations are now constrained. Datum feature C provides translation in x and y and datum feature D constrains translation in z. Translation in x was not necessary to locate the two .187 holes but was listed to complete the DRF for possible control of other features on the part.

Note: The ISO standards do not explicitly state the degrees of freedom in a feature control frame but rather states the overriding datum that each datum feature establishes. The terms PL for plane, L for line, and PT for point, will follow the datum feature letter in the feature control frame.

Multiple Datum Reference Frames

Datum reference frames (DRF) are selected based on function. In most cases, the establishment of one DRF on a part is sufficient to define the product. However, there are instances where additional DRF may be needed. This usually will occur when a part has multiple mating requirements, such as a gearbox housing or engine block. The location of certain features, such as oil pan mounting holes, have little to do with the main DRF on the engine block. In these cases, a secondary DRF may be established. In other cases, even though the part may have multiple mounting requirements, engineering will decide to stay with only one DRF. This is done when the tolerances are already relatively large and it certainly keeps the definition simple. The establishment of multiple DRF on a part should only be done where necessary.

The assembly below shows an adapter that may be considered for the establishment of multiple datum reference frames. Notice that the adapter mounts to the main housing on the bottom surface, aligned with the lower pilot, and fastened with four bolts. The electric motor mounts to the upper surface, aligned with the upper pilot, and fastened with four bolts. The relationship of the features on the bottom of the adaptor are important to each other. The relationship of the features on the top are also important to each other. But, the relationship of the two groups of features is relatively less important. The next pages will show how to establish multiple datum reference frames and relate the individual features while keeping the relationship between the groups of features.

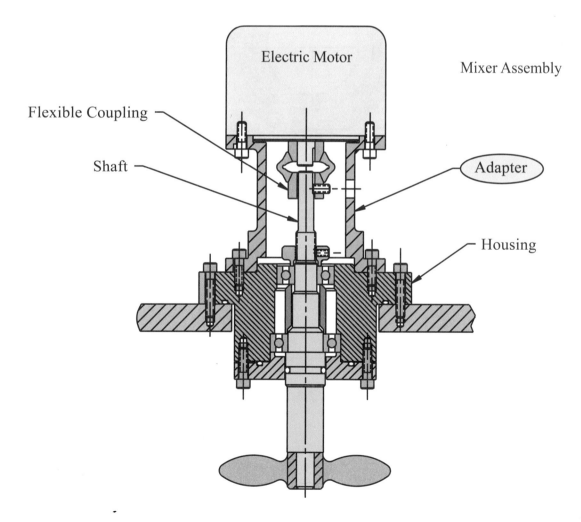

Establishing a Relationship Between Two DRF

The drawing below illustrates the adaptor which has two datum reference frames established. The reference frames also have an established relationship to each other. A DRF, we will call A, B, has been established by the bottom surface and the lower pilot. The datum features in this A, B DRF are qualified with a flatness and perpendicularity tolerance. The upper surface has been located up and down with a profile specification and then refined for orientation with a parallelism control. The upper pilot has been located with a position control. The upper surface and pilot are now related to the A, B DRF.

A secondary datum reference frame, C, D, is established from these upper features. The C, D DRF may move up and down within the profile zone, move diametrically within the position zone, and tilt within the parallelism zone all relative to the A, B DRF. The four .290 holes are then located to this new C, D DRF.

The next page illustrates the two datum reference frames and how the secondary C, D datum reference frame is related to the main A, B datum reference frame.

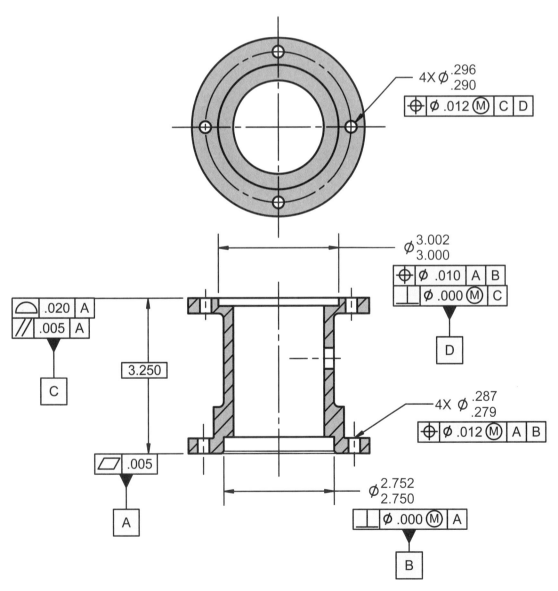

1. The upper face and pilot are located and oriented by the profile and position tolerance zones related to the A, B datum reference frame. The upper face orientation and flatness is further refined by the parallelism tolerance zone.

2. The upper face and pilot establish the secondary C, D datum reference frame. This C,D DRF may move up and down within the profile zone, move diametrically within the position zone, and tilt within the parallelism zone all relative to the A, B DRF.

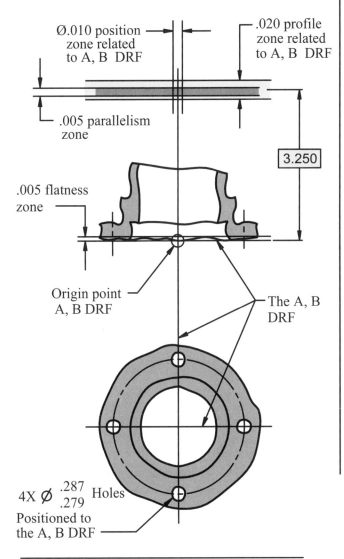

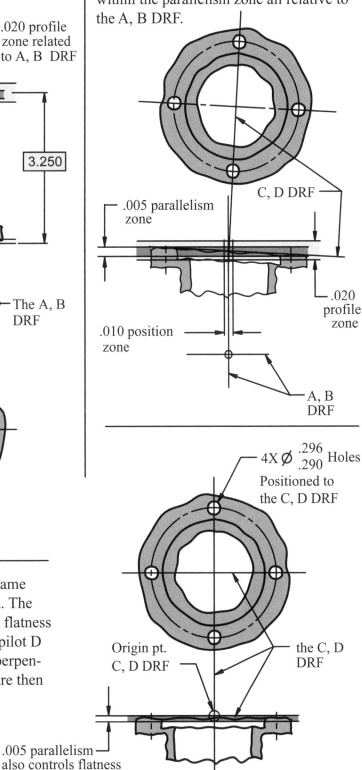

3. The secondary datum reference frame datum features C and D are qualified. The parallelism tolerance will control the flatness on the datum feature C surface. The pilot D is related to datum feature C with a perpendicularity tolerance. The four holes are then related to the secondary C, D DRF.

Workshop Exercise 9.2 - Customized Datum Reference Frame

The drawings below represents two parts that are identical in appearance but have different design requirements for the two .400 diameter holes. The upper drawing has a standard datum reference frame and references datum feature A as the primary datum feature at RMB. Because of the unique geometry of datum feature A (conical shape), it constrains 3 translational and 2 rotational degrees of freedom. The engineer feels the taper on the plug is too shallow to adaquetly constrain the translation along the axis and wants to use the large back face to arrest this translation.

On the drawing below create a customized datum reference frame to position the two .400 diameter holes and use datum feature A as a primary datum feature at RMB. Release the translational degree of freedom along the axis, and instead use the large face to constrain this translation. Make sure to create and label axes on the drawing to clearly identify your x, y and z directions. Study the other customized datum reference frames in this unit if you need help.

Standard Datum Reference Frame

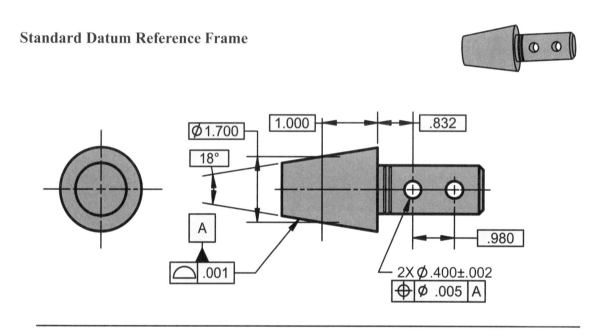

Customized Datum Reference Frame

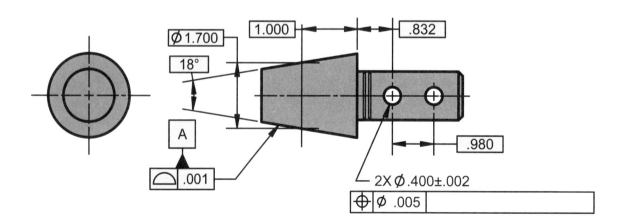

Unit 10

Form Tolerances

Form Tolerances

Form tolerances control the form of individual features.
No datums are allowed

Symbol	Form Tolerance	Common Shape of Tolerance Zone	2D / 3D	Application of Feature Modifier
—	Straightness	2 Parallel Lines	2D	**No** Controls Surface
⬭	Flatness	2 Parallel Planes	3D	
O	Circularity (Roundness)	2 Concentric Circles	2D	
⌭	Cylindricity	2 Concentric Cylinders	3D	
—	Straightness - Axis	Cylindrical	3D	**Yes** Controls Axis or Median Plane
⬭	Flatness - Median Plane	2 Parallel Planes	3D	

*

Overview:

There are six form tolerances shown in the chart above. Form tolerances control the form of an individual feature and there are no datum references allowed.

As we know, the Limits of Size (Rule #1) states that the size of an individual feature will control the size as well as the form of that feature. In effect, this means that the size tolerance of an individual feature will control how much the form will vary. Generally, in order to refine the form control on size features, a form tolerance is used.

Flatness and straightness-line elements are identical with the exception that flatness is a 3D control (tolerance zone is 2 parallel planes) and straightness - line elements is a 2D control (tolerance zone is 2 parallel lines). Cylindricity and circularity are also identical with the exception that cylindricity is a 3D control (tolerance zone is two concentric cylinders) and circularity is a 2D control (tolerance zone is two concentric circles).

* Flatness of a median plane is a new concept for the ASME Y14.5-2009 standard. This concept is an identical replacement of Straightness of a median plane found in the 1994 standard. It is merely a name change to make the standard more consistent.

Straightness of an axis and flatness of a median plane are the only form tolerances that control the axis or median plane, and therefore may have the material condition modifiers MMC, LMC, and RFS applied. All the other form tolerances control the surface of the feature.

The top four form tolerances in the table above (with the exception of the circularity average diameter concept) must be within the size and form requirements of Rule #1. The application of any of these four form tolerances to a feature will refine the form even further. In the case of straightness-axis or flatness-median plane, it will relax the form control provisions of Rule #1.

Flatness

Flatness is the condition of a surface having all elements in one plane. A flatness tolerance specifies a tolerance zone defined by two parallel planes within which the surface must lie.

This on the drawing

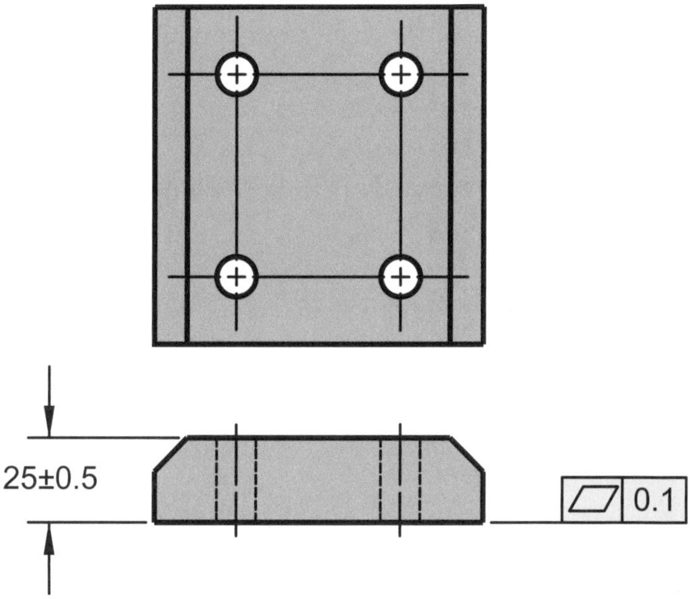
25±0.5
0.1

Application

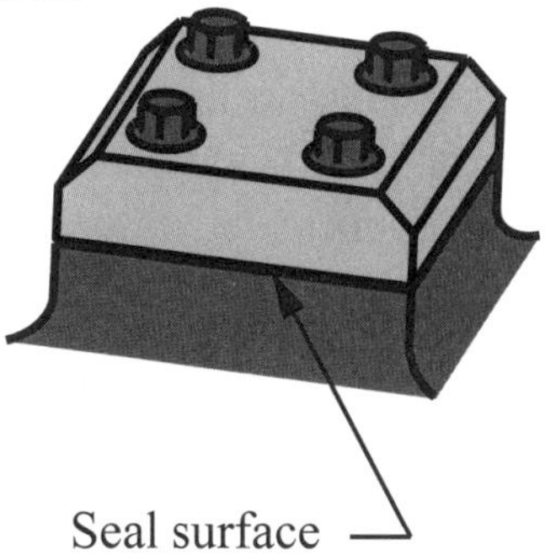
Seal surface

The surface must lie between two parallel planes 0.1 apart. In addition, the feature must be within the limits of size.

Means this

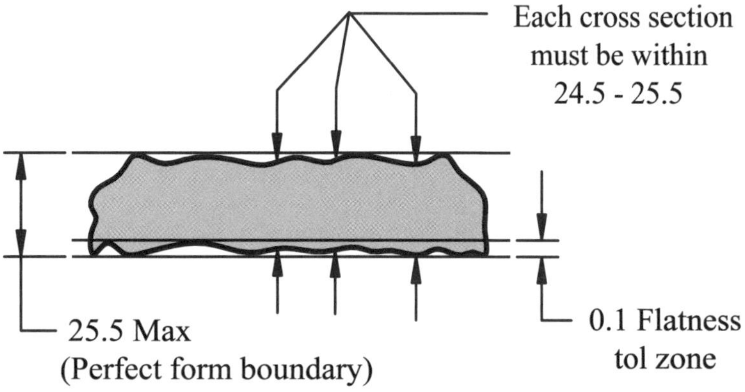
Each cross section must be within 24.5 - 25.5
25.5 Max (Perfect form boundary)
0.1 Flatness tol zone

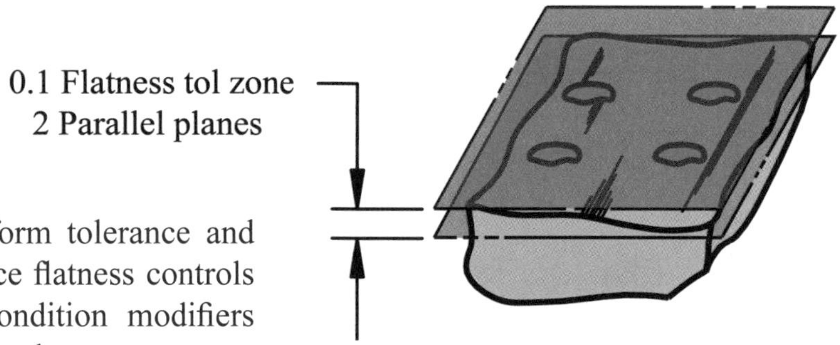
0.1 Flatness tol zone
2 Parallel planes

Flatness of a surface is a form tolerance and datums are not allowed. Since flatness controls the surface, the material condition modifiers MMC and LMC are not allowed.

Flatness Verification

There are many ways to check a flatness specification. Some are more precise than others. As with the verification for any geometric tolerance, the procedure used for verification will depend on many factors. How many parts are there to check? Is this the first part produced or the 1000th? Is the tolerance well within the process capability? Are statistical process controls being done? How tight is the tolerance? What kind of equipment is available. Is it an in-process check, or a final check?

All of these factors and many more may have an effect on how the part verification is developed by quality control. The procedures for verification should be recorded in the dimensional measurement plan and coordinated with anyone who is involved with the part.

The measurement techniques shown below are for illustration and background information. These procedures are intended to assist the reader in understanding the concepts. In addition to the methods shown below, there are many other ways to check flatness, including the use of a coordinate measuring machine (CMM).

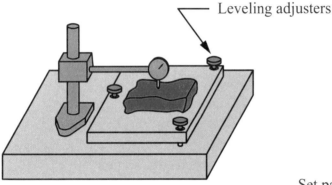

Leveling adjusters

Level part on surface plate using the three adjusters and move indicator over surface. The readings must not exceed flatness tolerance. This is a good check but it may be time consuming.

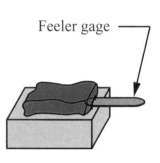

Feeler gage

Set part on surface plate and use feeler gage. This is a quick check used for larger tolerances. It may miss concave variations.

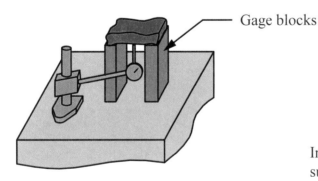

Gage blocks

Set part upside-down on gage blocks of equal height, then indicate underneath. This is also a precise check but cannot indicate on surface resting areas.

Indicate surface through a hole in the surface plate while sliding part. This is a good in process check. It may misread on convex parts.

Flatness Design Considerations

The manufacturability or producibility of a flatness tolerance, or any geometric tolerance for that matter, is not just dependent on the size of the tolerance zone. The size of the controlled surface is also a factor. A large geometric tolerance in conjunction with a large feature can be more difficult to produce than a small geometric tolerance on a small feature. When selecting geometric tolerances, the designer should consider the size of the controlled feature as well as the size of the tolerance zone.

Flatness is often used to qualify datum features. Specifying a flatness control on a primary datum surface will limit the amount of rock or instability of the surface relative to the datum reference frame. The rock or instability cannot be eliminated, it can only be minimized through a flatness control. There is an old "general rule" that states: "Datum features should be as good as the features that are related to it." This rule still holds good today with some clarification.

It stands to reason that if features are related to a datum feature that is unstable and can rock, then the feature that is related to the unstable datum surface will rock also. But, this does not mean that if the flatness tolerance is .005, that all the related features will rock .005. When considering the flatness tolerance, the designer should also consider the size, distance and relationship geometry of the datum feature relative to the related features.

Flatness is often applied to sheet metal, flexible or flanged surfaces that will be bolted, screwed or welded together. If the parts will bend or deform when assembled, the flatness tolerance may not control rock or instability, but rather how much the parts will be deformed or stressed in the assembly process.

Flatness is also used on gasket or seal surfaces. The flatness tolerance will control the size of the gap or crack when parts are assembled. A flatness on two mating surfaces of .005 each could result in a .010 total gap. Other examples of uses for a flatness control include surfaces used for heat transfer or electrical contacts, thrust bearings, clutches, and optics.

There is sometimes a question of how flatness affects the surface finish. The surface finish obviously must fall within the flatness tolerance zone, but it is usually on a much smaller scale. Surface finish is a localized check from an adjacent peak to valley to determine roughness on a surface. Flatness controls an entire surface at once. Because surface finish is a localized check, it may also be called out on round and curved surfaces whereas flatness can only be specified on flat surfaces.

In order to meet functional requirements and still meet producibility requirements, the designer may consider alternatives to specifying tight tolerances. The notation, MUST NOT BE CONCAVE or MUST NOT BE CONVEX, may be placed under the feature control frame as applicable. Another alternative is shown on the next page, using flatness per unit area.

Flatness Per Unit Area

Flatness may be applied on a unit basis as a means of preventing an abrupt surface variation within a relatively small area of the feature. This can be thought of as a rate of change. The unit variation can be used in combination with a total tolerance or alone. The zone for the unit variation can be specified as a square or diameter.

Note: This type of unit basis control may be used with straightness, perpendicularity, profile, and other geometric controls to obtain similar effects.

This on the drawing

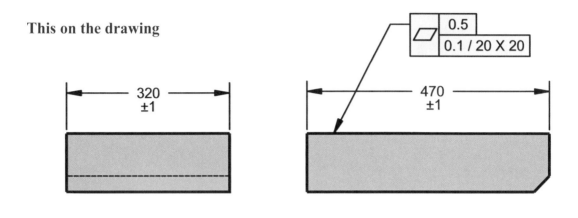

The entire surface must lie within two parallel planes 0.5 apart. In addition, each 20 X 20 unit of the feature must lie within two parallel planes 0.1 apart.

Means this

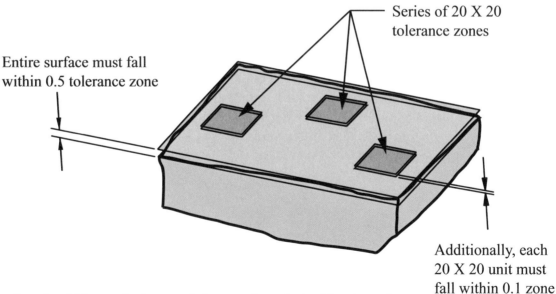

Entire surface must fall within 0.5 tolerance zone

Series of 20 X 20 tolerance zones

Additionally, each 20 X 20 unit must fall within 0.1 zone

Caution should be used when applying a unit control without specifying a maximum limit. Each unit of the feature must be within the specified unit tolerance, but can produce a cumulative effect if allowed to continue over the entire length of the feature without a specified maximum value.

Two Types of Straightness and Two Types of Flatness

Straightness:

There are two types of straightness controls. The two controls are different and will be discussed separately. The two types of straightness use the same geometric characteristic symbol. The manner in which the feature control frame is placed determines which control applies.

1. Straightness - line elements: This is a 2D specification and controls line elements of surfaces. No material condition modifiers are allowed. This control refines and further restricts the form control provisions of the Limits of Size (Rule#1) if applicable. This control may also be applied to features without a size tolerance. The feature control frame is directed to the surface.

2. Straightness - axis: This is a 3D specification and controls the derived median line of features of size (usually holes, pins, or shafts). This control relaxes the form control provisions of the Limits of Size (Rule#1) and the perfect form boundary at MMC may be violated. Material condition modifiers are applicable. The feature control frame is attached to, placed under, or associated with, the size tolerance.

Flatness:

Like Straightness, there are also two types of flatness controls. The two controls are different and will be discussed separately. The two types of flatness use the same geometric characteristic symbol. The manner in which the feature control frame is placed determines which control applies.

1. Flatness - surface: This control is by far the most common. It is a 3D specification and controls surfaces. No material condition modifiers are allowed. This control refines and further restricts the form control provisions of the Limits of Size (Rule#1) if applicable. This control may also be applied to features without a size tolerance. The feature control frame is directed to the surface.

2. Flatness - median plane: This is a 3D specification and controls the derived median plane of features of size (usually tabs, slots, or widths). This control relaxes the form control provisions of the Limits of Size (Rule#1) and the perfect form boundary at MMC may be violated. Material condition modifiers are applicable. The feature control frame is attached to, placed under, or associated with, the size tolerance.

Note: Straightness of a median plane is a concept found in the past ASME Y14.5M-1994 standard. The concept is still usable but it has been renamed flatness of a median plane in the current ASME Y14.5-2009 standard. See Flatness of a median plane later in this unit.

Straightness - Line Elements Design Considerations

Straightness - line elements is a surface control and when applied to a feature, it will tighten or further restrict the straightness control provisions of Rule #1. This type of control might be used on dowel pins or shafts where the straightness must be better than provided with the size control.

It can also be used in conjunction with, and as a refinement to other geometric tolerances. It can be used on corrugated material where control in one direction is more important than another.

It might also be specified on features that are not controlled by the form provisions of Rule #1, such as: stock sizes or commercially purchased parts.

Caution: As we know, the straightness 2D line elements apply in the view in which shown. Manufactured parts are not perfect and will have inherent variations on all the surfaces. In some instances, when the orientation of straightness tolerances or specifications are critical, it may be unclear on how to orient the part in order to make the straightness line element checks. Since straightness cannot be applied with datums, the designer might consider the profile of a line control with datums.

Straightness-Axis / Flatness-Median Plane Design Considerations

Straightness of an axis and Flatness of a median plane are form controls that are often called "the renegades," as they react differently than the other form tolerances.

In most of the form tolerances (with the exception of circularity, average diameter concept) the feature must be within the size and form requirements of the Limits of Size (Rule #1). The application of a form tolerance to the surface of a feature will tighten or restrict the form even further. Straightness-axis and Flatness-median plane will relax the form control of Rule #1.

Straightness-axis and Flatness-median plane are the only form tolerances that are applied to the axis or median plane of a feature. They are also the only form controls that can have a modifier applied (MMC, RFS or LMC). The surface form tolerances are only surface controls and therefore, can not have material condition modifiers applied.

Straightness of an axis or median plane is used when the form requirements of Rule #1 are too restrictive. Examples of use for straightness of an axis would be on long shafts or rods and deep drilled holes. The local size of the feature may be important but the part may bend or bow within a relatively large straightness-axis tolerance.

Examples of use for flatness of a median plane would be on shims or large machined plates. The local size of the feature may be important but the part may bend or bow within a relatively large flatness-median plane tolerance.

Straightness Line Elements Applied on a Flat Surface

Straightness - line elements is a condition where an element of a surface is a straight line. This establishes a tolerance zone within which each considered line element must lie. A straightness tolerance is applied to the feature in the drawing view in which it is to apply. Note the feature control frame is directed to the surface because it is a surface control.

This on the drawing

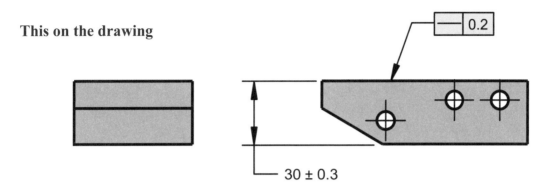

30 ± 0.3

The surface, in the view shown, must lie between a series of two parallel lines 0.2 apart. In addition, the feature must be within the limits of size. The straightness on the surface is a refinement of the form provisions defined within the size tolerance.

Means this

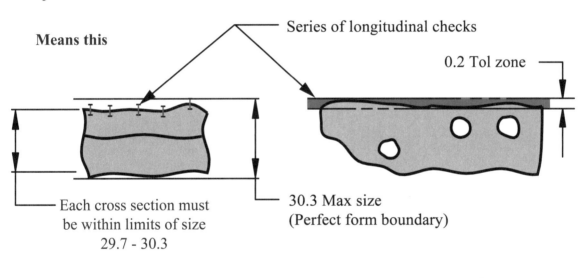

Series of longitudinal checks

0.2 Tol zone

Each cross section must be within limits of size 29.7 - 30.3

30.3 Max size (Perfect form boundary)

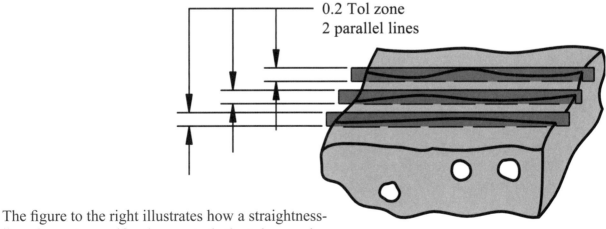

0.2 Tol zone
2 parallel lines

The figure to the right illustrates how a straightness-line elements specification controls the tolerance in a series of 2D line elements in the view shown.

Straightness Line Elements in Two Directions

Since straightness - line elements is a 2D control, a feature can have different straightness values specified in different views as shown below. The tolerance applies in the view in which it is shown.

This on the drawing

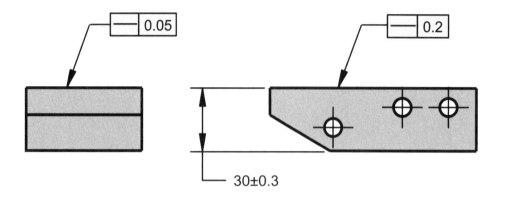

The surface, in the view shown, must lie between a series of two parallel lines 0.2 apart. In addition, the feature must be within the limits of size.

Means this

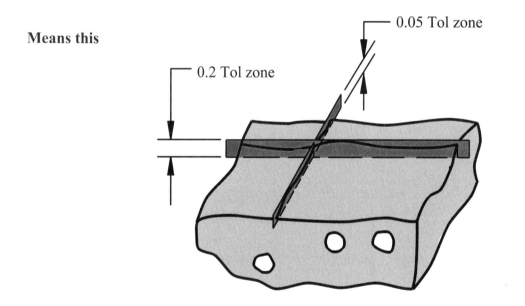

If the location or orientation of these tolerance zones is important, consider the use of profile of a line with datums.

Straightness Line Elements on a Pin

Straightness - line elements can be applied to the surface of a pin. The specification defines a tolerance zone in which the line elements on the surface must lie. Note the feature control frame is directed to the surface to make it a surface control.

This on the drawing

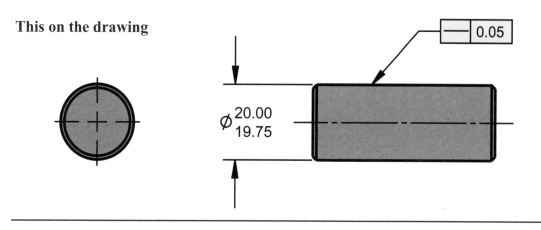

Each longitudinal element of the surface must lie between two parallel lines 0.05 apart where the two lines and the nominal axis of the part share a common plane. In addition, the feature must be within the limits of size, including perfect form at MMC. Straightness - line elements will control wasting, barrelling, and bending of the feature. It does not control taper.

Means this

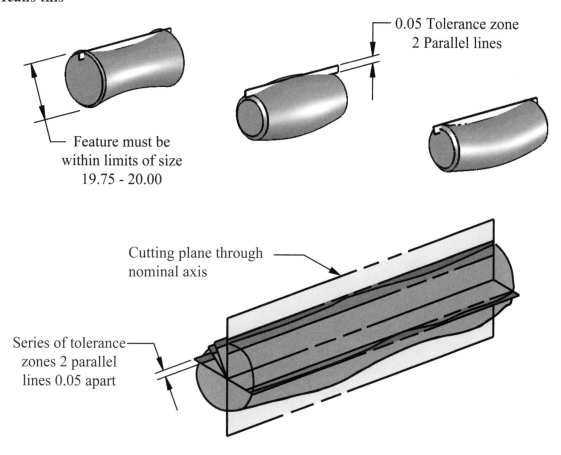

Straightness of an Axis

Straightness of an axis is a condition where an axis is a straight line. The specification defines a tolerance zone in which the derived median line must lie. This type of control is used where the local size of the pin is important, but the pin can bend or bow beyond the perfect form at MMC requirement defined by Rule #1. Note the feature control frame is associated with the size tolerance to make it an axis control.

This on the drawing

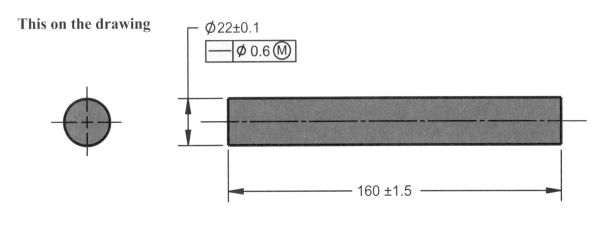

The derived median line of the feature's actual local sizes must be within a cylindrical tolerance zone of 0.6 at MMC. As the local size departs from MMC, an increase in the local diameter of the tolerance cylinder is allowed equal to the amount of such departure. In addition, each circular element must be within the limits of size.

Means this

The derived median line of the feature must lie within the specified dia tol zone.

A combination of the size and straightness tolerance requires that the entire feature must lie inside of a virtual boundary of 22.7

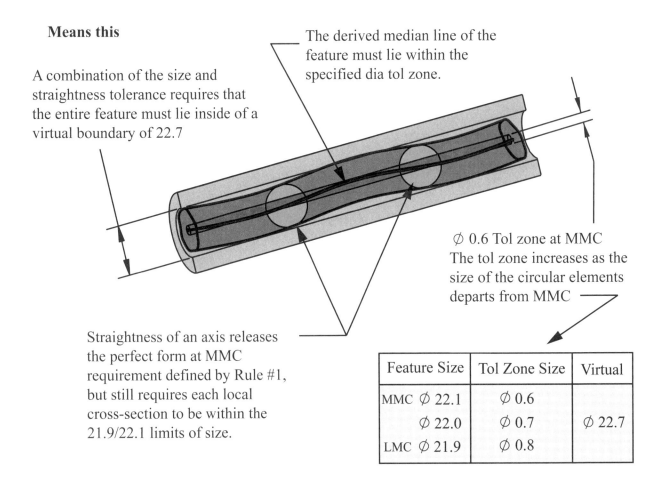

⌀ 0.6 Tol zone at MMC
The tol zone increases as the size of the circular elements departs from MMC

Straightness of an axis releases the perfect form at MMC requirement defined by Rule #1, but still requires each local cross-section to be within the 21.9/22.1 limits of size.

Feature Size	Tol Zone Size	Virtual
MMC ⌀ 22.1	⌀ 0.6	
⌀ 22.0	⌀ 0.7	⌀ 22.7
LMC ⌀ 21.9	⌀ 0.8	

Flatness of a Median Plane

Flatness can be applied to the median plane of a non-cylindrical feature. This is the same concept as straightness of an axis. Flatness of a median plane has replaced the past practice, straightness of a median plane. Since the specification is controlling the median plane, the modifiers RFS, MMC, and LMC may be applied. Note the feature control frame is associated with the size tolerance to make it a median plane control.

This on the drawing

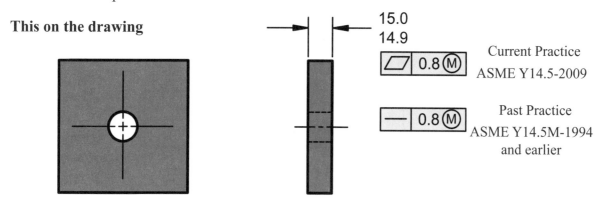

Current Practice
ASME Y14.5-2009

Past Practice
ASME Y14.5M-1994
and earlier

Application

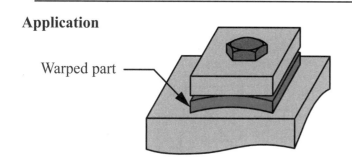

Warped part

The local size or thickness is important, but the part may be "bowed" or warped. It will be straightened out when bolted down in the assembly.

Means this

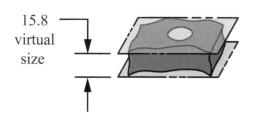

15.8 virtual size

A combination of size (exempted from Rule 1) and the flatness tolerance requires that the entire feature must be within a virtual boundary of 15.8. As the local size of the feature departs from MMC, additional flatness tolerance is allowed, equal to the departure from MMC, to a max of 0.9 flatness at LMC.

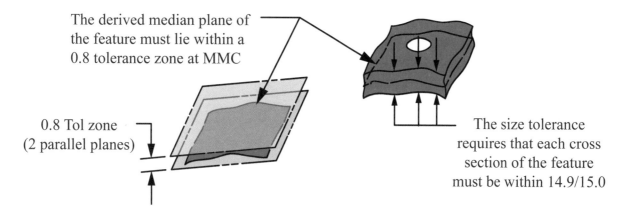

The derived median plane of the feature must lie within a 0.8 tolerance zone at MMC

0.8 Tol zone (2 parallel planes)

The size tolerance requires that each cross section of the feature must be within 14.9/15.0

Circularity (Roundness)

Circularity is a condition of a surface where:

    a.   for a feature other than a sphere, all points of the surface intersected by any plane perpendicular to an axis are equidistant to that axis.

    b.   for a sphere, all points of the surface intersected by any plane passing through a common center are equidistant from that center.

This on the drawing

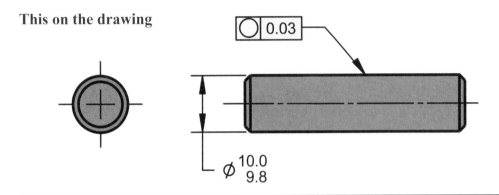

Each circular element of the surface in a plane perpendicular to the axis must lie between two concentric circles, one having a radius 0.03 larger than the other. In addition, the feature must be within the limits of size, including perfect from at MMC.

Means this

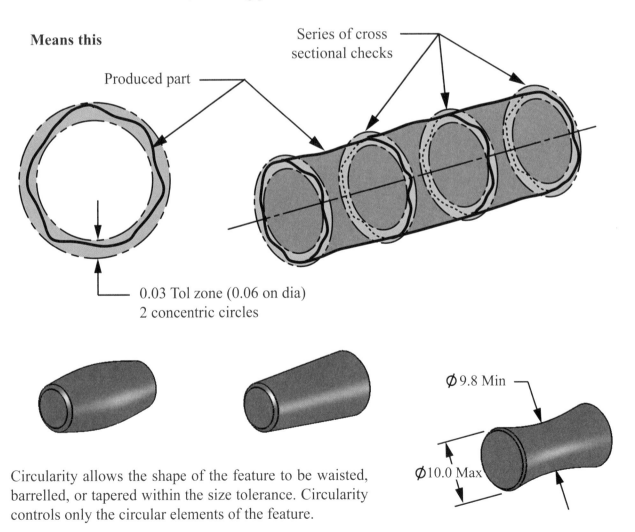

Circularity allows the shape of the feature to be waisted, barrelled, or tapered within the size tolerance. Circularity controls only the circular elements of the feature.

Cylindricity

Cylindricity is the condition of a surface of revolution within which all points of the surface are equidistant from a common axis. The tolerance zone is two concentric cylinders within which the surface must lie.

This on the drawing

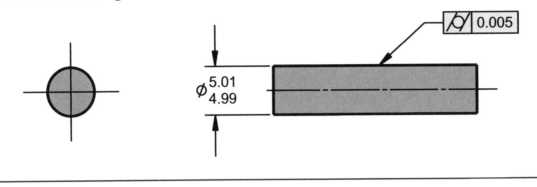

Application

The size of the bearings can vary within a greater tolerance relative to the cylindricity.

The cylindrical surface of the feature must lie between two concentric cylinders, one having a radius of 0.005 larger than the other. In addition, the feature must be within the limits of size, including perfect from at MMC.

Means this

0.005 Tol zone (0.01 on DIA)
2 Concentric cylinders

Produced part

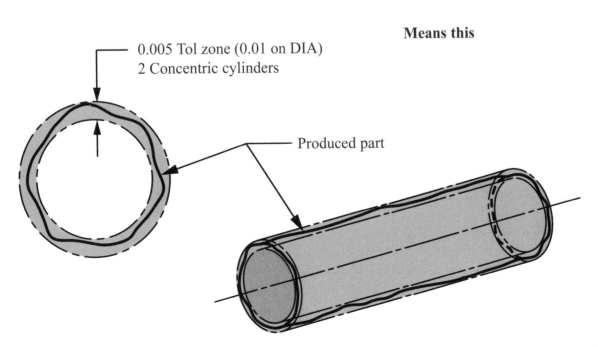

10.15

More on Circularity and Cylindricity

Circularity is a 2D tolerance. It controls circular elements only, not longitudinal elements. The value of the tolerance zone is the width of the band between the two concentric circles. Cylindricity is a 3D tolerance. It controls both the circular and longitudinal elements of the feature. It includes circularity, straightness, and taper of the surface.

Circularity and cylindricity are both form tolerances, therefore datums are not allowed. They are also surface controls; thus the modifiers MMC, LMC and RFS are not applicable. These controls are used to refine the size requirements and are always less than the size tolerance, except in the case of parts subject to free state variation. Circularity may be used on round or spherical features and cylindricity is used on cylindrical features. Usually the size and form requirements of Rule #1 are sufficient to control the form of the feature, but there are cases where the size tolerance may be large but the circularity or cylindricity must be refined.

Examples of use for circularity: ball bearings, tubes, hoses, pipes, and circular elements of tapered, barreled or waisted parts such as nose or tail cones, seals, valves, etc. It can also be used as a 2D refinement of cylindricity.

The verification of cylindricity and circularity are often difficult. The surface periphery of the part must be verified relative to a reference axis. Often an external feature is rotated in a vee bock, and the surface is verified with an indicator. This is a rough check. Depending on the varying number and arrangement of lobes on the part, plus the angle of the vee bock, this verification method can give varying results. In some cases, it can show a good part bad and other times it can show a bad part good.

A more precise method of verification is through the use of a roundness machine where the part is rotated utilizing a precision spindle. A stylus reads the surface and transcribes an enlarged profile of the part periphery on a polar graph or strip chart. A CMM could also be used.

The method of describing roundness and circularity in the ASME Y14.5 standard is called minimum radial separation. In the ANSI B89.3.1 standard on roundness, there are more sophisticated methods available to describe more specific needs. This includes methods such as the "least mean square" which determines the center of circular form based on a mathematical formula. This method is often used by coordinate measuring machines. There are also methods such as "minimum circumscribed circle (MCC) or maximum inscribed circle (MIC)" Consult the ANSI B89.3.1 standard for more specifics.

Average Diameter - Free State Variation

There are cases, especially with diameters subject to free state variation, where the circularity tolerance (and possibly other geometric tolerances) may exceed the size and form requirement of Rule #1. This is often the case of flexible aircraft rings that go "out of round" after fabrication.

If a diameter is qualified with the abbreviation AVG, the size of this diameter may exceed the form provisions of Rule #1. The size of the feature is the average of several diametric measurements taken across the feature. Usually enough measurements (at least four) are taken to assure the establishment of an average diameter. Specifying a circularity requirement on an average diameter of a nonrigid part is often necessary, as it will place a limit on the maximum out of roundness. This will ensure the actual diameter of the feature can be restrained to the desired shape at assembly and not be bent or deformed beyond its elastic limit. See page 12.30 in this text for an example drawing of average diameter with a circularity in the free state applied.

Workshop Exercise 10.1

1. Show the symbol and name for each of the four form tolerances.

Symbol	Name

2. Can any of the form tolerances have datum references in the feature control frame?

3. Can any of the form tolerances have feature modifiers applied? If so, which ones?

4. List the two dimensional (2D) form tolerances.

5. List the three dimensional (3D) form tolerances.

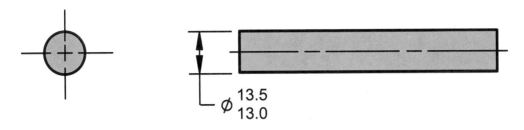

$\varnothing$ 13.5
13.0

6. On the drawing above, what is the maximum line element straightness of the pin before any geometric control is given?

7. What is the maximum "bow" of the pin before any geometric control is applied.

8. What rule in geometric tolerancing automatically controls the form of a feature within the specified limits of size?

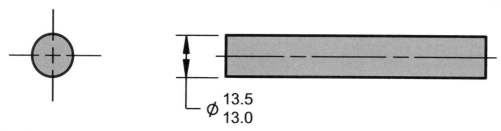

9. Apply a 0.15 straightness line element control to the surface of the part above and show the resultant tolerance zone on the produced part below.

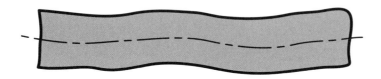

10. Can the MMC or LMC modifier be applied to this control? Why or Why not?

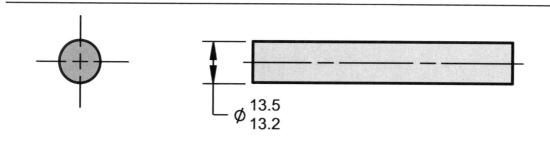

11. Apply a dia 0.5 at MMC straightness control to the axis of the part above and draw the resultant tolerance zone and virtual size on the sketch of the produced part below.

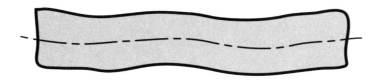

12. Complete the chart illustrating the effect of the MMC modifier applied to the straightness tolerance.

Feature Size	Tol Zone Size	Virtual
⌀ 13.5	⌀	
⌀ 13.4	⌀	⌀
⌀ 13.3	⌀	
⌀ 13.2	⌀	

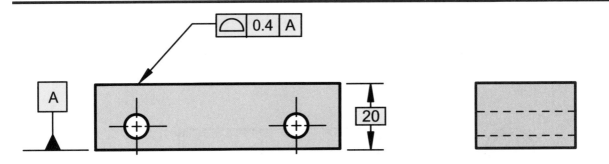

13. On the part above, what is the flatness on the top surface with only the profile applied?

_____

14. Apply a 0.1 flatness control to the top surface.

15. On the produced part below, show a simple sketch on how the flatness on the part may be verified and how the resultant flatness tolerance zone is defined.

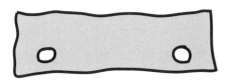

16. The entire top surface on the part below must be flat within 0.5 but each 25 X 25 patch must be flat within 0.1. Apply the necessary geometric controls to achieve this result.

17. What is the max circularity, on a radius, on the part below before a feature control frame is applied?

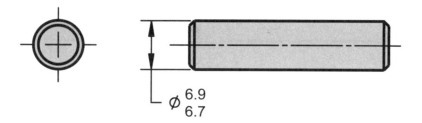

ϕ 6.9
6.7

18. Apply a 0.05 circularity on the part above and show the resultant tolerance zone on the produced part below.

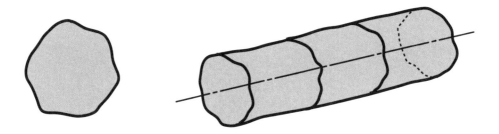

19. What is the max cylindricity, on a radius, on the part below before any geometric controls are applied?

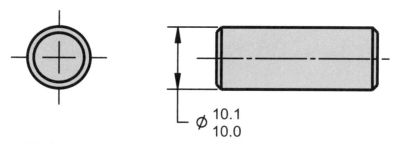

ϕ 10.1
10.0

20. Apply a 0.02 circularity on the part above and show the resultant tolerance zone on the produced part below.

Unit 11

Orientation Tolerances

Overview
Parallelism on a Surface
Parallelism Application
Angularity on a Surface
Angularity Verification
Perpendicularity on a Surface
Perpendicularity with Two Datum Feature References
Orientation of a Hole
Orientation to a Pattern of Holes
Perpendicularity Line Elements (2D)
Parallelism Tangent Plane
Perpendicularity of Zero at MMC
Workshop Exercise 11.1

Orientation Tolerances

Orientation tolerances control the orientation of features. Datum feature references are required.

Symbol	Orientation Tolerance	Common Shape of Tolerance Zone	2D / 3D	Application of Feature Modifier
∠	Angularity	2 Parallel Lines 2 Parallel Planes Cylindrical	3D or 2D*	Yes, if feature has size No, if feature is a surface
* ⊥	Perpendicularity			
* //	Parallelism			

* Note: As an option, the ASME Y14.5-2000 standard allows the use of the angularity symbol for all orientation tolerances.

Overview:

The three orientation tolerances, parallelism, angularity, and perpendicularity are shown in the chart above. Orientation tolerances control the orientation (tilt) of individual features. The orientation control of a surface will also control the form of that surface if no form tolerance is specified.

*The orientation tolerances are all 3D controls by default. If it is necessary to control a feature in 2D, a note, such as "EACH ELEMENT" or "EACH RADIAL ELEMENT," is placed under the feature control frame.

In some cases, it may be necessary to control only the tangent plane of a feature. In this situation, the T in a circle designation is placed in the feature control frame.

Datum feature reference(s) are required on all orientation tolerances. It is quite common to relate an orientation tolerance to one or more datum features. This requirement is specified if it is necessary to stabilize the tolerance zone relative to the datum reference frame.

The orientation tolerances will control features without size (planar surfaces) as well as features of size (holes, slots, tabs, pins etc.). If the orientation tolerances are applied to planar surfaces, then feature modifiers (MMC, RFS or LMC) are not allowed. If the orientation tolerances are controlling the axis or median plane of a feature of size, then feature modifiers apply.

As we know, the size requirements of Rule #1 will control the form and size of an individual feature but do not control the relationship between the individual features. In fact, the ASME Y14.5-2009 standard states, "Features shown perpendicular, coaxial, or symmetrical to each other must be controlled for location or orientation to avoid incomplete drawing requirements." The orientation tolerances are used to control the orientation of features to a datum reference frame.

The reason we still have the three symbols is simply because of past history. Mathematically the three orientation tolerances are identical. They all control orientation. The only difference between the three controls is the angle of orientation to the primary datum.

According to the past ASME Y14.5M-1994 standard, if the feature is 90 degrees from the primary datum, then the perpendicularity symbol is used. If the feature is 180 degrees from the primary datum, then the parallelism symbol is used. If the feature is defined at any other angle than 90 or 180 degrees from the primary datum, then the angularity symbol is used. In reality, the angularity symbol could be used in place of the parallelism and perpendicularity symbols and would have the same meaning.

In fact, this is why in the current ASME Y14.5-2009 standard, the angularity symbol may be used in place of perpendicularity and parallelism to simply mean orientation. In many places in this text, the word orientation is used in place of parallelism, angularity and perpendicularity to help you think generically about orientation tolerances.

As you review the orientation tolerances in this chapter, you will find that in some complex examples it is difficult to determine if the specification is perpendicularity, parallelism or angularity. They are all orientation tolerances. An angularity tolerance will control the orientation of an individual feature at the specified angle(s) to the datum reference frame.

Orientation tolerances do not locate features. They will only control the angle of orientation. In order to locate features, a profile or position tolerance is used, and the orientation tolerances refine the orientation.

Parallelism is the condition of a surface, axis, or centerplane, oriented at 180° to the datum reference frame.

This on the drawing

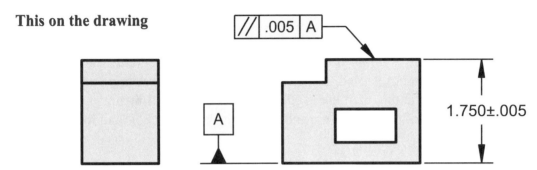

All points of the surface must lie within two parallel planes .005 apart which are parallel to datum plane A. In addition, the surface must be within the limits of size or profile. Since parallelism is a refinement of the size or location, the parallelism tolerance must be tighter in order to be effective.

Means this

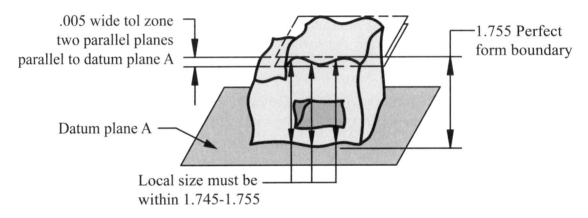

The above parallelism specification is a surface control. The MMC, LMC, and RFS modifiers are not applicable. Parallelism on a plane surface also controls the flatness. Thus, if no flatness is specified, the flatness tolerance requirement will be equal to the parallelism requirement.

The plus/minus tolerance on this part is for illustrative purposes only and could be replaced with a basic dimension and a profile tolerance. If parallelism is applied in conjunction with a profile tolerance, the parallelism tolerance zone must be smaller than the profile zone.

Sample Inspection for Parallelism

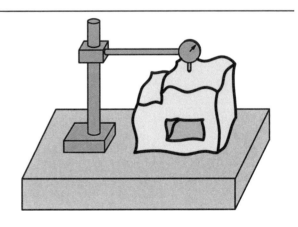

Part is mounted on datum A and an indicator is placed on the surface. As the indicator rides on the surface, the full indicator movement (FIM) can be no more than .005

Parallelism Application

This bearing spacer is a good application for a loose location tolerance with a tighter parallelism requirement.

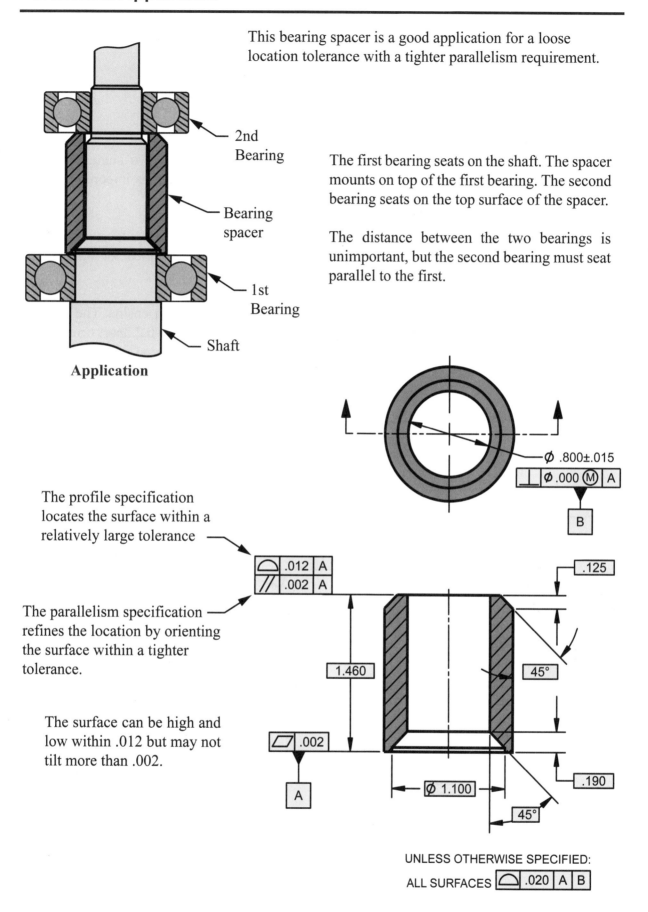

Application

2nd Bearing

Bearing spacer

1st Bearing

Shaft

The first bearing seats on the shaft. The spacer mounts on top of the first bearing. The second bearing seats on the top surface of the spacer.

The distance between the two bearings is unimportant, but the second bearing must seat parallel to the first.

The profile specification locates the surface within a relatively large tolerance

The parallelism specification refines the location by orienting the surface within a tighter tolerance.

The surface can be high and low within .012 but may not tilt more than .002.

Ø .800±.015

B

.125

1.460

45°

Ø 1.100

45°

.190

A

UNLESS OTHERWISE SPECIFIED:
ALL SURFACES

11.5

Angularity is the condition of a surface, axis, or center plane oriented at a specified angle to the DRF.

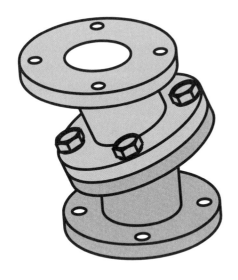

Application

The application on the left shows two angled fittings. The location of the surfaces are not as important as the angles of orientation.

The drawing below shows the surface is located with a 1 mm profile tolerance and then refined for orientation with an angularity tolerance.

The angle of 20° must be a basic dimension. The surface must lie within two parallel planes 0.2 apart oriented at 20° to the DRF. Because of the shape of this tolerance zone, the flatness is also controlled within 0.2.

This on the drawing

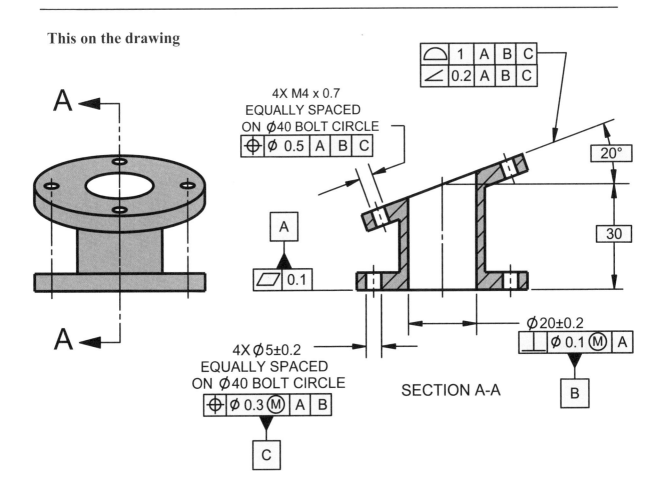

Means this

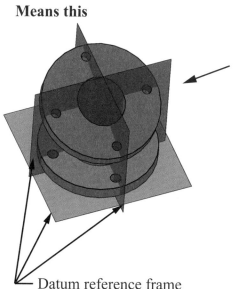

⌓	1	A	B	C
∠	0.2	A	B	C

Datum feature A establishes a datum plane on the bottom surface. Datum features B and C create an axis and two orientation planes. The angularity tolerance zone is defined at basic 20 degrees to datum plane A and 90 degrees to the orientation planes established by B and C.

Without references to datum features B and C in the angularity feature control frame, the tolerance zone would not have to be rotationally oriented to the pattern of holes.

Datum reference frame established in order by datum features A, B, & C.

1 mm profile zone basically **located** and **oriented** to the DRF established by datum features A, B, & C

0.2 angularity zone basically **oriented** to the DRF. The zone may translate back/forth and up/down, but the zone must remain at the basic angles (oriented) to the DRF.

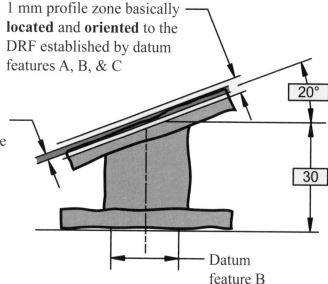

20°

30

Datum feature B

Sample Inspection

To verify the angularity requirement, the part is mounted on datum feature A, centered on datum feature B, and oriented with datum feature C, the pattern of four holes. The part is then inclined at 20 degrees using a sine plate. The surface to be measured should now be near parallel to the surface plate. The full indicator movement on the surface should be no more than 0.2.

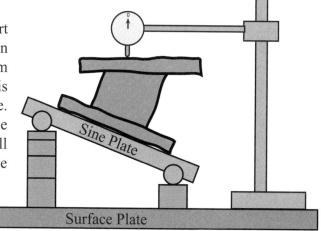

Sine Plate

Surface Plate

Angularity is a 3D control; the tolerance zone is two parallel planes. Angularity tolerance applied to a plane surface also controls the flatness on the surface.

Perpendicularity on a Surface

Perpendicularity is the condition of a surface, median plane, or axis oriented at 90° to the datum reference frame.

This on the drawing

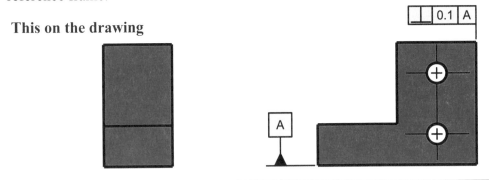

Application

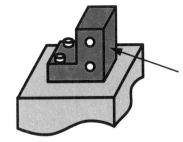

Surface must be perpendicular when part is mounted on bottom surface.

Means this

The surface must lie between two parallel planes 0.1 apart. In addition, the feature must be within the applicable limits of size or location.

Datum plane A
(Datum reference frame)

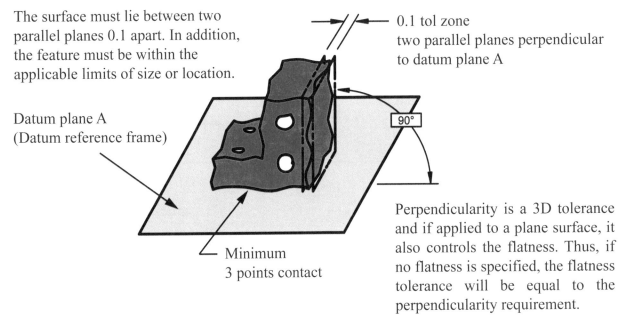

0.1 tol zone
two parallel planes perpendicular to datum plane A

90°

Minimum
3 points contact

Perpendicularity is a 3D tolerance and if applied to a plane surface, it also controls the flatness. Thus, if no flatness is specified, the flatness tolerance will be equal to the perpendicularity requirement.

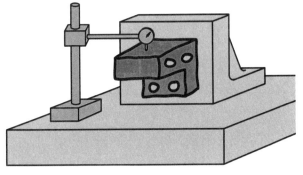

Sample Inspection

The part is mounted on datum feature A and the surface to be verified is leveled. The full indicator movement must not exceed 0.1.

Perpendicularity with two Datum Feature References

Perpendicularity is an orientation tolerance and can be applied with multiple datum feature references. There is an example of this specification below.

The current ASME Y14.5-2009 standard allows the use of the angularity symbol for all orientation tolerances. This practice makes the specification easier to understand. It is not interpreted as "the surface must be perpendicular to A and perpendicular to B," but rather "the surface must be oriented to the datum reference frame established by datum features A primary and B secondary." The tolerance zone is oriented at the angle shown (in this case implied 90° basic to A and implied 180° basic to B). See the 3D interpretation on the next page.

Note: Separate feature control frames on the drawing with a perpendicularity to A and a parallelism to B will yield entirely different results. This will create two different primary datums and as a result have different datum reference frame setups. This will not match the mounting conditions shown below.

This on the drawing

Alternate practice allowed by current ASME Y14.5-2009

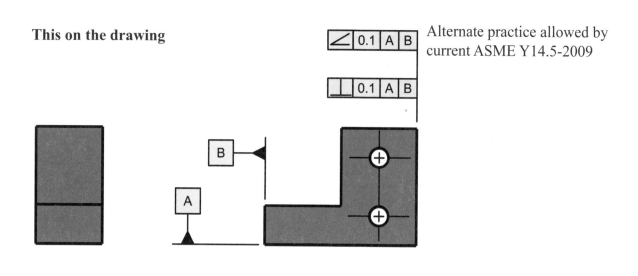

Application

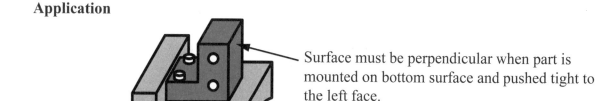

Surface must be perpendicular when part is mounted on bottom surface and pushed tight to the left face.

The graphic below illustrates in 3D how the orientation (perpendicularity) requirement is related to two datum features. Notice how the datum selection and the function of the part shown on the previous page correlate with the datum set up and interpretation below.

Perpendicularity or any other orientation tolerance with multiple datum references, can be confusing if you do not clearly understand the datum reference frame concept. It is important to remember that the datum features create a datum reference frame. The order in the feature control frame is the order of the set up procedure. In this case, 3 points on A and 2 points on B. The orientation tolerance zones are oriented to the datum reference frame and not the datum features themselves.

Means this

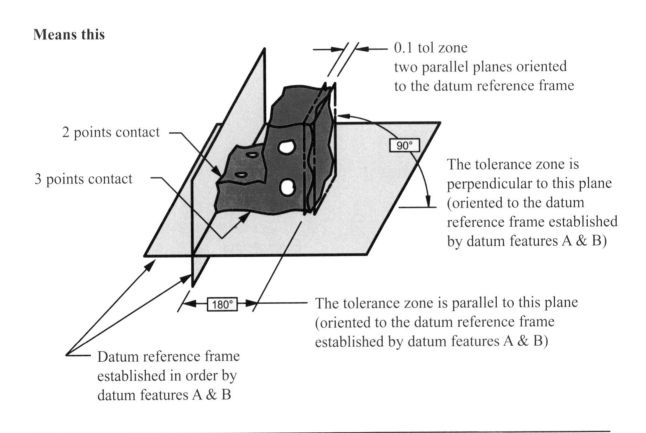

0.1 tol zone
two parallel planes oriented
to the datum reference frame

2 points contact

3 points contact

90°

The tolerance zone is perpendicular to this plane (oriented to the datum reference frame established by datum features A & B)

180°

The tolerance zone is parallel to this plane (oriented to the datum reference frame established by datum features A & B)

Datum reference frame established in order by datum features A & B

Sample Inspection

The part is mounted on datum feature A and leveled to datum feature B. The surface must indicate within 0.1.

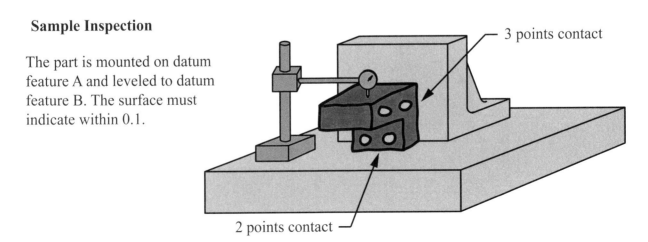

3 points contact

2 points contact

11.10

Orientation of a Hole

The part in this example is fully toleranced according to its assembly requirements. All locating dimensions are basic and have been left off for clarity. The purpose of this example is to show the application of an orientation tolerance applied to a hole with multiple datum feature references.

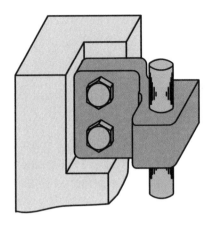

Application

This part is mounted by aligning the three surfaces shown and fastened with the two screws.

The three aligned mounting surfaces are selected as the datum features. The location of the .313 hole is not important and is given a relatively large .010 position tolerance. If no other controls were applied, the position tolerance would also control the orientation (tilt) of the hole. However, the orientation of the hole is more critical, so an orientation tolerance was added to control this variation within a smaller tolerance zone.

This on the drawing

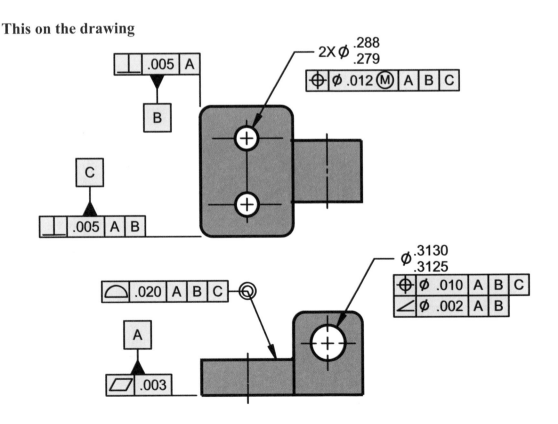

The 3D figure below illustrates how the orientation tolerance zone is related to the datum reference frame. The order of datum features specified in the feature control frame define the establishment of the DRF.

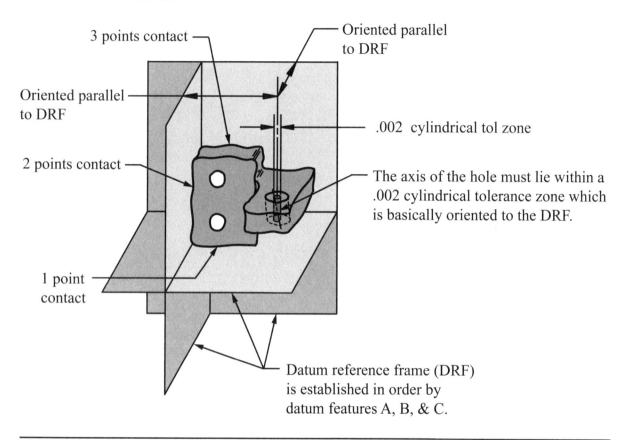

3 points contact

Oriented parallel to DRF

Oriented parallel to DRF

2 points contact

1 point contact

.002 cylindrical tol zone

The axis of the hole must lie within a .002 cylindrical tolerance zone which is basically oriented to the DRF.

Datum reference frame (DRF) is established in order by datum features A, B, & C.

The graphics below show the resulting tolerance zones created by the two feature control frames to the right.

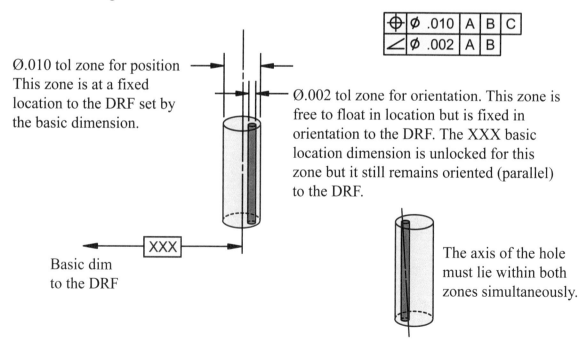

Ø.010 tol zone for position
This zone is at a fixed location to the DRF set by the basic dimension.

Ø.002 tol zone for orientation. This zone is free to float in location but is fixed in orientation to the DRF. The XXX basic location dimension is unlocked for this zone but it still remains oriented (parallel) to the DRF.

Basic dim to the DRF

XXX

The axis of the hole must lie within both zones simultaneously.

11.12

Orientation to a Pattern of Holes

The example part below illustrates an orientation tolerance to a pattern of holes. This example is an extension of the principle shown in the last problem with parallelism to multiple datum references. Again, the basic dimensions were left off the drawing for clarity.

The reason for this example is to reinforce the idea that tolerances are related to the datum reference frame established by the datum features, not the datum features themselves. This example will concentrate on the position and orientation specification on the .313 hole.

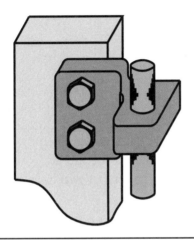

Application

This part is mounted and aligned with the two screws. The location of the large hole is not important but the orientation is critical.

Datums and geometric tolerances are selected based on the functional requirements. For this example, the mounting face is selected as datum feature A and the two mounting holes are selected as datum feature B. Just as in the last example, the location of the .313 hole is not important and is given a relatively large .010 position tolerance. If no other controls were given, this position tolerance would also control the orientation (tilt) of the hole. However, the orientation of the hole is more critical so an orientation tolerance was added to control this variation within a smaller tolerance zone.

This on the drawing

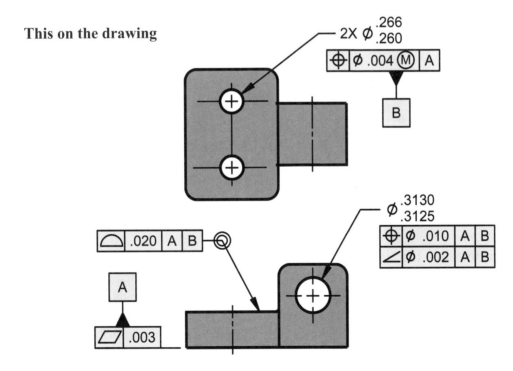

The back face establishes datum plane A. The axes of the 2 mounting holes must lie within two position tolerance cylinders set at basic and then establish datum B at RMB. The 6 degrees of freedom on the part have been constrained.

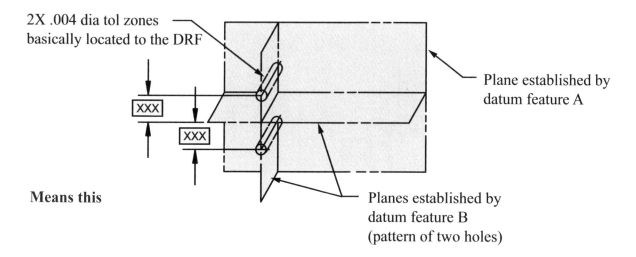

2X .004 dia tol zones basically located to the DRF

XXX

XXX

Means this

Plane established by datum feature A

Planes established by datum feature B (pattern of two holes)

The orientation tolerance zone is related to the DRF established by the datum features as shown below. Note; Even though the two mounting holes establish datum B and are referenced at RMB, they have size and possible clearance with their mating components. The part may shift and twist on the clearance holes in the mated condition. The designer should also factor this condition into calculating the over all orientation (parallelism) on the .312 diameter hole.

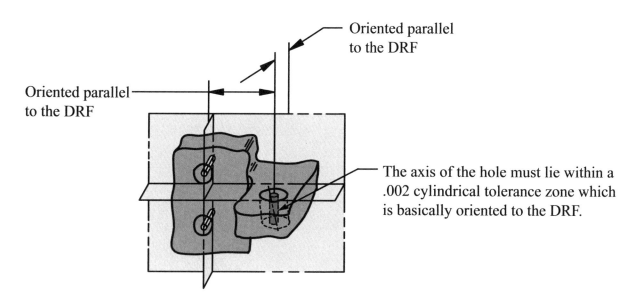

Oriented parallel to the DRF

Oriented parallel to the DRF

The axis of the hole must lie within a .002 cylindrical tolerance zone which is basically oriented to the DRF.

This orientation specification also reinforces the discussion in the overview in the beginning of this unit. As the orientation requirements on the part becomes more complex and is related to more than one datum, it is difficult to decide whether to use a parallelism, perpendicularity, or angularity control. In the final analysis, it probably does not matter which symbol is used. This is why in the current ASME Y14.5-2009 standard, the angularity symbol may be used to control the orientation of a feature at any angle specified.

Perpendicularity Line Elements (2D)

All of the orientation tolerances are by default 3D. If there is a need to specify an orientation tolerance in 2D, the words "EACH ELEMENT" are placed under the feature control frame. The direction of the tolerance zone applies in the view where it is specified.

This on the drawing

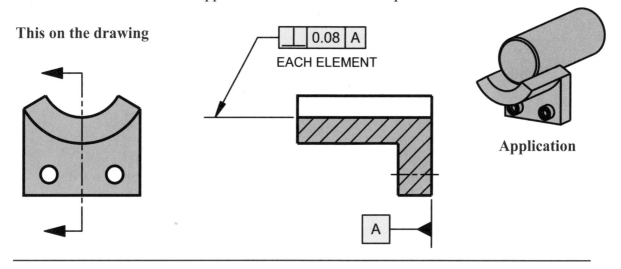

Application

Each line element of the surface must lie between two parallel lines 0.08 apart which are oriented perpendicular to the specified datum reference frame. In addition, the surface must lie within the applicable size or profile tolerance.

Means this

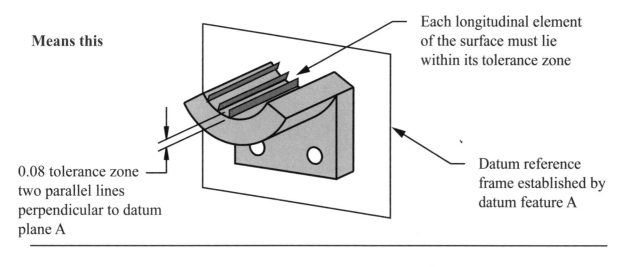

Each longitudinal element of the surface must lie within its tolerance zone

0.08 tolerance zone two parallel lines perpendicular to datum plane A

Datum reference frame established by datum feature A

Sample Inspection

Part is mounted on datum feature A and the indicator is placed on the surface traveling in a series of line element checks that are perpendicular to the DRF.

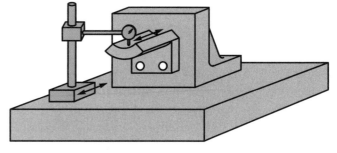

Line element tolerance applied to a surface also controls the straightness of that surface. The line element concept may be expanded to control other 2D requirements. Notations such as, "EACH RADIAL ELEMENT" OR "EACH CIRCULAR ELEMENT" may also be used.

Parallelism Tangent Plane

The tangent plane symbol can be applied to many of the geometric characteristics. The tangent plane symbol, T in a circle, is shown below applied to a parallelism specification. A plane contacting the high points of the surface must lie within the parallelism tolerance zone. A tangent plane specification will not control the flatness on the surface.

This on the drawing

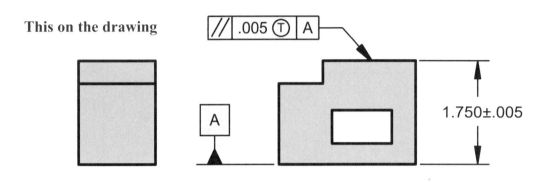

| // | .005 Ⓣ | A |

1.750±.005

The tangent plane contacting the high points of the surface must lie within two parallel planes .005 apart which are parallel to datum plane A. In addition, the surface must be within the limits of size or profile.

Means this

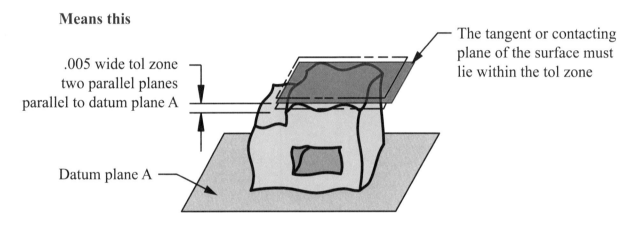

.005 wide tol zone
two parallel planes
parallel to datum plane A

The tangent or contacting plane of the surface must lie within the tol zone

Datum plane A

The tangent plane symbol may also be used with profile of a surface. Profile of a surface with a tangent plane modifier is often used when locating parts in a weldment. The flatness of the surface is then controlled on the individual piece part drawing.

Sample Inspection

Part is mounted on datum feature A with a parallel bar placed covering the surface. As the indicator moves on the parallel bar, the full indicator movement (FIM) can be no more than .005

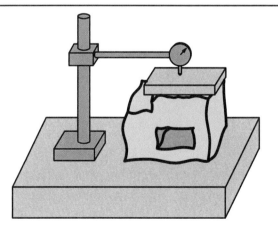

Perpendicularity of Zero at MMC

The two examples shown each represent a mating assembly. Example 1 specifies a .002 at MMC perpendicularity requirement. Example 2 specifies a .000 at MMC requirement. Notice example 2 provides a greater variation on size while still maintaining the same virtual size as example 1. The zero tolerancing at MMC or LMC may be applied to other geometric controls as well.

Example 1

Use this method if it is necessary to control the size and orientation with separate specs.

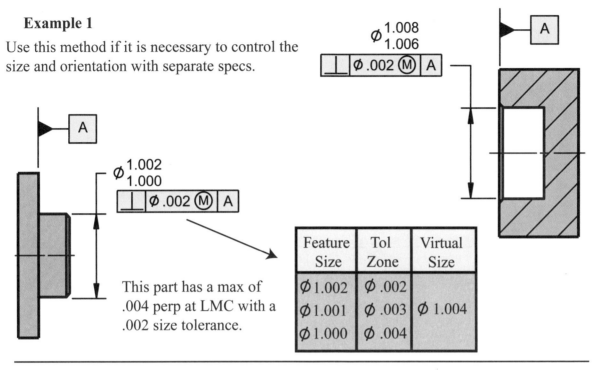

This part has a max of .004 perp at LMC with a .002 size tolerance.

Feature Size	Tol Zone	Virtual Size
⌀ 1.002	⌀ .002	
⌀ 1.001	⌀ .003	⌀ 1.004
⌀ 1.000	⌀ .004	

Example 2

Use this method if it is not necessary to control the size and perpendicularity with two values. This method combines the size and orientation and simply states the virtual size of the feature. It is a common specification on pilots. It maintains the same perpendicularity but provides more size tolerance than example 1.

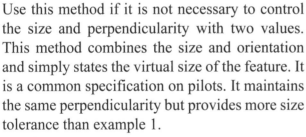

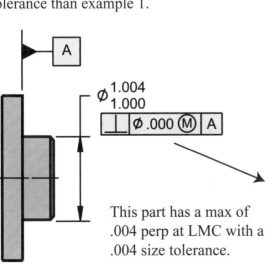

This part has a max of .004 perp at LMC with a .004 size tolerance.

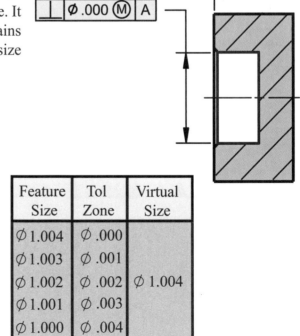

Feature Size	Tol Zone	Virtual Size
⌀ 1.004	⌀ .000	
⌀ 1.003	⌀ .001	
⌀ 1.002	⌀ .002	⌀ 1.004
⌀ 1.001	⌀ .003	
⌀ 1.000	⌀ .004	

1. Name the three orientation tolerances. _____

2. Are datums always required with orientation tolerances? _____

3. Which of the orientation symbols may be used in place of the other two symbols? _____

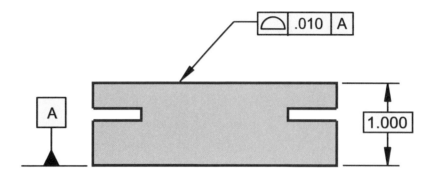

4. On the part above, what is the maximum variation in parallelism with only the profile applied? _____

5. On the part above, specify a parallelism tolerance to make the top surface parallel to datum A within a total of .002.

6. Can the MMC modifier be applied to the feature tolerance? Why or why not?

_____

7. What can you expect the flatness tolerance to be on the top surface? _____

8. With both the profile and parallelism applied to the part above, what is the maximum height of the top surface to datum A. _____

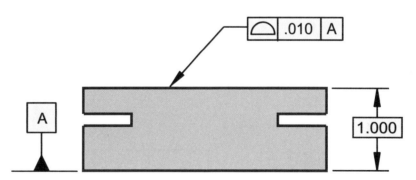

9. On the part above, make the tangent plane of the top surface parallel to the bottom surface within a total of .002.

10. Explain the difference between the parallelism in problem 4 and the control above.

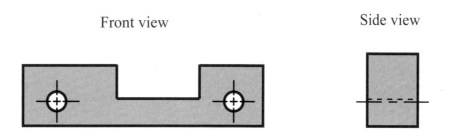

Front view   Side view

ALL ANGLES ± 1°

11. On the part above, specify a geometric control to make the left surface in the front view perpendicular to the bottom surface within .005 total.

12. Can a datum or feature modifier be applied to this feature control frame? Why or why not?

_____

13. What is the flatness on the left surface of the part above? _____

14. On the part above, there is no perpendicularity requirement specified between the right surface and bottom surface. What perpendicularity is implied? _____

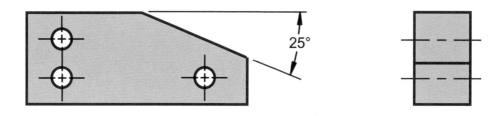

15. On the part above, show an angularity specification to orient the angled surface in relation to the right surface in the side view and top surface in the front view within .005 total. Make sure to label the datum features.

16. Should the 25 degree angle above have a plus/minus tolerance applied or should it be basic? Complete this requirement on the drawing.

17. What is the flatness tolerance on the angled surface above? _____

18. Draw a simple sketch below of the shape of the resultant angularity tolerance zone.

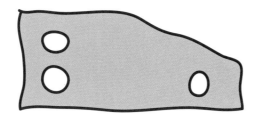

11.19

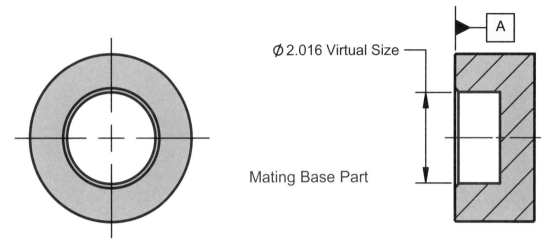

Ø 2.016 Virtual Size

Mating Base Part

A

19. The two pilot parts below mate with the part above. There is a design clearance of .002 between the virtual size fit of the two parts. Complete the two charts below showing the perpendicularity tolerance for the different pilot sizes. Also fill in the virtual size for each of the parts.

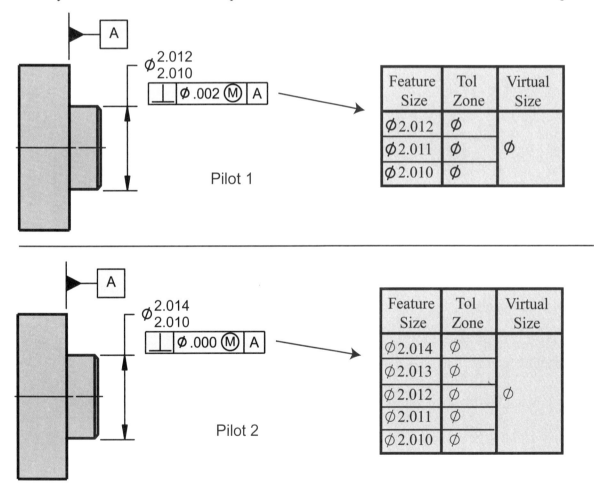

Feature Size	Tol Zone	Virtual Size
Ø 2.012	Ø	
Ø 2.011	Ø	Ø
Ø 2.010	Ø	

Pilot 1

Feature Size	Tol Zone	Virtual Size
Ø 2.014	Ø	
Ø 2.013	Ø	
Ø 2.012	Ø	Ø
Ø 2.011	Ø	
Ø 2.010	Ø	

Pilot 2

20. Which part provides the most manufacturing tolerance and is easiest to work with, part 1 or part 2? Explain why.

_____

_____

Unit 12

Profile Tolerancing

Profile Tolerancing

Profile tolerances control the location, and/or orientation, and/or form, and/or size of a feature and may or may not have datum references.

Symbol	Type of Tolerance	Common Shape of Tolerance Zone		2D / 3D	Application of Feature Modifier
⌒	Profile of a Line	≋	 2 Dimensional Uniform Boundary	2D	No
⌓	Profile of a Surface	≋	 3 Dimensional Uniform Boundary	3D	No

Overview:

There are two types of profile tolerances: Profile of a surface and profile of a line. These profile tolerances along with their associated symbol are shown in the chart above. Profile of a line is a 2D control and profile of a surface is a 3D control. Profile tolerances may or may not have datum feature references.

Profile of a line is often used in conjunction with profile of a surface. Profile of a line refines the cross sectional elements of a feature and is commonly used on extruded type parts where control in one direction is more important than another. Since profile of a line is a 2D control, the tolerance applies in the view shown.

Profile of a surface is probably the most powerful control in the geometric tolerancing system. Depending on datum feature references, profile of a surface can be used to control size, form, orientation and the location of feature surfaces. Profile of a surface is primarily used to replace the directly toleranced or plus/minus type tolerancing on parts. As we have seen in earlier units, parts that are defined with plus/minus type tolerances are not clear and are subject to many interpretations.

Profile tolerance can be related to a datum reference frame and provides a clear uniform boundary within which the feature surface must lie. In industry today, most companies use Computer Aided Design (CAD) to design their product and Coordinate Measuring Machines (CMM) to verify their product. These systems require that the parts are mathematically defined relative to a 3D coordinate system.

Profile tolerance is easy to verify. It provides clear boundaries relative to a defined datum reference frame. Of course, it is impossible to check all the points in a profile tolerance, but it is also impossible to check all the points with a plus/minus tolerance and just like it is impossible to check all the points with a surface finish. How many points are checked is based on the risk taken and is defined in the measurement plan. Defining a part with profile tolerance does not mean it is more important or more difficult. The size of the profile tolerance zone is what will determine if it is a tight tolerance.

The use of profile tolerancing does not require any new or exotic type of verification procedures. Depending on the tolerance and situation, profile may be verified with a CMM, micrometer, calipers, gages, optical comparator, indicators or visually. Verification procedures depend on many variables, such as how many parts need to be checked, is it a first article or the ten thousandth part, or is it a final check or an in process check? The tightness of the tolerance, the method of manufacture, and acceptable level of risk and other factors are all taken into account. The method of verification and procedure is usually determined by the quality engineer and defined in the dimensional measurement plan.

In using profile tolerances, the basic dimensions define the geometry of the part and the location of the profile tolerance zone. Manufacturing will aim for the basic dimensions and inspection will check points to identify the deviation from basic. The inspection report will define the deviation from basic and not the variation of the basic dimension. Options for the reporting of the inspection results can be found in the ASME Y14.45 Measurement Data Reporting standard expected to be released in 2010. There are also some examples shown later in this unit.

Many companies are beginning to tolerance the solid model according to the ASME Y14.41 standard. Profile tolerancing works well to tolerance the model. It is not possible to apply plus/minus tolerances to the model. In some industries, especially automotive sheet metal and aircraft parts, profile tolerancing is used exclusively. Plus/minus tolerances are used only for the size of features. If you are using some profile tolerances on your drawing now, you may wish to gradually begin using more in the future. Profile tolerancing is certainly the future direction of industry.

In the past, parts were dimensioned and toleranced with plus/minus tolerancing with the general default tolerance shown in the title block such as: 2 place dimensions having one tolerance and 3 place dimensions having another tolerance. If profile tolerances are used for unspecified tolerances, all surface defining dimensions must be basic. This can be accomplished by placing all the dimensions in a box or by a note on the drawing or in a standard referenced on the drawing stating: ALL UNTOLERANCED DIMENSIONS ARE BASIC.

Profile tolerances are applied to the features with a feature control frame and datum feature references. General default profile tolerances can be defined in a note or in the title block. There are examples of this concept shown later in this unit.

Non-uniform profile is new for the ASME Y14.5-2009 standard. This concept allows the designer to define the zone of tolerance to any shape necessary. Also, if you work with ISO drawings and ASME drawings, pay particular attention to the differences in interpretations shown later in this unit.

Profile tolerancing specifies a uniform boundary along the true profile defined by the basic dimensions within which the elements of the surface must lie. Profile tolerancing can be applied on a bilateral, unilateral, or unequal distribution basis. The example below shows a bilateral distribution. A single arrow from the feature control frame points directly to the surface.

Profile Application - Bilateral

This on the drawing

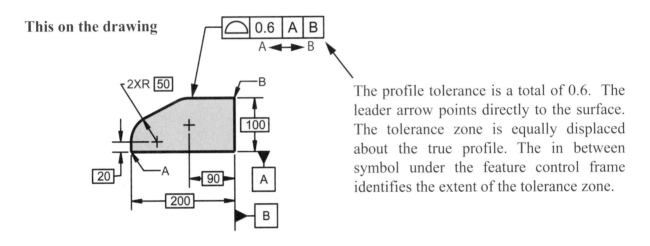

The profile tolerance is a total of 0.6. The leader arrow points directly to the surface. The tolerance zone is equally displaced about the true profile. The in between symbol under the feature control frame identifies the extent of the tolerance zone.

The tolerance zone established by the profile of a surface control is three dimensional and extends along the full length and width of the considered feature. The in between symbol under the feature control frame states the tolerance extends between points A and B.

Means this

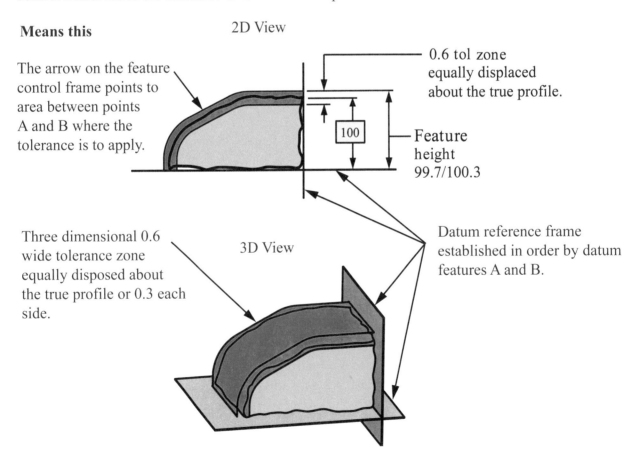

The arrow on the feature control frame points to area between points A and B where the tolerance is to apply.

2D View

0.6 tol zone equally displaced about the true profile.

Feature height 99.7/100.3

Three dimensional 0.6 wide tolerance zone equally disposed about the true profile or 0.3 each side.

3D View

Datum reference frame established in order by datum features A and B.

Profile of a Surface - Unilateral Out

Profile tolerance by default is a bilateral control that specifies a tolerance zone that is equally disposed about the true profile. In some cases, an unequal distribution of the tolerance zone may be desired. If an unequal distribution is required, there are two methods available to designate this specification on the drawing or model.

1. Option 1 *(New for ASME Y14.5-2009 and recommended practice)*
   Use the circle U symbology in the feature control frame.

2. Option 2 Use two arrows on the drawing to illustrate the width and/or direction of the tolerance zone. *(Alternate earlier practice to eventually be phased out.)*

Profile Application - Unilateral Out

This on the drawing

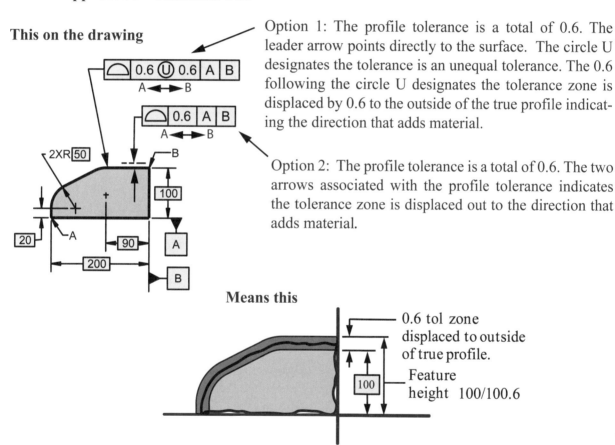

Option 1: The profile tolerance is a total of 0.6. The leader arrow points directly to the surface. The circle U designates the tolerance is an unequal tolerance. The 0.6 following the circle U designates the tolerance zone is displaced by 0.6 to the outside of the true profile indicating the direction that adds material.

Option 2: The profile tolerance is a total of 0.6. The two arrows associated with the profile tolerance indicates the tolerance zone is displaced out to the direction that adds material.

Means this

Unilateral and bilateral unequal tolerances are certainly a tool that can be used on engineering drawings. However, it should be used with some caution. Engineering, manufacturing and inspection often share the same math database. If the tolerance zones are artificially pushed to one side by a unilateral profile tolerance zone, manufacturing also needs to shift their data so they manufacture to the middle of the tolerance zone. This can be easily missed and can cause misunderstandings. It is often more prudent for an engineer to simply adjust the basic dimensions and allow the tolerance zone to be displaced equally about the math data.

Profile Application - Unilateral In

This on the drawing

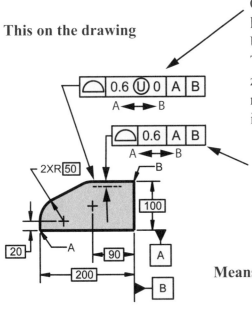

Option 1: The profile tolerance is a total of 0.6. The leader arrow points directly to the surface. The circle U designates the tolerance is an unequal tolerance. The 0 following the circle U designates the tolerance zone is displaced by 0 to the outside that adds material. Since 0 is displaced out to add material, then 0.6 is displaced in the direction that removes material.

Option 2: The profile tolerance is a total of 0.6. The two arrows associated with the profile tolerance indicate the tolerance zone is displaced in the direction that removes material. Since 0 is displaced out then 0.6 is displaced in the direction that removes material.

Means this

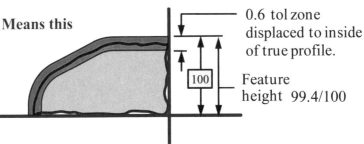

0.6 tol zone displaced to inside of true profile.

Feature height 99.4/100

Profile Application- Unequal Bilateral

This on the drawing

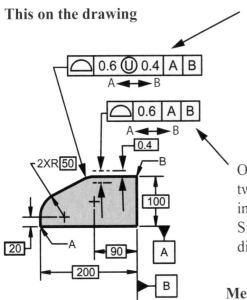

Option 1 The profile tolerance is a total of 0.6. The leader arrow points directly to the surface. The circle U designates the tolerance is a unequal tolerance. The 0.4 following the circle U designates the tolerance zone is displaced by 0.4 to the outside that adds material. Since 0.4 is displaced out to add material then 0.2 is displaced in the direction that removes material.

Option 2 The profile tolerance is a total of 0.6. The two arrows associated with the 0.4 profile tolerance indicate the tolerance zone is unequally displaced. Since 0.4 is displaced to add material, then 0.2 is displaced in the direction that removes material.

Means this

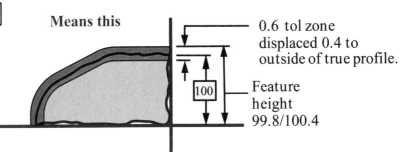

0.6 tol zone displaced 0.4 to outside of true profile.

Feature height 99.8/100.4

Profile of a Surface - Unilateral Clarification

Profile can be applied on a unilateral or unequal basis. The circle U following the tolerance defines an unequal tolerance and the value behind the symbol specifies how much the tolerance applies out or adds material. The example below illustrates a unilateral profile applied to an internal feature and a external feature. The unilateral out tolerance allows an external feature to increase in size from basic and the internal feature to decrease in size from basic.

This on the drawing

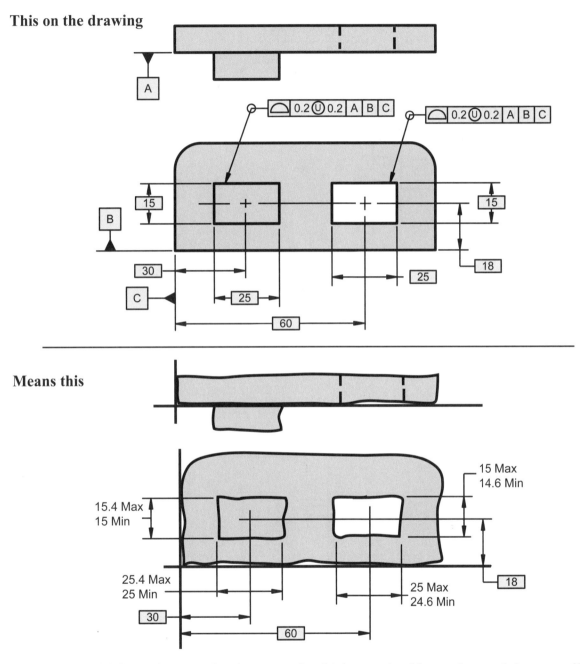

Unilateral and bilateral unequal tolerances should be used with caution and is generally not recommended. Engineering, manufacturing and inspection often share the same math database. If the tolerance zones are artificially pushed to one side by a unilateral profile tolerance zone, manufacturing also needs to shift their data so they manufacture to the middle of the tolerance zone. This can be easily missed and can cause misunderstandings. It is often more prudent for an engineer to simply adjust the basic dimensions and apply a bilateral profile tolerance zone that is displaced equally about the math data.

12.7

Profile tolerancing specifies a uniform boundary along the true profile defined by the basic dimensions within which the elements of the surface must lie. In some cases, it may be necessary to refine a unit area or control the rate of change of the feature with a more restrictive tolerance. This can be accomplished by using profile per unit area.

This on the drawing

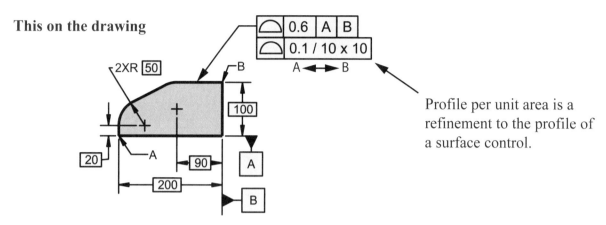

Profile per unit area is a refinement to the profile of a surface control.

The tolerance zone established by the 0.6 profile of a surface control is three dimensional and extends along the full length and width of the considered feature. The profile per unit area is a series of 10 x 10 tolerance zones that are 0.1 wide and float within the 0.6 profile of a surface zone.

The profile per unit area is used to refine the requirements and control rate of change within the larger 0.6 zone profile zone. It limits abrupt changes such as bumps, wrinkles, dimples, machine mis-match etc. Profile per unit area can be used in conjunction with a customized datum reference frame to further control the orientation, and/or location of the zones relative to a DRF. If required, the unit area can also be specified in a diameter zone by using the diameter symbol.

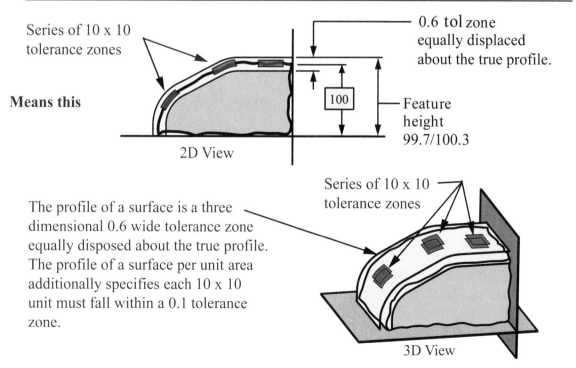

Series of 10 x 10 tolerance zones

Means this

0.6 tol zone equally displaced about the true profile.

Feature height 99.7/100.3

2D View

The profile of a surface is a three dimensional 0.6 wide tolerance zone equally disposed about the true profile. The profile of a surface per unit area additionally specifies each 10 x 10 unit must fall within a 0.1 tolerance zone.

Series of 10 x 10 tolerance zones

3D View

Profile of a Line

Profile of a line specifies a series of cross sectional uniform tolerance boundaries along the true profile, and along the full length and width of the considered feature, within which the elements of the surface must lie. Profile of a line can be applied on a bilateral, unilateral, or unequal distribution basis. It may be specified with or without a datum feature reference and applies in the view shown.

In the example below, profile of a line is used in conjunction with profile of a surface to further refine the cross sectional elements of the feature in the view shown. To clarify that the cross sections are only oriented to the referenced DRF, a customized datum reference frame was used. The letters following the datum reference define only the u, v and w rotational degrees of freedom constrained.

**Profile of a Line - Bilateral
Customized Datum Reference Frame**

This on the drawing

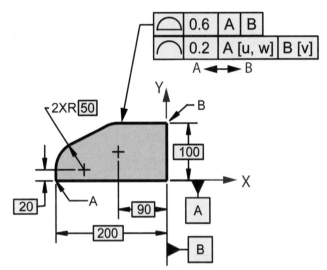

A 0.6 total wide profile of a surface zone equally disposed about the true profile within which lies a series of 2 dimensional 0.2 wide profile of a line cross sectional tolerance zones each of which are oriented to the DRF. The surface elements of the feature must lie in the profile of a surface zone and series of refining profile of a line zones simultaneously.

The profile of a line feature control frame was used to insure the series of profile of a line cross sectional tolerance zones are oriented only, and not located to the referenced DRF.

Means this

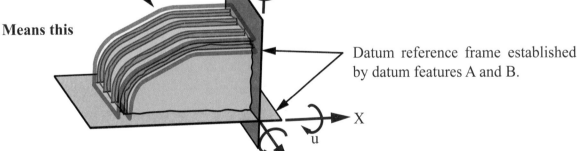

Datum reference frame established by datum features A and B.

Profile of a line is often used on extruded type parts where only cross sectional zones are necessary as in use for a flexible extruded rubber moulding. Profile of a line may or may not have any datum feature references. If referencing datum features, a customized datum reference frame can be used to specifically state the arresting degrees of freedom for each datum feature. This will relate the orientation and/or location of the series of cross sectional tolerance zones. See the customized datum reference frame explained earlier in unit 9.

Profile tolerance is used to locate surfaces. It can be used to locate contoured surfaces as well as flat surfaces. The cam wheel below has two profile controls applied. The upper 0.4 profile specification is by default, a bilateral control which specifies a tolerance zone that is equally disposed about the basic true profile.

The lower 0.2 profile specification is a unequal control because of the circle U following the feature tolerance in the feature control frame. The 0 following the circle U designates the amount the tolerance zone is displaced to the outside of the basic true profile (adds material). Since zero is displaced to the outside, the profile zone is all shifted to the inside of the true profile by 0.2. The surface of the part must fall within the profile zones.

Profile application cam wheel

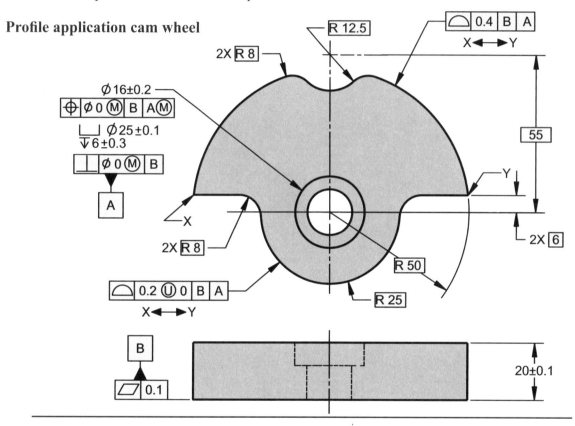

Profile tolerance zones

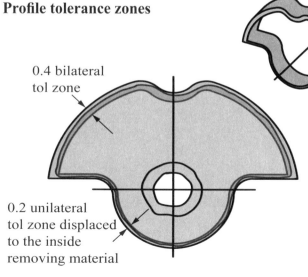

0.4 bilateral
tol zone

0.2 unilateral
tol zone displaced
to the inside
removing material

Profile tolerance is a 3D specification and the tolerance zone extends the depth of the feature.

This figure shows the imperfect part in the designated profile tolerance zones. The between symbol under the feature control frame designates the upper 0.4 profile bilateral zone extends from point X to Y on the top half and the lower 0.2 profile unilateral tolerance zone extends from point X to Y on the bottom half. The surface of the part must fall within the profile tolerance zones.

A sample inspection report is shown for the two profile specifications. Inspection bubbles are put next to the callouts so they may be referenced in the chart. An electronic height gage and a surface plate are used to check the two profile specifications. Four points were selected for feature 2 because of its relatively small size and large tolerance. Six points were selected for feature 1 because of its larger surface area and relatively small tolerance.

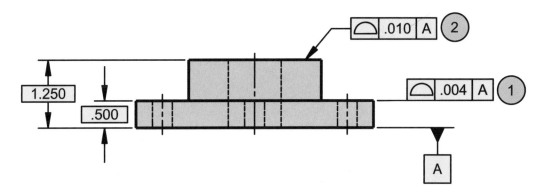

Feature	Allowed Profile	# of Pts Measured	Min/Max Deviation	Measured Profile	Accept Reject
1	.004	6	-.0005/ +.0015	.003	A
2	.010	4	+.001/ +.003	.006	A

Depending on the complexity of the measurement plan, the value of each measured point may be included in the report. When reporting, all values are converted to deviations from basic. They are reported as plus material or minus material.

The measured profile is recorded as a total zone about basic. The extreme deviation in either direction is doubled for this value. This allows it to be compared to the allowed profile.

Measured Values   Reported Values

 1

1	.5010
2	.5002
3	.5015
4	.4998
5	.5004
6	.4995

Pt.	Deviation from Basic
1	+ .0010
2	+ .0002
3	+ .0015
4	- .0002
5	+ .0004
6	- .0005

2

1	1.253
2	1.252
3	1.253
4	1.251

Pt.	Deviation from Basic
1	+ .003
2	+ .002
3	+ .003
4	+ .001

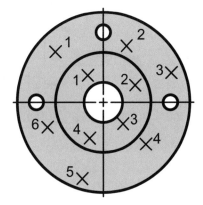

The locations of the points may also be defined. They may be shown in a simple picture or with exact x, y, z coordinates included in a chart.

Profile of a Surface - Fuel Line Bracket

The assembly below is a bracket and clamp that secures a fuel line in location relative to the engine. As you can see from the assembly drawing, the bracket mounts to the engine on the back face and by screws in the two holes. The periphery of the bracket must clear the engine in one area but the remaining contour is unimportant. The elongated holes on the bracket are for adjustment and must line up with the clamp holes. The clamp contour is relatively unimportant except where it mounts to the bracket and the fuel line.

The drawings of the bracket and the clamp illustrate a simple application of establishing a datum reference frame and applying geometric tolerancing that reflects functional requirements.

Fuel line bracket assembly

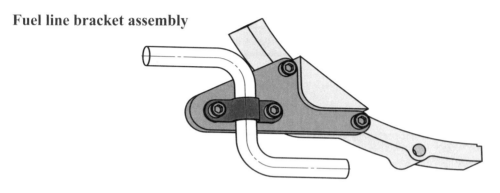

Fuel line bracket drawing

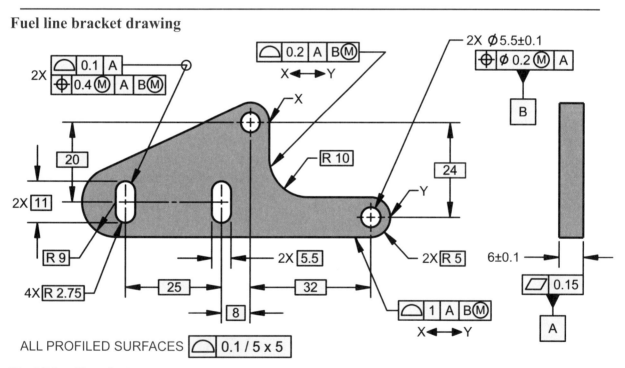

Fuel Line Bracket

A datum reference frame is established on the bracket using the face and two holes. The face is controlled with flatness and the two holes are positioned to each other relative to the face. Profile tolerance is applied to the remaining features. Notice the profile tolerance is held closer on the right upper side area of the bracket to clear the engine housing. The remaining outside contour has a larger profile tolerance. The slotted holes are profiled for size and then positioned using the boundary concept. The profile per unit area was applied to control abrupt changes or mismatch on the profiled surfaces.

12.12

Fuel line clamp drawing

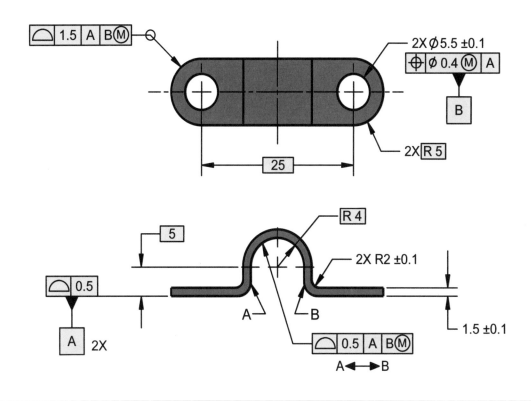

Fuel Line Clamp

The datum reference frame on the clamp is established by the two mounting surfaces and the two holes. The two surfaces are controlled for coplanarity with a profile tolerance. A profile tolerance was selected instead of flatness because flatness is a form tolerance and controls each surface individually. Flatness does not control the relationship between features, where as profile does. The two holes are positioned to each other and relative to datum feature A. The two surfaces and the pattern of two holes establish the DRF. The contour where it clamps to the tube was important so it did not squash the tube. The outside contour of the clamp was relatively unimportant so a large profile tolerance was applied.

Workshop Exercise 12.1

Often in the verification process, it is only necessary to monitor a few simple dimensions on the part. On the three parts below, calculate the minimum and maximum of the designated distances. See the drawings earlier in the unit for the dimensions and tolerances.

Cam Wheel
See page 12.10

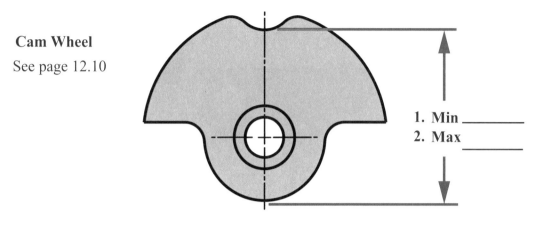

1. Min _____
2. Max _____

Fuel Line Bracket
See page 12.12

3. Min _____
4. Max _____

5. Min _____
6. Max _____

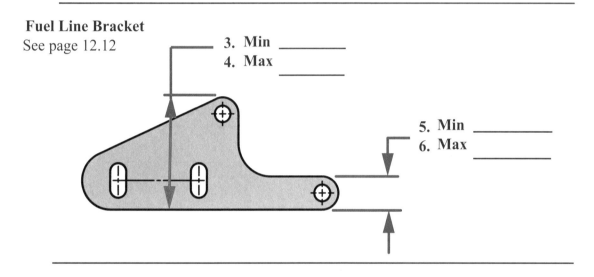

Fuel Line Clamp
See page 12.13

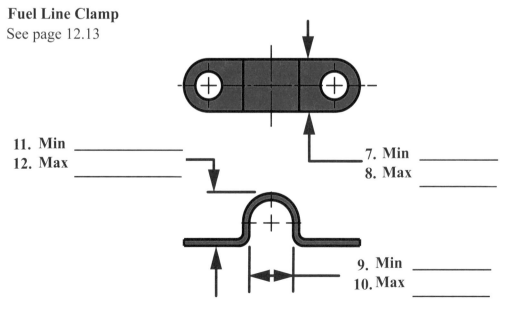

11. Min _____
12. Max _____

7. Min _____
8. Max _____

9. Min _____
10. Max _____

12.14

Profile of a Surface - Hat Bracket

The sheet metal bracket below has geometric tolerancing applied to the drawing. The datum features were selected based on the mounting conditions. Profile tolerancing was applied to the surfaces, and position tolerancing was applied to the holes.

Hat bracket assembly

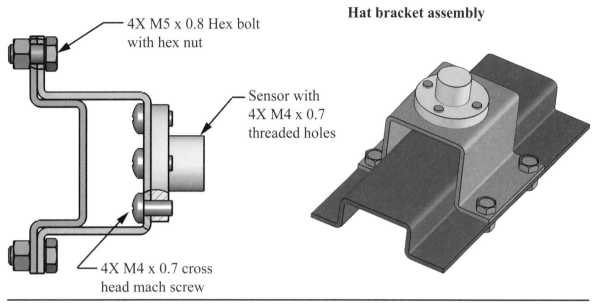

4X M5 x 0.8 Hex bolt with hex nut

Sensor with 4X M4 x 0.7 threaded holes

4X M4 x 0.7 cross head mach screw

Hat bracket drawing

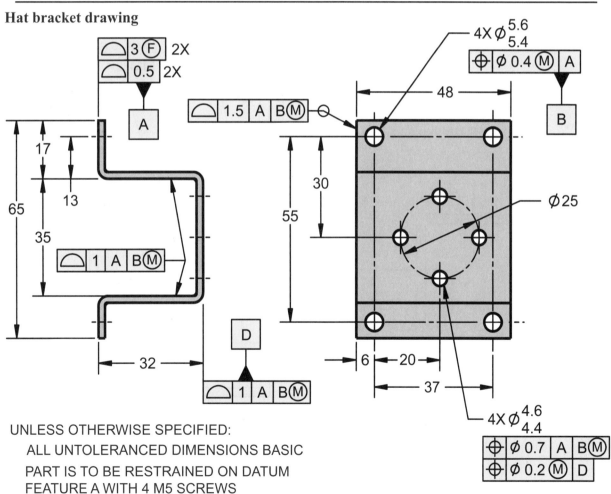

UNLESS OTHERWISE SPECIFIED:
ALL UNTOLERANCED DIMENSIONS BASIC
PART IS TO BE RESTRAINED ON DATUM
FEATURE A WITH 4 M5 SCREWS

Profile of a Surface - Hat Bracket

The hat bracket mounts on the bottom two flanges and the four hole pattern. The inside width clears the mating part. The bottom two flanges and the pattern of four holes establish the DRF. The default condition for all parts is the free state condition. This means all measurements are taken in the free state without any external forces applied. The designer had concerns about the flexibility of the part and applied a note that changes the default free state condition to a restrained condition. This requires all tolerances, except those noted with the free state symbol, circle F, to be verified in the restrained condition based on the restrictions in the note. See Non-Rigid Parts on page 12.29.

The 3 mm profile tolerance applied to the flanges with the free state symbol is verified with the part in the free state to insure the part is not twisted beyond a predetermined limit. The second 0.5 mm profile tolerance and all the remaining tolerances are verified in the restrained condition per the requirements dictated in the note.

The figure below illustrates the maximum material and least material boundaries of the profile and position tolerances in the restrained condition. Note how simple it is to evaluate the tolerances between the features. The verification procedure is equally as simple. The pattern of holes can be verified in an open set-up using the paper gage concept shown in unit 9 or a CMM can be used with the "best fit" option. If the holes conform, then retain this set-up as the DRF to verify all other features from this DRF setup.

Hat bracket tolerance zones

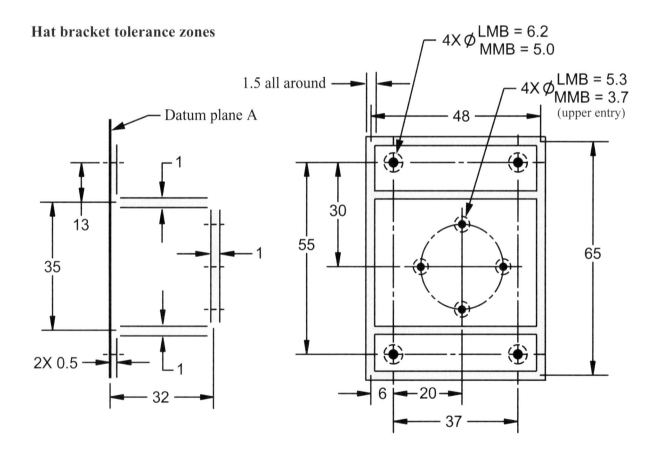

Workshop Exercise 12.2

Calculate the min and max distances in the restrained condition on the hat bracket below. Enter your answers in the table.

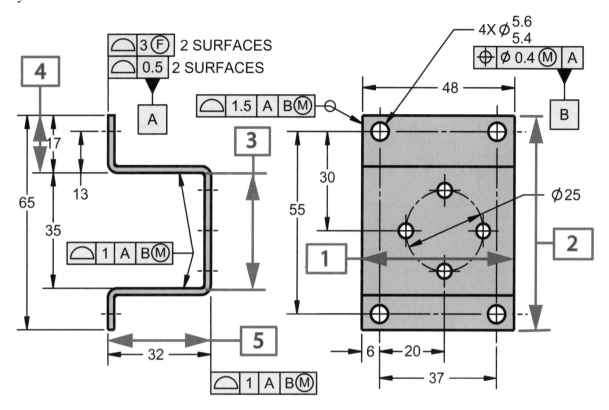

UNLESS OTHERWISE SPECIFIED:
  ALL UNTOLERANCED DIMENSIONS BASIC

  PART IS TO BE RESTRAINED ON DATUM
  FEATURE A WITH 4 M5 SCREWS

Calculate the Min and Max Dimensions		
No.	Min	Max
1		
2		
3		
4		
5		

Profile Interpretation - ASME Y14.5-2009

The shape of the profile tolerance zone has a different interpretation when applied per the ASME Y14.5-2009 standard vs. the ISO 1660-1987 standard. The figures and text below and throughout this book are per the ASME Y14.5-2009 standard. See the following page for the ISO 1660 interpretation.

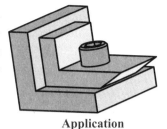

Application

This on the drawing

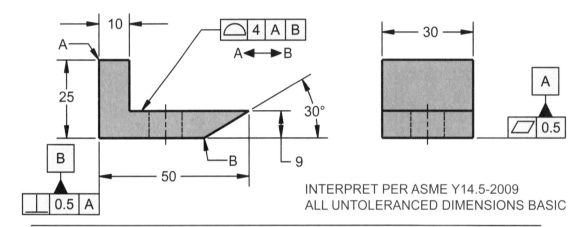

INTERPRET PER ASME Y14.5-2009
ALL UNTOLERANCED DIMENSIONS BASIC

Means this

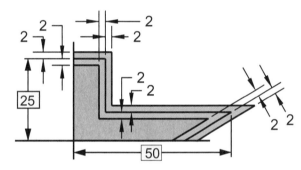

Profile tolerances apply normal (perpendicular) to the true profile at all points along the surface. The boundaries of the tolerance zone follow the geometric shape of the true profile.

Where a profile tolerance encompasses a sharp corner, the tolerance zone extends to the intersection of the boundary lines. The actual surface or line element must be within the specified tolerance zone.

Calculations to sharp corner

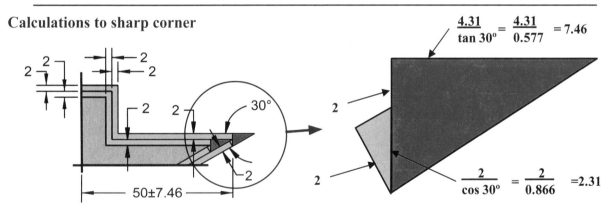

$$\frac{4.31}{\tan 30^{\circ}} = \frac{4.31}{0.577} = 7.46$$

$$\frac{2}{\cos 30^{\circ}} = \frac{2}{0.866} = 2.31$$

Since the profile tolerance zone extends to the intersection of the boundary lines. The overall length of the part measured from the datum to a sharp corner can vary more than the specified 4 mm profile zone. As seen from the right angle geometry calculations, the over all length to the corner can vary as much as 50±7.46. Compare the overall length of the ASME drawing with the ISO drawing on the next page. If necessary, the corner zone can be modified by applying a min or max radius or using a nonuniform profile tolerance.

Profile Interpretation - ISO

The shape of the profile tolerance zone has a different interpretation when applied per the ASME Y14.5-2009 standard vs. the ISO standard. The figures and text below represent the ISO 1660 - 1987 interpretation of profile tolerance zones.

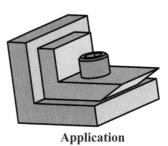

Application

Profile tolerances in the ISO 1660-1987 standard are defined with respect to the "true" profile which is itself defined by theoretically exact (basic) dimensions. The tolerance zone shall be equally disposed on either side of the true profile. The width of the tolerance zone is uniform when measured normal to the true profile at any point.

The tolerance zone may be explained by dragging the center point of a 4 mm sphere around the true profile. Notice that depending on the geometry of the true profile, the tolerance zone can have rounded corners. The actual surface or line element must be within the specified tolerance zone.

This on the drawing

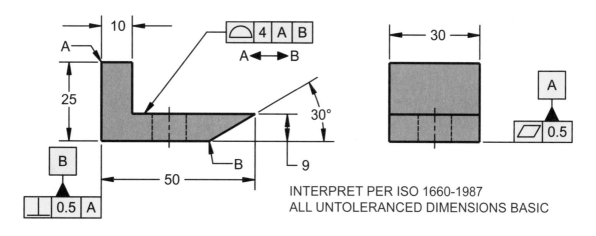

INTERPRET PER ISO 1660-1987
ALL UNTOLERANCED DIMENSIONS BASIC

Means this

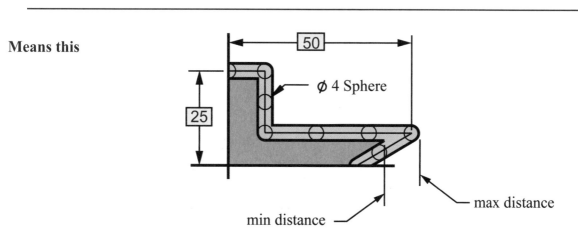

The profile tolerance zone is shown in the drawing above. Notice the external corners of the tolerance zone have a radii of 2mm established by the passing of a 4mm ball traveling along the true profile. The surface of the part must fall within the tolerance zone. Compare the overall length of the ASME drawing with the ISO drawing on the next page to see the difference.

Profile Tolerance - Non-Uniform Zone

The profile non-uniform tolerance zone is new for the ASME Y14.5-2009 standard. The profile non-uniform zone can be used where the standard default profile zone does not adequately define functional requirements. The term "NON-UNIFORM" replaces the tolerance value within the feature control frame. The leader line from the non-uniform feature control frame is directed to the true profile.

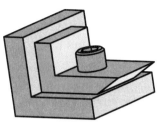

Application

A non-uniform profile tolerance zone is represented by a maximum material boundary and a least material boundary of unique shape that encompasses the true profile. These boundaries are defined in a CAD file or by basic dimensions on a drawing with phantom lines to indicate the tolerance zone.

On the drawing below, the MMB and LMB boundaries of the non-uniform profile zone are shown in phantom lines and defined with basic dimensions. The tolerance zone is defined as offset from the true profile by 2 mm with 5 intersecting corners that have a radii of 2 mm. The end point has a radii of 4mm resulting in a tapered tolerance on the top surface. The non-uniform profile tolerance zone applies between points A and B.

This on the drawing

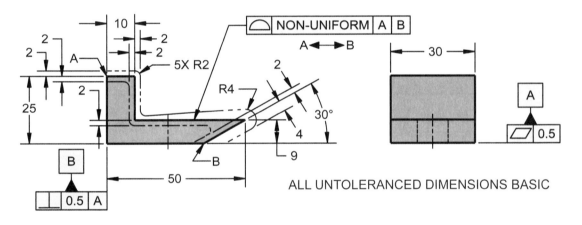

ALL UNTOLERANCED DIMENSIONS BASIC

Means this

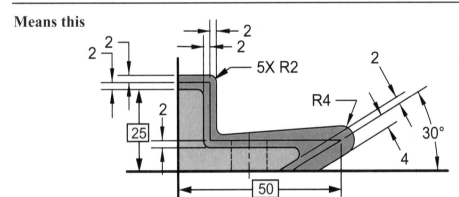

Basic dimensions define the size and shape of the tolerance zones.

This figure shows the non-uniform tolerance zone defined in phantom lines and dimensioned on the drawing above. The zone is tapered and has a larger radii on the end point. The zones can be defined in any geometric shape to meet functional requirements. The actual surface of the part must lie within the specified tolerance zone.

Composite Profile Tolerancing

Composite tolerancing is special. It can be used with both profile and position tolerances. The symbol is entered once and it is applicable to both horizontal entries. It is different than two single segmented feature control frames.

The upper segment on a composite profile feature control frame controls location, orientation, form and in some cases, size to the referenced datums. The lower segment does not control location to the specified datums, but does control orientation, form and in some cases, the size to the referenced DRF.

In contrast, all single segmented profile feature control frames control the location, orientation, form and in some cases, size of the feature to the referenced datums. This type of control is interpreted simply as two profile requirements.

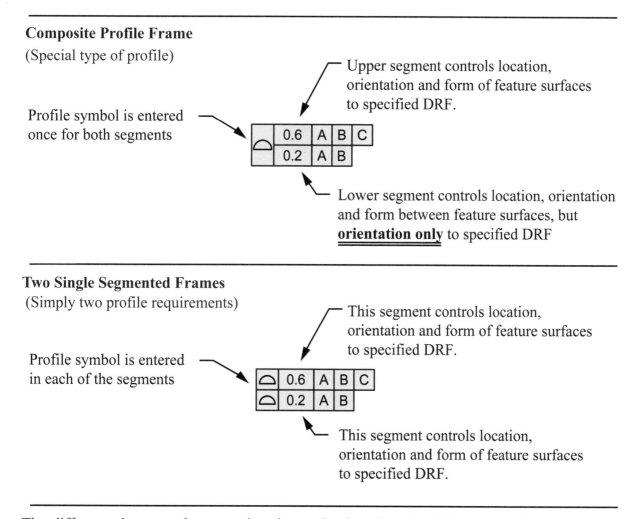

Composite Profile Frame

(Special type of profile)

Profile symbol is entered once for both segments

Upper segment controls location, orientation and form of feature surfaces to specified DRF.

| 0.6 | A | B | C |
| 0.2 | A | B |

Lower segment controls location, orientation and form between feature surfaces, but **orientation only** to specified DRF

Two Single Segmented Frames

(Simply two profile requirements)

Profile symbol is entered in each of the segments

This segment controls location, orientation and form of feature surfaces to specified DRF.

This segment controls location, orientation and form of feature surfaces to specified DRF.

The difference between the terms location and orientation should be clear. Location locates features and is associated with basic linear dimensions. It can also include orientation. Orientation however, is not associated with location or with basic linear dimensions, only basic angles. Orientation is thought of as parallelism, perpendicularity or angularity.

The composite tolerancing concepts above apply to both position and profile tolerances. See the position unit of this text for a more complete explanation of composite tolerancing.

Composite Profile - One Datum Feature

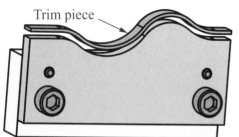

Trim piece

Application: The part mounts on the face and 2 dowels. These features are used to establish the DRF. The top surface contour mates with a trim piece. The location of the top contoured surface to the A, B DRF may vary within 0.5. The shape and orientation of the surface to the back mounting face must remain within 0.1.

Composite profile was applied to the surface. The upper specification locates and orients the contour to the DRF and the lower specification controls the profile of the shape and orients it only to datum A. The remaining surfaces are unimportant and are profiled within 1.5 to the DRF.

This on the drawing

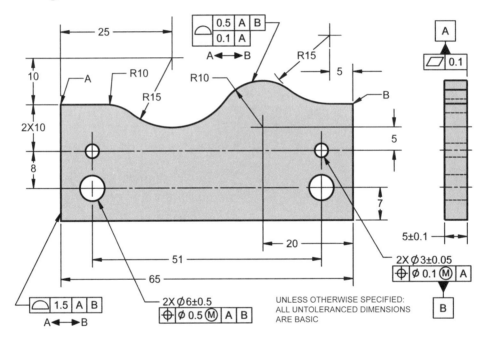

Means this

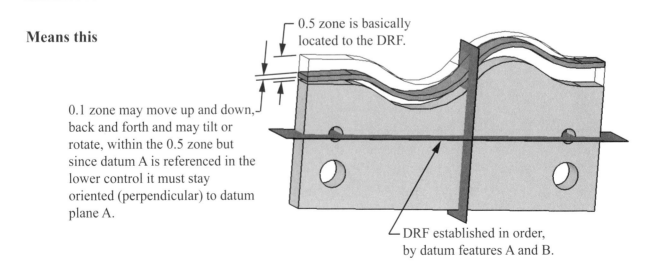

0.5 zone is basically located to the DRF.

0.1 zone may move up and down, back and forth and may tilt or rotate, within the 0.5 zone but since datum A is referenced in the lower control it must stay oriented (perpendicular) to datum plane A.

DRF established in order, by datum features A and B.

The contoured surface of the part must lie in both tolerance zones simultaneously.

12.22

Composite Profile - Two Datum Features

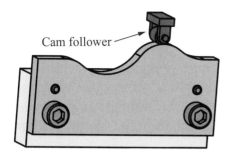

Cam follower

Application: The part mounts on the face and 2 dowels. These features are used to establish the DRF. A cam follower rides on the surface and may adjust up and down but has no provisions to allow it to tilt or rotate. The location of the top contoured surface to the A, B DRF may vary within 0.5 as long as the shape and orientation of the surface to the DRF remains within 0.1.

Composite profile was applied to the surface. The 0.5 profile upper specification references datum features A and B and locates and orients the contour to the DRF. The 0.1 profile lower specification references datum features A and B and refines the shape of the contour to itself and refines only the orientation (not location) to the DRF. The remaining surfaces are unimportant and are profiled within 1.5 to the DRF.

This on the drawing

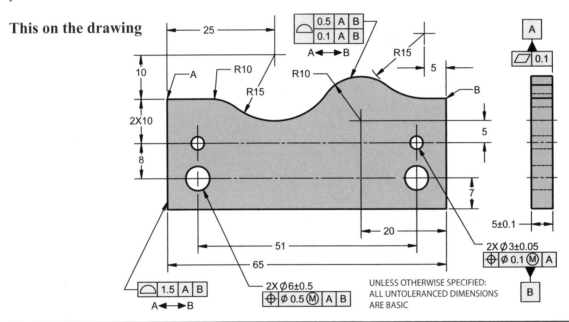

UNLESS OTHERWISE SPECIFIED:
ALL UNTOLERANCED DIMENSIONS
ARE BASIC

Means this

0.5 zone is basically located to the DRF.

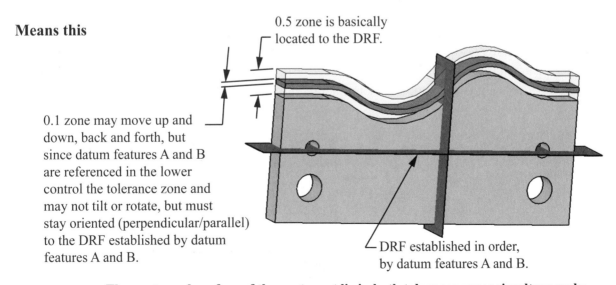

0.1 zone may move up and down, back and forth, but since datum features A and B are referenced in the lower control the tolerance zone and may not tilt or rotate, but must stay oriented (perpendicular/parallel) to the DRF established by datum features A and B.

DRF established in order, by datum features A and B.

The contoured surface of the part must lie in both tolerance zones simultaneously.

Composite Profile of an Irregular Feature

Composite profile may be used to locate irregular shaped features as shown below. The irregular feature is defined with basic dimensions. A composite profile is applied to the feature.

The profile symbol is entered once and is applicable to both the upper and lower segments. The upper segment of the feature control frame, in effect, locates the feature to the specified datum features. The lower segment of the feature control frame specifies the size, form (or shape) of the feature and the orientation (perpendicularity in this example) of the feature to the specified datum features.

Composite profile can be used define size, form, orientation and location of an Irregular feature.

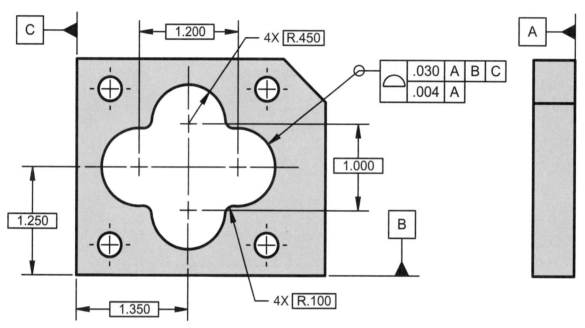

Composite profile used for locating irregular features is similar to the position boundary concept for locating irregular features. The difference is that position boundary concept applied at MMC or LMC establishes a single MMB or LMB boundary that the feature must clear.

The composite profile concept can not be applied with the feature modifiers MMC or LMC. Composite profile establishes both an MMB and LMB in which the feature must lie. Using composite profile to locate an irregular feature is a more restrictive control than position boundary. The position boundary concept is explained later in the position section, unit 13.

Calculate the min/max dimensions.

1. Min _____
2. Max _____

3. Min _____
4. Max _____

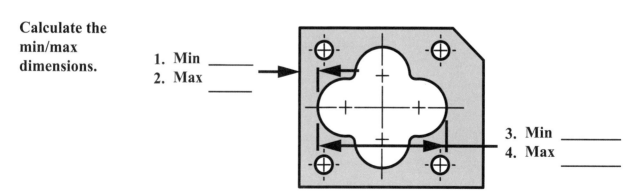

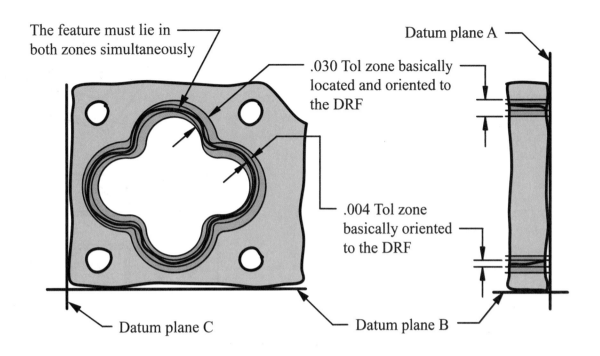

The feature must lie in both zones simultaneously

.030 Tol zone basically located and oriented to the DRF

Datum plane A

.004 Tol zone basically oriented to the DRF

Datum plane C

Datum plane B

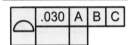

The upper segment specifies the location of the feature to the specified datums.

DRF is established in order by datum features A, B, & C

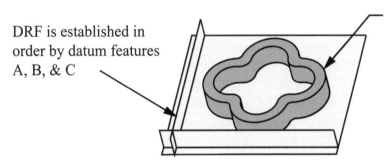

First:
The feature must lie within a .030 zone basically located and oriented to the datum reference frame.

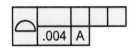

The lower segment specifies the size, shape, and the orientation of the feature to the stated datums.

DRF is established by datum feature A. This is an orientation plane only.

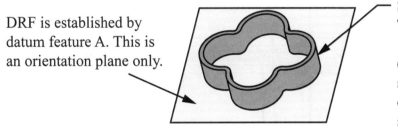

Second:
The feature must lie within a .004 zone that is only oriented (perpendicular) to the lower segment DRF. This zone also controls the size, form and shape of the feature.

The two segments in the composite profile feature control frame create two tolerance zones. The .030 zone locates the feature to the datum reference frame. The .004 zone refines the size, shape, and orientation of the feature. This .004 zone may float up, down, left, right and rotate inside the .030 zone to accept the feature.

12.25

Profile - Coplanarity

The part on the drawing below has multiple geometric controls applied to the four pads. Multiple controls are not common, but in certain cases there are design requirements that require more exacting control on the features. Each applied feature control frame defines a specific location, orientation and/or form requirement for the pads. This is an illustration to show the requirements of the applied feature control frames below.

The upper segment of the composite profile locates, orients, and controls form of the 4 pads within .040 to datum A.

The lower segment of the composite profile controls the location between the pads (coplanarity) within .020, while also refining orientation (parallelism) to datum A.

The profile of .010 has no datum reference so it refines the coplanarity between the 4 pads (to each other).

This on the drawing

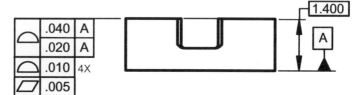

The flatness of .005 controls the surface of each individual pad. Note: Flatness and parallelism are individual controls and do not control the relationship between features.

The feature control frames are basically stating that the height of the pads must be located within .040. However the pads may not tilt, they must remain parallel within .020. The pads to themselves or coplanarity must be within .010. Also, each pad individually must be flat within .005.

Means this

The .040 upper segment composite profile locates the 4 pads to datum A. This zone is centered about the basic dimension of 1.400.

Tolerance zones do not locate tolerance zones. The features must lie in the required zones.

The .010 profile locates the 4 pads to each other (coplanarity) without a datum feature reference. All surfaces must lie in this zone, this zone may float and tilt within the confines of the other specifications because it has no datum feature reference.

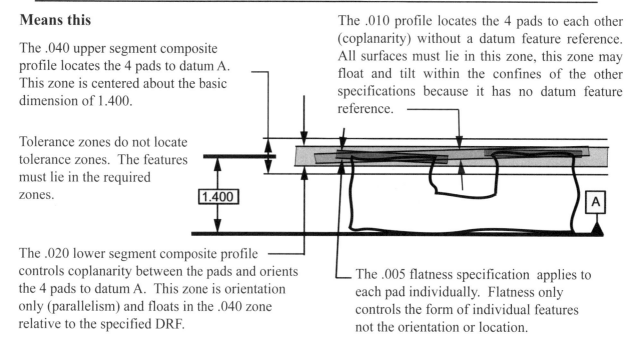

The .020 lower segment composite profile controls coplanarity between the pads and orients the 4 pads to datum A. This zone is orientation only (parallelism) and floats in the .040 zone relative to the specified DRF.

The .005 flatness specification applies to each pad individually. Flatness only controls the form of individual features not the orientation or location.

Profile - Multiple Controls

The part below has multiple profile controls applied to illustrate how adding and removing datum feature references in the feature control frames can effect the profile (or position) specification. Notice when datum references are added or removed the specification can be changed from location, and/or orientation, and/or form and/or size.

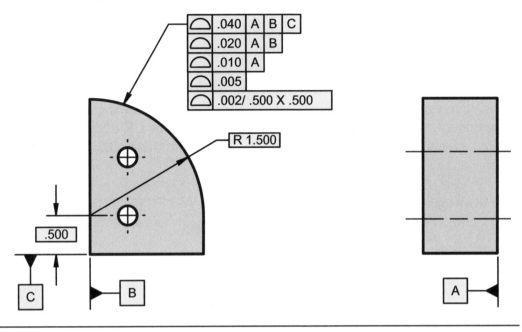

.002 zone controls form for a series of specified areas of .500 X .500 or rate of change. This zone floats in the .005 zone. (no datums) Often used to control machine mis-match.

.005 zone controls form and shape only. This zone floats in the .010 zone. (no datums)

.010 zone controls orientation to datum A. It includes the form. This zone floats in the .020 zone.

.020 zone controls location to datums A and B. It includes orientation and form. This zone floats only up and down in the .040 zone.

.040 zone controls location to datums A, B, C. It includes orientation and form. This zone is fixed to the basic dimensions.

The part must meet all requirements. Each control starting from the top refines the tolerance.

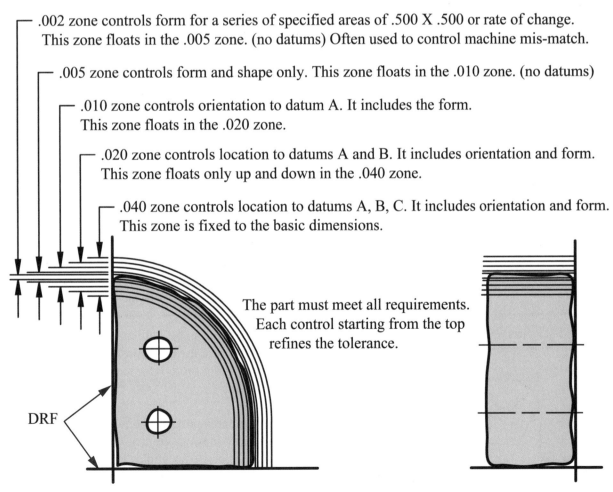

There are two types of directly toleranced radii that can be applied to parts, the radius and controlled radius. The radius (R) is for general applications. The controlled radius (CR) is used when it is necessary to place further restrictions on the shape of the radius, as in high stress applications.

In the more general R application, a crescent shaped tolerance zone is established and the feature surface must lie within the tolerance zone. In the more restrictive CR application, a crescent shaped tolerance zone is established along with an additional requirement that the feature surface contour must be a fair curve with no flats or reversals. The intent is to have the radii free of sharp corners in order to avoid stress concentrations. The verification that the contour must be a fair curve with no flats or reversals can be subjective. When using the CR designation, the engineer may need to refer to some supporting company document that will identify the detailed verification procedure or use profile tolerancing.

This on the drawing   **Means this**

Radius

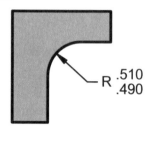

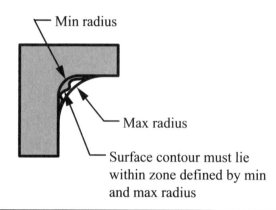

Min radius

Max radius

Surface contour must lie within zone defined by min and max radius

Controlled Radius

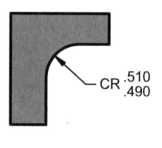

Min radius

Max radius

Surface contour must lie within zone defined by min and max radius. The contour must also be a fair curve with no flats or reversals.

Profiled Radius

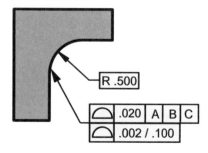

Profile tolerances can also be applied very effectively, to a radii. On this part, the upper profile control locates, orients and controls the form of the surface to the datum feature references.

The lower per unit area profile specification provides a series of smaller zones that controls the rate of change of the surface. See earlier examples in this unit for additional information on controlling surfaces.

Non-Rigid Parts - Restrained Condition

Unless otherwise specified, all dimensioning and tolerancing applies in a free state condition with no restraint. Some parts, such as sheet metal, thin metal, plastics and rubber, are nonrigid in nature. These parts are naturally somewhat flexible or wobbly. Commonly, it may be necessary to specify design requirements on the part in a natural or free state as well as in a restrained condition. The restrained condition usually requires some restraint or force to insure functional requirements. The restraint or force on the nonrigid parts is usually applied in such a manner to resemble or approximate the functional or mating requirements. Sometimes, it may also be necessary to specify the direction of the force of gravity.

Nonrigid parts can fall into two categories:

1. Inherently flexible parts - These parts are very flexible and wobbly. The force of gravity alone can cause distortion in inspection and lead to the misreadings. Parts fitting in this category are thin metal rings, rubber, gasket material, etc. To ensure design intent, these inherently flexible parts will require the application of forces or restraints to hold or bring them into shape. It might also be necessary to specify the direction or force of gravity.

2. Inherently rigid parts - These parts are rigid, but due to the imposition or release of internal stresses resulting from the manufacturing process, these parts can distort beyond the specified design requirements. Parts fitting in this category are sheet metal stampings, flexible welded or riveted assemblies, plastic parts etc. These inherently rigid parts may require the application of certain restraints to force them into the functional design requirements.

The dimensioning and tolerancing of the inherently flexible parts and the inherently rigid parts are usually similar. To ensure design requirements, it may be necessary to assess these parts both in the natural free state and in a restrained condition.

Example: An inherently rigid part such as a sheet metal stamping may have some surfaces that have spring-back or draft due to the stamping operation. These surfaces may be used as datum features. These datum features are bolted, screwed or otherwise restrained in the assembly process, but in their free state, the parts may be in or out of tolerance. The designer may specify the datum surfaces restrained in such a way to resemble the bolted or screwed assembly condition. All of the characteristics related to the datums are verified in the specified restrained condition.

As you might imagine, the method of restraining the datum features can vary widely depending on the particular part. There is no standard note or symbol available that will cover all the possible restraint conditions. The usual method for specification of restrained conditions is done with a note on the drawing. If all the parts are similar or there is a standard restraint procedure, this information may be specified in a company standard specification. This company standard specification may be referenced on the drawing.

This note or specification should explain how the part is restrained and the force required to facilitate the restraint. A sample note may be found on the drawing below.

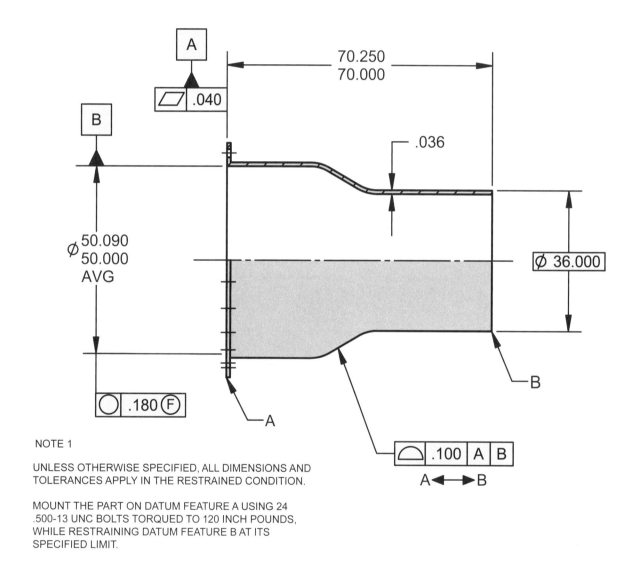

NOTE 1

UNLESS OTHERWISE SPECIFIED, ALL DIMENSIONS AND
TOLERANCES APPLY IN THE RESTRAINED CONDITION.

MOUNT THE PART ON DATUM FEATURE A USING 24
.500-13 UNC BOLTS TORQUED TO 120 INCH POUNDS,
WHILE RESTRAINING DATUM FEATURE B AT ITS
SPECIFIED LIMIT.

12.30

In some cases it may be necessary for a designer to specify some dimensions and tolerances to the part in the free state condition. This may be required because even though the part may meet the design requirements in the restrained condition, there are certain design requirements that must also be met in the free state condition. The free state requirements may be required before the part is restrained, it should be checked to make sure it is not bent beyond its elastic limit or that it is not bent so much that the assembly process will require too much force as to restrict assembly or insert screws. In another case, the part may be bent so much that the forces required to restrain it may be great enough that it will deform the part to which it is attached. These are but a few of the many reasons that a designer may require the part to be checked both in the free and restrained state.

If any of the part specifications are to be verified in the free state, the designer may specify this requirement with the words FREE STATE or the free state symbol. The free state symbol is an F in a circle. This free state symbol means that dimensions and tolerances that have the free state symbol applied are checked in the free state and not the restrained condition. The free state condition is the condition of a part when it is only subjected to the forces of gravity. The free state symbol is applied by placing it next to or associating it with the required dimensions and tolerances. If it is applied in a feature control frame, it always follows the feature tolerance and any modifiers.

In theory, there should not be any assumptions. It is important to specify all the design requirements on the drawing to ensure functional requirements. But, is there a point where we can apply too many specifications? Nonrigid parts have been built for a long time. There are many established procedures that manufacturing and quality already use to ensure product integrity. It is generally understood that certain nonrigid or flexible parts are subject to variation and must be clamped or restrained. If an organization is in the business of working with these parts on a regular basis, there are often some written or unwritten rules than will determine how a part shall be verified.

There may be some instances where, rather than specifying restraint requirements, the designer might let standard shop procedures or "common sense" apply. It should be understood that there is some risk when it is assumed that everyone understands all the parameters in which the tolerance is applied and verified. It might be wise for a designer to consult with manufacturing and quality to determine what, if any, standard procedures may apply or are already in place.

In many cases, the part may be of significant importance to require the specification of restraint forces to insure functional requirements. In this case, it may be absolutely necessary that the restraint forces and methods of application are specified in great detail.

Where the line is drawn to specify or not specify restraints will depend on the part, the number built, common sense, tolerance limits, part size, material, stock size and past history, to name a few. The designer must make the final decision on the specifications.

There is additional information on nonrigid parts in the ASME Y14.5-2009 standard and the ISO 10579, 1993 (E) standard. If the designation ISO 10579 is found on ISO drawings in or near the title block, the conditions under which the part shall be restrained may be found in the notes.

See the Hat Bracket in this unit for an example of a sheet metal bracket in which the tolerances were applied in the restrained condition with a feature control frame and free state symbol.

Workshop Exercise 12.3 - Profile Stack

Calculate the minimum and maximum values for the indicated dimensions below. All minimum and maximum dimensions indicated with an asterisk (*) should be calculated as originating from the DRF established by the datum targets. All other minimum and maximum dimensions should be calculated as the relationship between the features.

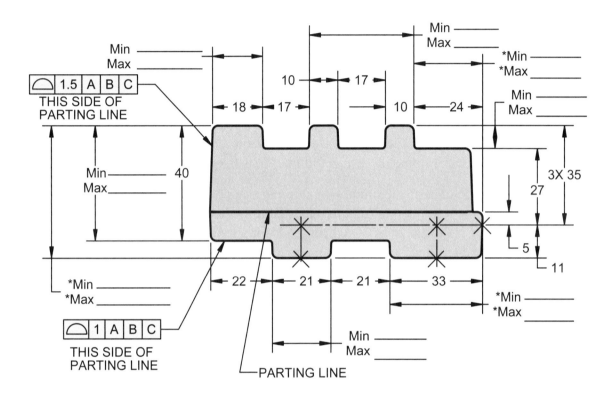

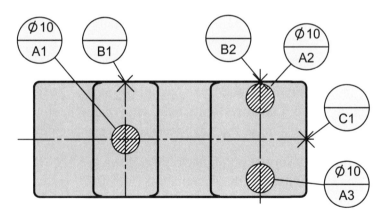

NOTES:
ALL UNTOLERANCED DIMENSIONS BASIC
2° BASIC DRAFT ANGLE
FILLET AND CORNER RADII = 2

Calculate the minimum and maximum values for the indicated dimensions below. All minimum and maximum dimensions indicated with an asterisk (*) should be calculated as originating from the DRF established by the datum targets. All other minimum and maximum dimensions should be calculated as the relationship between the features.

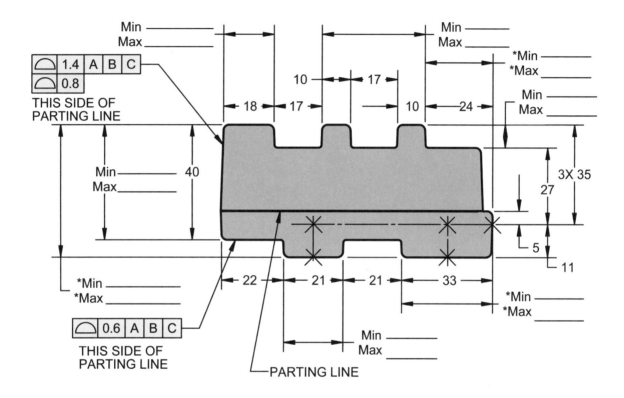

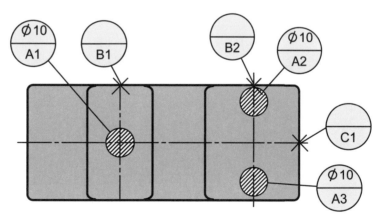

NOTES:
ALL UNTOLERANCED DIMENSIONS BASIC
2° BASIC DRAFT ANGLE
FILLET AND CORNER RADII = 2

Unit 13

Position Tolerances

Position Tolerancing

Position tolerance controls the location of a center point, axis, median plane, or boundary of a feature of size.

Symbol	Tolerance	Common Shape of Tolerance Zone	2D / 3D	Application of Feature Modifier
⊕	Position	2 Parallel planes Cylindrical Spherical Conical Boundary	3D	Yes

Overview:

Position tolerancing is used for locating features of size. It defines a zone within which the center point, axis, median plane or boundary of a feature of size is permitted to vary from a true (theoretical exact) position. The true position is established with basic dimensions from a datum reference frame or from other features that are related to the datum reference frame. All dimensions locating the features must be basic and are considered theoretically exact. The tolerance for the location deviation of the feature from basic is found in the position feature control frame.

Position tolerancing is a three dimensional control. The common shapes of the tolerance zones are shown in the above chart. If the diameter symbol is used, the position tolerance is a cylindrical shape. If no symbol is used, the tolerance zone defaults to a total wide zone as in parallel planes for a slot. An "S" preceding the diameter symbol, designates a spherical diameter zone.

Position tolerancing can also be used for locating irregular shaped features that have no discernible axis or median plane. This will create a boundary tolerance zone. The surface of the feature must not enter this theoretical boundary of identical shape located at true position.

In the past ASME Y14.5M-1994 standard, If a boundary shaped tolerance zone was needed, the term BOUNDARY was placed under the feature control frame. In the current ASME Y14.5-2009 standard, this concept is implied and the term BOUNDARY is no longer required under the feature control frame.

Per the ASME Y14.5-2009 standard, position tolerancing is applied on an MMC, LMC or RFS basis. Where MMC or LMC are required, the appropriate feature modifier symbol is applied in the feature control frame following the feature tolerance. If no modifier is specified, the default condition is implied at RFS. See modifier rules and effect of modifiers shown earlier in Unit 3.

Position tolerance is usually associated with datums and the datum reference frame. Although there are cases, where a group of features have an interrelationship to each other and are used as datums. In this case, datums may not be necessary for the interrelationship of those features.

Axis Versus Surface Interpretation

Position tolerancing is often described as locating the axis or center plane of a feature. The axis or center plane of a feature is derived by using the actual mating envelope which contacts the high points of a feature. The location of the axis or center plane of the actual mating envelope explanation is often used because of past history and because it is easy to explain. In a functional assembly, it is really the surface of the feature that comes in contact with the mating part. This is especially evident when a functional gage is used to verify a position call out. The feature's axis or center plane is not located, but rather the surface of the feature must not enter a virtual condition boundary.

The difference in the axis and surface interpretations is usually a minor concern. The problem occurs when a hole is out of round and only the axis of the mating envelope is controlled.

In the verification procedure, the actual value of position tolerancing can be defined in terms of the axis or median plane or in terms of the surface of a feature. The axis and median plane inspection procedure is explained in unit 3. The surface interpretation is reported in terms of the actual virtual size (MMB or LMB) of a feature.

This concept of axis versus surface was first recognized in the ANSI Y14.5M-1982 standard. The current ASME Y14.5-2009 and ASME Y14.5.1M-1994 standards recognize that the axis versus surface interpretation can yield minor but different results. The standards state that if there is a conflict between the two procedures, the surface interpretation shall prevail.

ISO Interpretation

In ISO, the default condition for establishing an axis, point or center plane of a feature is different than the ASME Y14.5-2009 standard. ISO 14660, part 1 and 2 defines the default extraction process for an axis, point or center plane as a least squares center process. If another method is desired, such as minimum inscribed, maximum circumscribed or surface is desired, it can be referenced on the drawing. See ISO standards for additional information.

Note: A common default fitting routine generally used by a CMM is least squares.

Position - Cylindrical Tolerance Zone

Position tolerancing is used for locating features of size. It defines a zone within which the feature axis or centerplane must lie. The size of this tolerance zone is equal to the amount of variation allowed from the true (theoretically exact) position. Basic dimensions establish the true position from the specified datums as well as the interrelationship between the features.

This on the drawing

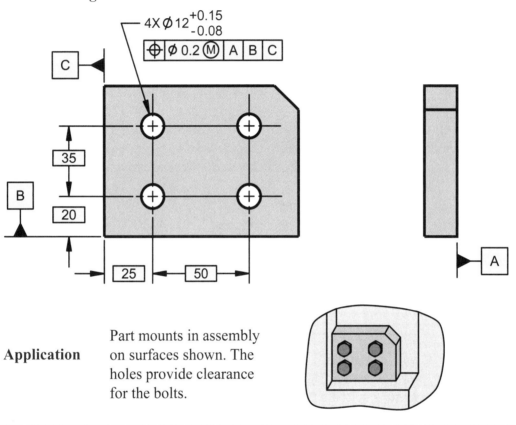

Application   Part mounts in assembly on surfaces shown. The holes provide clearance for the bolts.

Means this

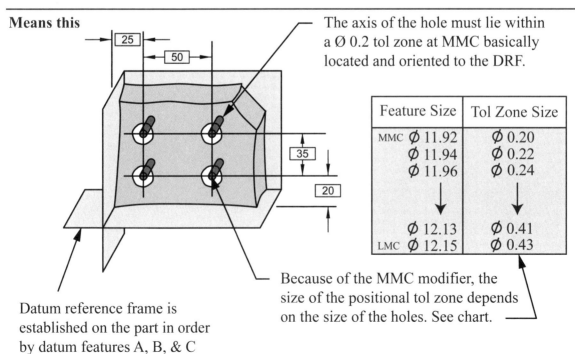

The axis of the hole must lie within a Ø 0.2 tol zone at MMC basically located and oriented to the DRF.

Feature Size	Tol Zone Size
MMC Ø 11.92	Ø 0.20
Ø 11.94	Ø 0.22
Ø 11.96	Ø 0.24
↓	↓
Ø 12.13	Ø 0.41
LMC Ø 12.15	Ø 0.43

Because of the MMC modifier, the size of the positional tol zone depends on the size of the holes. See chart.

Datum reference frame is established on the part in order by datum features A, B, & C

13.4

The graphic below shows the sectioned imperfect part from the previous page. The theoretical 3D cylindrical tolerance zones control the location and orientation of the hole axes. These axes can be simulated with a gage pin, CMM, or other methods.

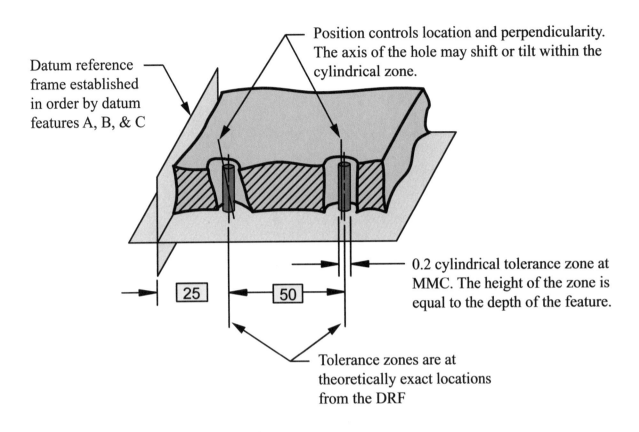

Position controls location and perpendicularity. The axis of the hole may shift or tilt within the cylindrical zone.

Datum reference frame established in order by datum features A, B, & C

0.2 cylindrical tolerance zone at MMC. The height of the zone is equal to the depth of the feature.

25

50

Tolerance zones are at theoretically exact locations from the DRF

Functional gage for position call out

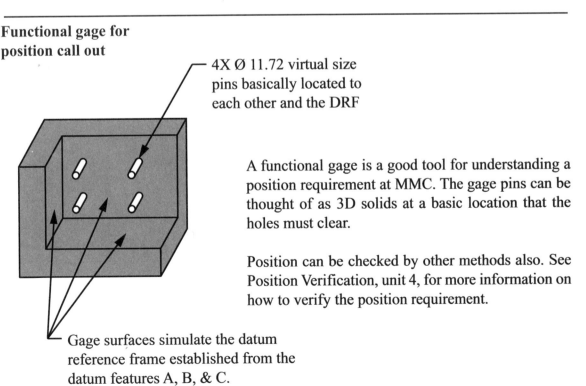

4X Ø 11.72 virtual size pins basically located to each other and the DRF

A functional gage is a good tool for understanding a position requirement at MMC. The gage pins can be thought of as 3D solids at a basic location that the holes must clear.

Position can be checked by other methods also. See Position Verification, unit 4, for more information on how to verify the position requirement.

Gage surfaces simulate the datum reference frame established from the datum features A, B, & C.

Position - Rectangular Tolerance Zone

In some cases, it may be desirable to locate or orient a feature within a tighter tolerance in one direction than another. This may be accomplished by the method shown below. Separate feature control frames are used to indicate the direction and magnitude of the variation. Note that the diameter symbol is not present in the feature control frames indicating a distance between two parallel planes. This is a common way to locate slotted holes.

This on the drawing

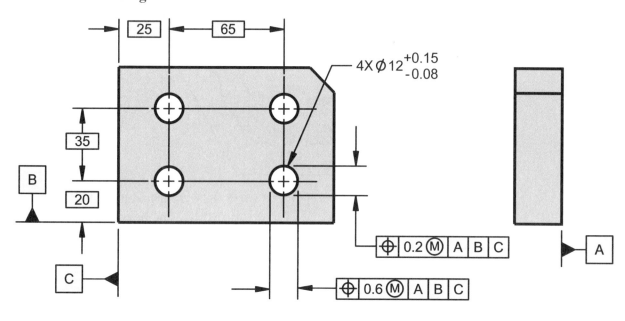

Means this

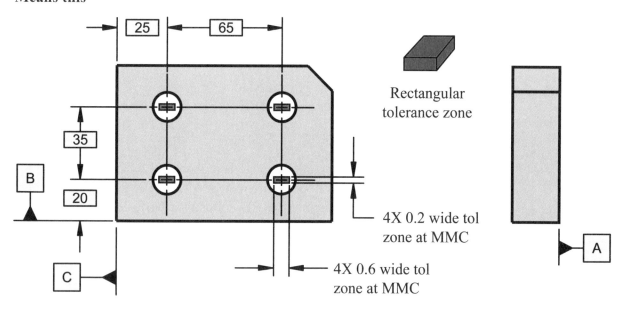

The axes of the holes must lie within the 0.2 X 0.6 rectangular tolerance zones basically located to the specified datum reference frame.

13.6

Position - Spherical Tolerance Zone

Position tolerancing may be used to control the location of a spherical feature. The symbol for spherical diameter precedes the size dimension to indicate a spherical feature. The SØ is also entered before the feature tolerance in the feature control frame to indicate the shape of the position tolerance zone is a sphere.

This on the drawing

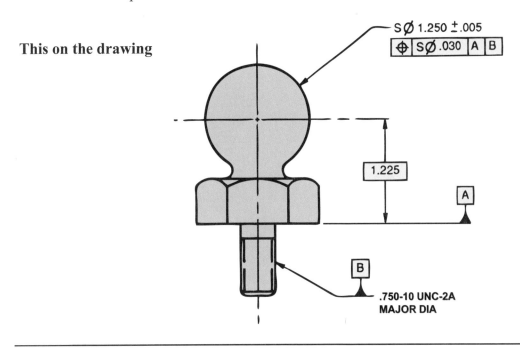

Means this

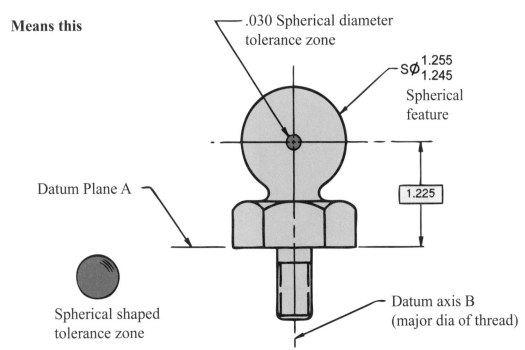

The center point of the spherical diameter must lie within a spherical diameter zone of .030 RFS, which is basically located to the datum reference frame (DRF).

Position - Conical Tolerance Zone

Position tolerance can be used to control a feature such as a deep drilled hole. The center hole in the manifold needs to be closely located where it meets a port but can be allowed more location tolerance at the left surface. Two feature control frames are used. One defines a close tolerance zone where it meets the port, the other feature control frame defines a larger tolerance at the left surface. The result is a conical shaped tolerance zone within which the feature axis must lie.

Application

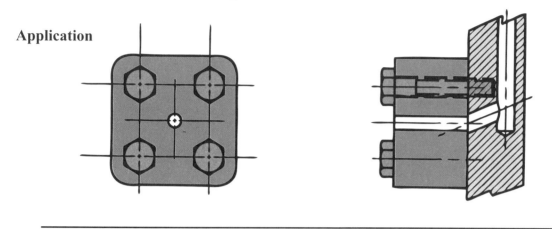

This on the drawing

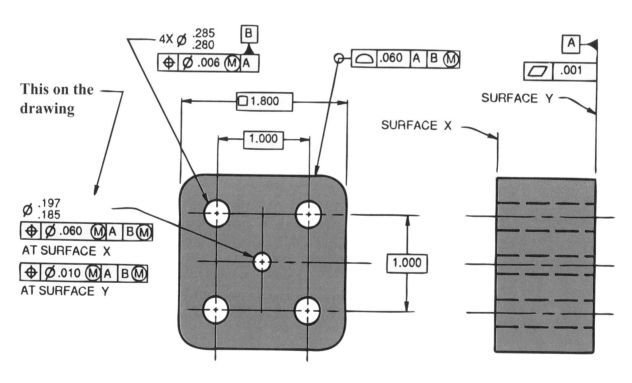

Means This

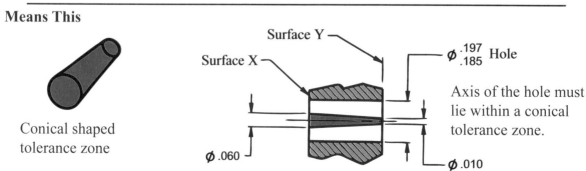

Conical shaped
tolerance zone

Axis of the hole must
lie within a conical
tolerance zone.

Position Boundary

Position may be used to locate irregular features of size. Rather than locating the axis or median plane of the feature, a single maximum material boundary is established. In the illustration below, a datum reference frame is established by the back surface, height, and width. The profile tolerance on the irregular opening defines the size, shape, and orientation to datum A. The position tolerance controls the location of the feature. It defines a boundary in which no element of the feature must lie. The feature modifiers, MMC or LMC, may be applied.

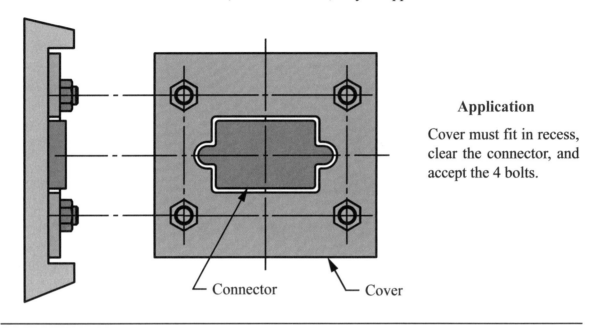

Application

Cover must fit in recess, clear the connector, and accept the 4 bolts.

Connector   Cover

This on the drawing

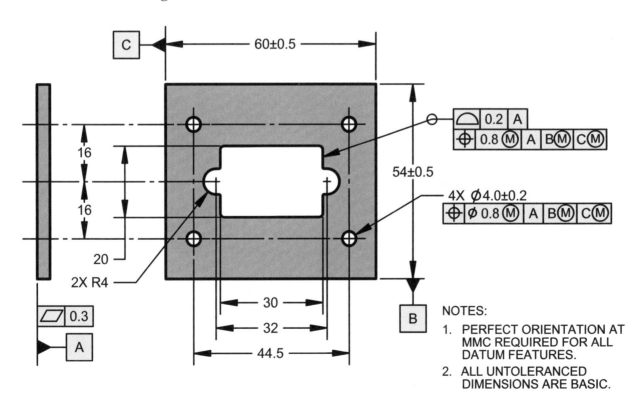

NOTES:

1. PERFECT ORIENTATION AT MMC REQUIRED FOR ALL DATUM FEATURES.

2. ALL UNTOLERANCED DIMENSIONS ARE BASIC.

Means this

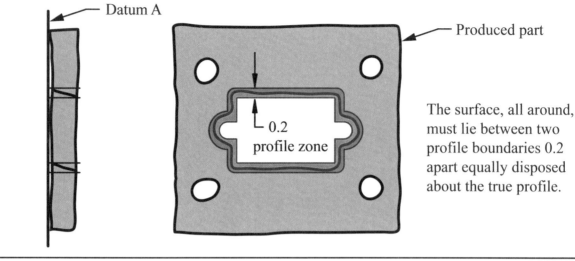

The profile requirement controls the size, shape, and orientation of the feature.

Datum A

Produced part

0.2
profile zone

The surface, all around, must lie between two profile boundaries 0.2 apart equally disposed about the true profile.

The position requirement locates the feature to the datum reference frame.

No portion of the surface may lie inside the virtual boundary created by the smallest profile boundary minus the position tolerance. This maximum material boundary is basically located to the datums A, B at MMB and C at MMB.

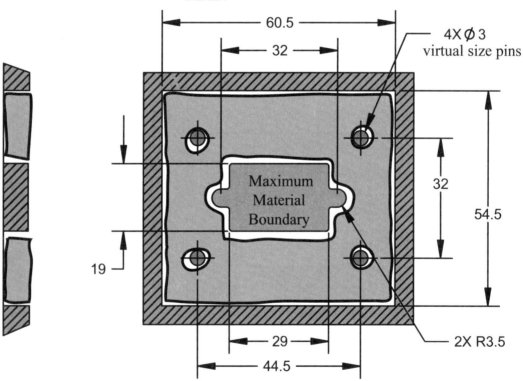

Position boundary for locating irregular features of size is a similar concept to composite profile. The difference is position boundary, when applied with MMC and LMC modifiers, defines only a MMB or LMB in which no part of the surface may lie. Composite profile creates both a MMB and a LMB in which the feature must lie. See composite profile in unit 12.

13.10

Composite Tolerancing is Special

Composite positional tolerancing provides for the location of feature patterns as well as the features within these patterns. These requirements are stated by the use of a composite feature control frame. The position tolerance symbol is entered once and is applicable to all horizontal segments. Each complete horizontal segment in the feature control frame may be verified separately, but the lower segment is always a subset of the upper segment. The composite tolerancing concepts can also be applied to profile.

Composite Position Feature Control Frame

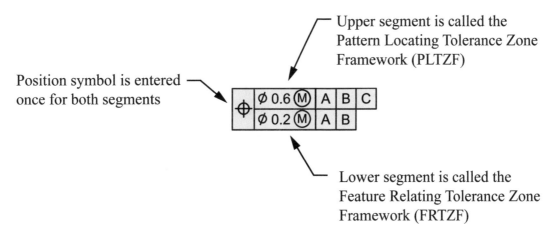

The upper segment of a composite position feature control frame is called the Pattern Locating Tolerance Zone Framework (PLTZF). The PLTZF is located to applicable datums with basic dimensions. It specifies the larger positional tolerance for the location of the pattern of features as a group. Applicable datums are entered in the upper segment in their order of precedence. The upper segment locates and orients the features to each other as well as the specified datums.

The lower segment of a composite feature control frame is called the Feature Relating Tolerance Zone Framework (FRTZF). The FRTZF specifies a smaller position tolerance for the feature to feature relationship within the pattern. Basic dimensions apply between the features but do not apply to the datums. The lower entry is orientation only to the specified datums and not location.

If datums are not specified in the lower segment, the FRTZF is allowed to tilt, rotate and/or shift within the confines of the PLTZF. If datums are specified in the lower segment, they govern the orientation (not location) of the FRTZF relative to the specified datums.

Note: Different datum features, different datum feature modifiers, or the same datums in a different order of precedence, constitutes a different datum reference frame. This requirement is not to be specified using the composite positional tolerancing method, since such a requirement no longer represents a liberation within the given limits of the FRTZF. A separately specified feature relating tolerance, using a second single segment feature control frame, should be used, including applicable datum feature references as an independent requirement.

Composite Tolerancing vs. Two Single Segments

Composite position tolerancing is special. The position symbol is entered once for both horizontal entries. It is different than two single segmented feature control frames.

The upper segment on a composite feature control frame controls location and orientation to the DRF. The lower segment on a composite controls location between the features but controls orientation only (parallelism, perpendicularity, angularity) of the features to the specified DRF.

In contrast, all single segment position frames control both the location and orientation between the features and the specified DRF. This type of control is interpreted simply as two position requirements.

Composite Position Frame
(Special type of position)

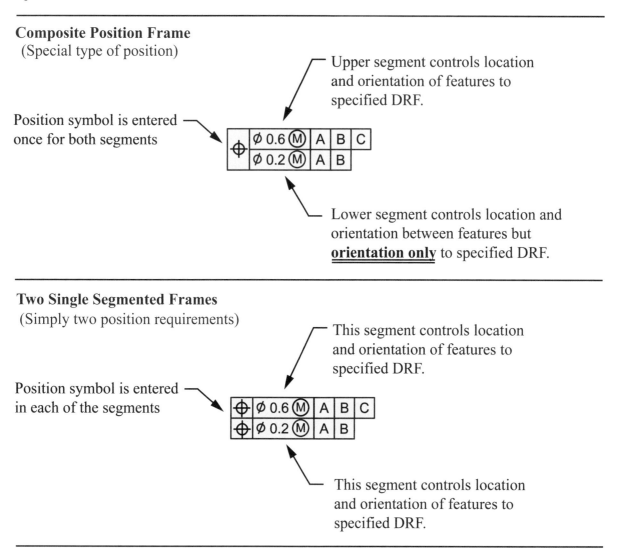

Upper segment controls location and orientation of features to specified DRF.

Position symbol is entered once for both segments

Lower segment controls location and orientation between features but **orientation only** to specified DRF.

Two Single Segmented Frames
(Simply two position requirements)

This segment controls location and orientation of features to specified DRF.

Position symbol is entered in each of the segments

This segment controls location and orientation of features to specified DRF.

The difference between the terms position and orientation should be clear. Position locates features and is associated with basic linear dimensions. It also includes orientation. Orientation, on the other hand, is not associated with location or with basic linear dimensions, only basic angles. Orientation is usually thought of as parallelism, perpendicularity or angularity. The composite tolerancing concepts explained above apply to both position and profile tolerances.

In some particular cases, either the application of a composite tolerance or two single segmented frames to a group of features can yield the same result. This is especially true when only a primary datum plane is established perpendicular to the features and is entered in the lower segment. The lower entry on a composite is only orientation to the datums. Although all the entries on single segmented frames are location, only orientation can be established to the perpendicular plane. See the four hole plate composite position examples in this text for an illustration.

The concepts concerning composite tolerancing (both position and profile) was enhanced and clarified in the ASME Y14.5M-1994 standard and reaffirmed in the 2009 standard. The composite enhancement happens when more than one datum is entered in the lower segment of a composite tolerance or the primary datum is established from a feature other than perpendicular to the positioned features. The ANSI Y14.5M-1982 standard did not cover this topic in detail. This expanded interpretation of composite tolerancing is applicable to both position and profile tolerancing.

Composite tolerancing is very useful for providing a loose location but a more restrictive orientation of the features. A simple example might be a specification for a pattern of holes that locate an emblem or name plate. The holes are important to each other but not important to the datums. The orientation of the pattern, however, is important as the emblem or name plate is not allowed to tilt or mount crooked.

Composite position tolerancing is used and appropriate datums features are specified in the lower segment to control orientation. Applying orientation tolerances such as parallelism, perpendicularity and angularity will not work as they can not control location between the group of features, only the orientation of individual features.

The name plate is only a simple example. There are many more examples where a designer may allow a relaxation for the location of a group of features, if the orientation of the features can be more restrictive. Composite tolerancing increases manufacturing tolerance by providing the designer with the tools to specify only the requirements needed. If location is required to the specified datums, use single segmented feature control frames. If a large location to the datums, but close location between the features and orientation is required, use composite tolerancing.

The following examples in the text illustrate composite position tolerancing. These examples can be compared to the examples of single segmented feature control frames for additional clarification of principles. See also composite profile tolerancing.

13.13

Composite Position - One Datum Feature Reference

Composite tolerancing is a special method of locating features of size. The position symbol is entered once and is applicable to both horizontal segments. With composite tolerancing, the upper segment locates and orients the pattern of holes to the specified DRF. The lower segment locates the holes to each other and controls the orientation of the pattern to datum A.

This on the drawing

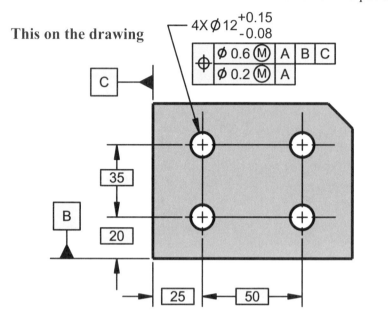

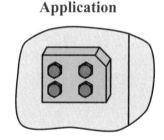

Application

Composite positional tolerancing may be used where the location of the holes to each other is important, but the location of the holes relative to the DRF is less important. This might be an application where a bracket or switch attaches with the holes but the location of the bracket to the edges is not as essential.

Means this

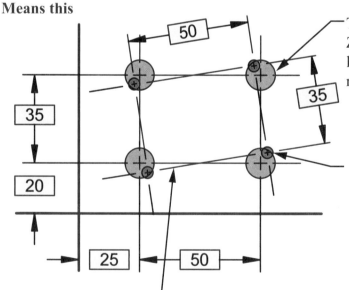

The 0.6 dia Pattern Locating Tolerance Zone Framework (PLTZF) is basically located and oriented to the datum reference frame (planes A,B,C).

The 0.2 dia Feature Relating Tolerance Zone Framework (FRTZF) is basically located and oriented between the features and basically oriented (perpendicular) to the datum reference frame (plane A).

The FRTZF may skew, rotate, and/or be displaced within the confines of the PLTZF. The axes of the holes must lie in both zones simultaneously.

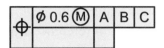

The upper segment specifies the pattern location to the stated DRF.

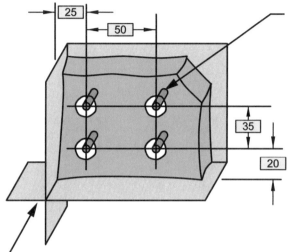

First:
The axes of the holes must lie within a Ø 0.6 at MMC pattern locating zone that is basically located and oriented to the stated datum reference frame.

4X Ø 11.32 virtual size pins basically located and oriented to each other and the DRF. ⌐

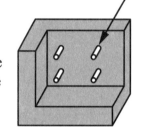

Datum reference frame is established in order by datum features A, B, & C.

Sample functional gage for the upper segment. The size of the holes must also be verified.

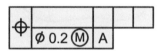

The lower segment specifies the hole to hole requirement and the <u>orientation</u> of this requirement to the stated DRF.

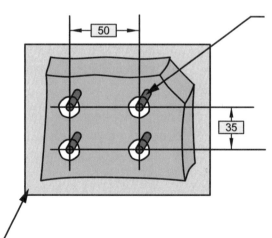

Second:
The axes of the holes must lie within a Ø 0.2 at MMC feature relating zone (hole to hole) that is only oriented (perpendicular) to the lower segment datum reference frame.

4X Ø 11.72 virtual size pins basically located to each other and perpendicular to the face. ⌐

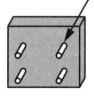

DRF is established by datum feature A. This is an orientation plane only.

Sample functional gage for the lower segment. The holes must meet both requirements.

13.15

Composite position tolerancing may be applied with two or more datum features in the lower segment. This type of control is useful for locating a pattern of holes to a datum reference frame with a large tolerance while providing a smaller hole to hole location tolerance. Because this is composite tolerancing, the lower segment controls orientation only to the specified DRF.

This on the drawing

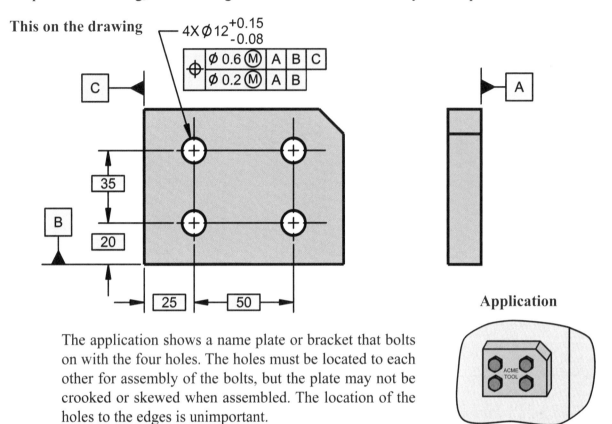

The application shows a name plate or bracket that bolts on with the four holes. The holes must be located to each other for assembly of the bolts, but the plate may not be crooked or skewed when assembled. The location of the holes to the edges is unimportant.

Application

Means this

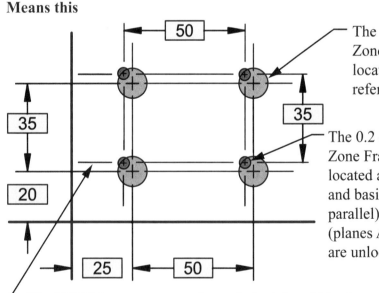

The 0.6 dia Pattern Locating Tolerance Zone Framework (PLTZF) is basically located and oriented to the datum reference frame (planes A,B,C).

The 0.2 dia Feature Relating Tolerance Zone Framework (FRTZF) is basically located and oriented between the features and basically oriented (perpendicular & parallel) to the datum reference frame (planes A & B). The basic dimensions are unlocked to the datums.

The FRTZF may move up, down, left, right within the confines of the PLTZF, but it may not skew or rotate. The axes of the holes must lie in both zones simultaneously. The axes of the holes may only rotate within the confines of the FRTZF.

13.16

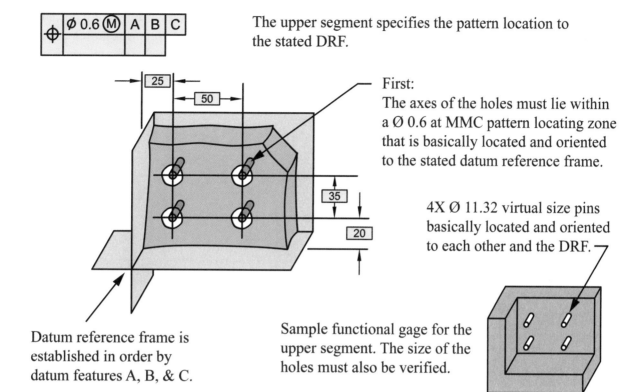

The upper segment specifies the pattern location to the stated DRF.

First:
The axes of the holes must lie within a Ø 0.6 at MMC pattern locating zone that is basically located and oriented to the stated datum reference frame.

4X Ø 11.32 virtual size pins basically located and oriented to each other and the DRF.

Datum reference frame is established in order by datum features A, B, & C.

Sample functional gage for the upper segment. The size of the holes must also be verified.

The lower segment specifies the hole to hole requirement and the <u>orientation</u> of this requirement to the stated DRF.

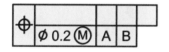

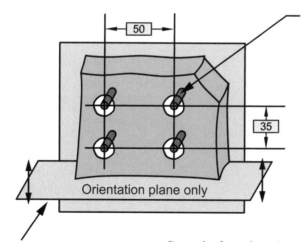

Orientation plane only

Second:
The axes of the holes must lie within a Ø 0.2 at MMC feature relating zone (hole to hole) that is only oriented (perpendicular & parallel) to the lower segment datum reference frame.

4X Ø 11.72 virtual size pins basically located to each other and perpendicular to the face.

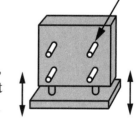

DRF is established by datum features A and B. These are orientation planes only. Basic dimensions to the datums are unlocked.

Sample functional gage for the lower segment. The sliding gage rail allows part movement up, down, left, and right, keeping the hole to hole requirement but only orienting the part to datums A & B.

Note: If datum C were also entered in the lower segment, the interpretation would be the same. Since the lower segment controls orientation only to the DRF, for this example, datums A and B are enough to establish complete orientation.

Position - Two Single Segments

This drawing is not using composite tolerancing. It is using 2 single position callouts. These position callouts are read separately and simply control the location between the holes and the location and orientation to the specified DRF. Multiple frames may be used with datum feature references added or deleted as required.

This on the drawing

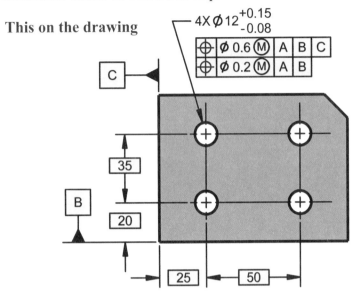

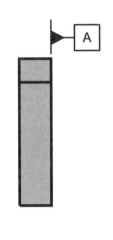

Two single segmented feature control frames can be used where the holes are important to each other. In this application, they are also more important to only two of the datums. The holes can move more left and right as a group but up and down to a lesser tolerance.

Application

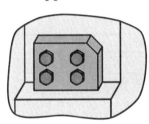

Means this

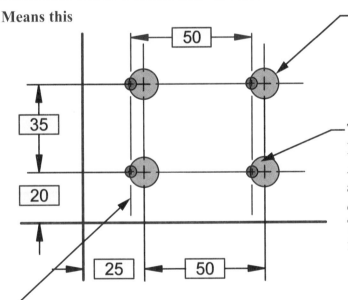

The 0.6 dia tolerance zones are basically located and oriented to the datum reference frame (planes A,B,C).

The 0.2 dia tolerance zones basically located and oriented to the DRF (Planes A, B). They are also basically located and oriented to the lower segment datum reference frame (planes A & B). There is no requirement to Datum feature C.

The 0.2 tolerance zones may move right or left within the confines of the 0.6 tolerance zones. The axes of the holes must lie in both zones simultaneously.

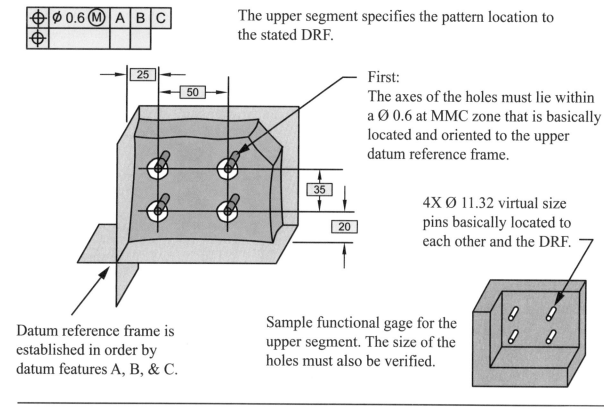

The upper segment specifies the pattern location to the stated DRF.

First:
The axes of the holes must lie within a Ø 0.6 at MMC zone that is basically located and oriented to the upper datum reference frame.

4X Ø 11.32 virtual size pins basically located to each other and the DRF.

Datum reference frame is established in order by datum features A, B, & C.

Sample functional gage for the upper segment. The size of the holes must also be verified.

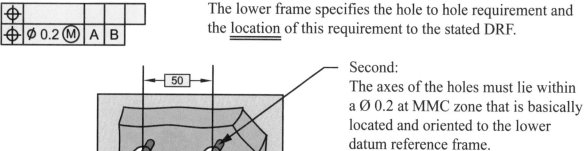

The lower frame specifies the hole to hole requirement and the <u>location</u> of this requirement to the stated DRF.

Second:
The axes of the holes must lie within a Ø 0.2 at MMC zone that is basically located and oriented to the lower datum reference frame.

4X Ø 11.72 virtual size pins located to each other and the specified DRF.

DRF is established by datum features A and B. Basic dimensions to datum B are locked.

Sample functional gage with a fixed gage rail for the lower callout. This gage will check the hole to hole requirement, the orientation to datum A, and the location to datum B. The part is still free to move left and right to accept the four pins.

Note: If datum C were also entered in the lower segment, there would be a conflict. Both the upper and lower callouts would control location to all three datums, there would be no need for the larger tolerance in the upper feature control frame.

Composite Position - Paper Gage Verification

The top drawing is a plate with composite position tolerancing applied to the three holes. The lower drawing is the produced part. The example below and following pages explain how to use the paper gage to evaluate the location tolerance for the 3 holes. The upper segment of the composite can be determined by the x/y position table. The lower segment can be evaluated by using the paper gage, functional gage or appropriate CMM software.

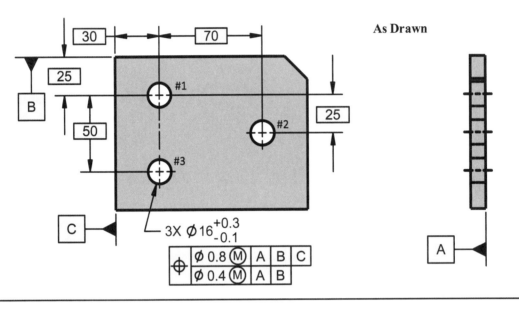

As Drawn

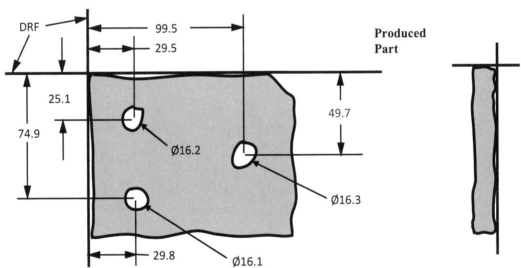

Produced Part

Upper segment calculations | Lower segment calculations

Hole No.	MMC Size	Act Size	Pattern Locating Allowed	"X" Dev	"Y" Dev	Pattern Relating Actual	Acc Rej	Feature Relating Allowed	Feature Relating Actual	Acc Rej
1	15.9	16.2	1.1	-0.5	-0.1	1.02	A	0.7	0.6	A
2	15.9	16.3	1.2	-0.5	+0.3	1.17	A	0.8	0.6	A
3	15.9	16.1	1	-0.2	+0.1	0.45	A	0.6	0.6	A

The upper entry on the composite position can be calculated using the x/y position chart or with the paper gage as shown below. The deviation from basic for each hole is plotted out on the graph, taking care to note positive and negative direction. The polar coordinate represents the position tolerance zones. The axis of the holes must lie within the allowable pattern tolerance. The axis of the polar coordinate lies on the center of the Cartesian coordinate.

The axes of the 3 holes fall within their respective pattern locating tolerance zones located to the DRF.

The axes of the 3 holes are related to each other and fall within their respective feature relating tolerance zones oriented to the DRF.

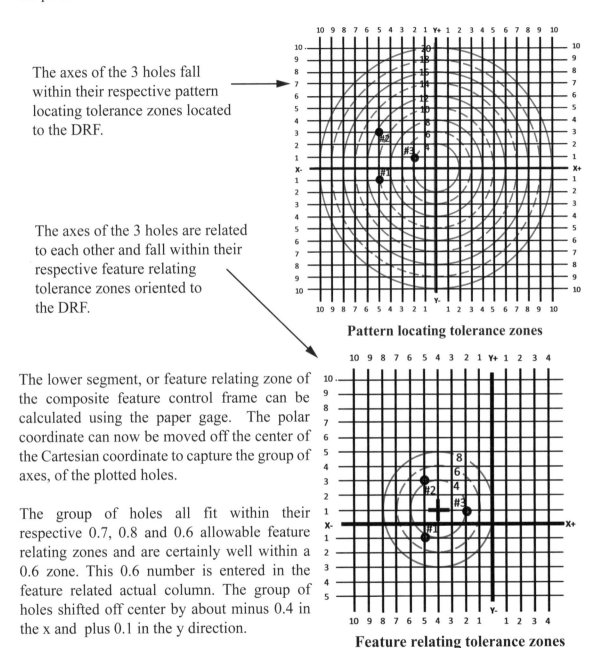

Pattern locating tolerance zones

Feature relating tolerance zones

The lower segment, or feature relating zone of the composite feature control frame can be calculated using the paper gage. The polar coordinate can now be moved off the center of the Cartesian coordinate to capture the group of axes, of the plotted holes.

The group of holes all fit within their respective 0.7, 0.8 and 0.6 allowable feature relating zones and are certainly well within a 0.6 zone. This 0.6 number is entered in the feature related actual column. The group of holes shifted off center by about minus 0.4 in the x and plus 0.1 in the y direction.

Two datum features were referenced in the lower segment of the composite feature control frame. Since this is composite tolerancing, and the lower entry is orientation only to the DRF, the collection of data and verification was accomplished while the part was set-up and oriented to the DRF, established by datum features A and B. If datum feature B were eliminated from the lower entry of the feature control frame, the DRF set-up to verify would change. The collection of data and verification would take place while the 3 holes are only balanced to each other in the initial set up without orientation to datum feature B.

13.21

Composite Position Application - Coaxial Holes on a Hinge

Composite position tolerancing may be used to control coaxial holes. Shown below are four examples of composite and segmented position frames to help illustrate the difference and effect of adding and releasing datum references. The explanation of scenario #1 is shown at the bottom of the page. Scenarios #2 - #4 are shown on the next page.

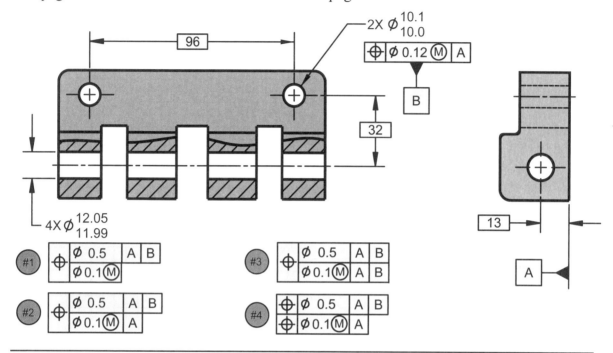

Scenario #1- A composite frame is created by entering the position symbol once for both segments. The upper segment controls the location and orientation of the four holes to the A, B reference frame within 4 basically located 0.5 dia tol zones. The lower segment controls the location and orientation between the 4 holes within the tighter 0.1 dia tol zones. Since there are no datum features called off in the lower segment, these zones are free to float, tilt, or skew together within the confines of the upper tol zone. The hole axes must lie within both zones simultaneously.

Views show tolerance zones only.
The holes have been omitted for clarity.

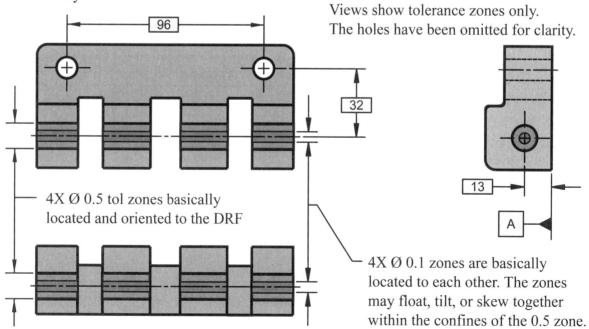

4X Ø 0.5 tol zones basically located and oriented to the DRF

4X Ø 0.1 zones are basically located to each other. The zones may float, tilt, or skew together within the confines of the 0.5 zone.

It is important to understand the differences between location and orientation when interpreting composite position and multiple position frames. These factors, plus the application or elimination of datum feature references, can have a substantial effect on the interpretation of specifications. It is important that the designer select a control that properly reflects design requirements. The following scenarios apply to the coaxial holes in the hinge illustration on the previous page.

Scenario #2 - In the accompanying hinge figure, the composite position tolerance is revised and datum feature A is added to the lower segment. The upper segment did not change and locates and orients the holes to the specified datum reference frame.

The lower segment or 0.1 tolerance zone controls the coaxiality of the holes. In addition, since datum A is also entered in the lower segment, the orientation (parallelism) of this tolerance zone is controlled in only one direction, relative to the face or datum A.

The basic dimensions to the datums are unlocked for the lower segment.

Scenario #3 - In the accompanying hinge figure, the composite position tolerance is revised and both datum feature A and B are added to the lower segment. The upper segment did not change and locates and orients the holes to the specified datum reference frame.

The lower segment or 0.1 tolerance zone controls the coaxiality of the holes. In addition, since datum A and datum B are also entered in the lower segment, the orientation (parallelism) of this tolerance zone is controlled in both directions, (up and down as well as back and forth) relative to the datum reference frame established by datum features A and B.

The basic dimensions to the datums are unlocked for the lower segment.

> **Note:** A parallelism tolerance would not yield the same results as the lower segment of the composite position tolerance. Parallelism is an orientation tolerance for individual features. It will not control the location between the holes as it does in the lower segment of the composite position tolerance. A parallelism specification will require each hole to be parallel but will not control the location between the holes. If there were only one hole, as shown in the example in Unit 11 of this text, a parallelism control would be used.

Scenario #4 - In the accompanying hinge figure, the composite position tolerancing is replaced with two single segmented position controls. The upper segment has the same interpretation as the composite control and locates and orients the holes to the specified datum reference frame.

The lower segment, or 0.1 tolerance zone, controls the coaxiality of the holes. In addition, since datum A is referenced in the lower segment, both the location and orientation of this zone is controlled relative to datum A. The 0.1 tolerance zone is located and oriented to datum A but may slide up and down relative to the reference planes established by B.

The basic dimension to datum A remains locked, but the basic dimension to datum B is unlocked for the lower segment.

Continuous Feature

Size tolerance applies to individual features and does not control the relationship of features. There is no relation between features unless specified. Witness lines between the features define the size of each individual feature but do not control the location between them.

The location between the features may be controlled with a position tolerance or with a continuous feature symbol. This makes the group of features act like an individual feature of size.

When using a continuous feature symbol, it is not required to state the number of features. However, in certain cases it might be helpful to specify the number of diameters and clarify design intent. The continuous feature symbol is new for the ASME Y14.5-2009 standard.

The continuous feature symbol is used for features of size. It is not applicable to coplanar surfaces. Profile of a surface is used to control the relationship between coplanar surfaces.

This on the drawing

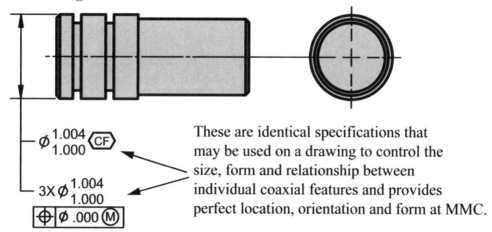

These are identical specifications that may be used on a drawing to control the size, form and relationship between individual coaxial features and provides perfect location, orientation and form at MMC.

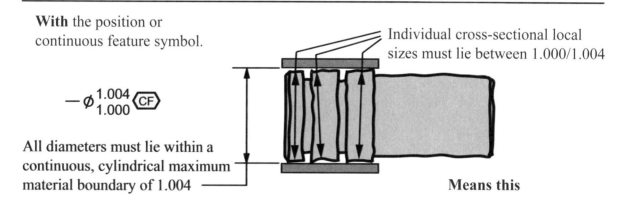

With the position or continuous feature symbol.

All diameters must lie within a continuous, cylindrical maximum material boundary of 1.004

Individual cross-sectional local sizes must lie between 1.000/1.004

Means this

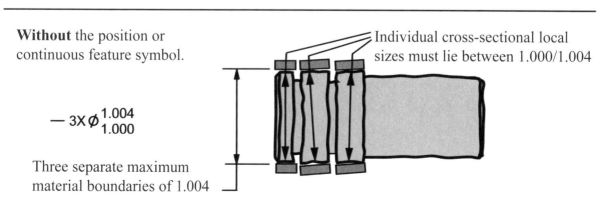

Without the position or continuous feature symbol.

Three separate maximum material boundaries of 1.004

Individual cross-sectional local sizes must lie between 1.000/1.004

Datum Features Referenced Individually

The assembly to the right includes a plate with three separate connectors. Each connector pilots on the large center hole and is fastened through the side screw holes. Each set of screw holes must follow its respective pilot hole but has little relationship to the edges of the plate.

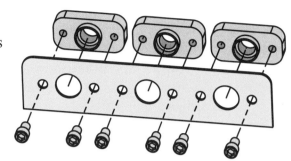

This on the drawing

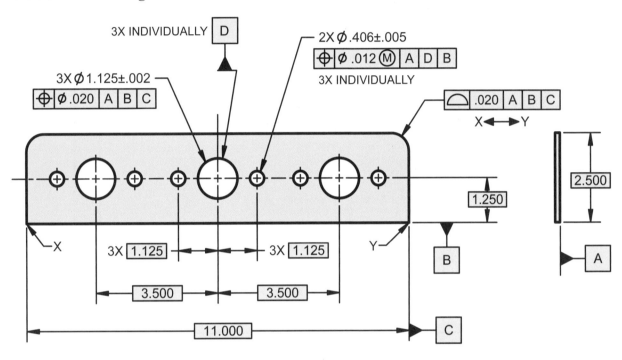

The plate mounts on the back face, bottom edge and outside width. A DRF is established from these datum features. The position of .020 on the 3 large holes locates the mating connectors to this DRF. The position of .012 on the screw holes makes sure the sets of screw holes follow their pilot hole and ensure the screws will fasten through. Also, because B is referenced for orientation in this frame, the position of .012 will control the rotation of the connectors.

The notation INDIVIDUALLY is applied to the datum feature D on the drawing to identify the three pilot holes as individual datum features. This creates three separate datum D axes. The INDIVIDUALLY notation is also applied to the position tolerance for the three sets of two screw holes. The datum reference for each set of two holes relates them to the primary datum plane A, and the center of each datum feature D hole individually. The orientation of the 2 holes are established by the bottom surface datum feature B. The use of datum features with INDIVIDUALLY is designed to make the drawing less complicated and cluttered.

13.25

Three Single Segments - Rotational Control

Position tolerancing is flexible as it allows the designer to add and release datums to state exacting requirements while allowing maximum manufacturing tolerance. The example below illustrates a pattern of holes with three specifications. It is sometimes easier to read these from the bottom up. The bottom specification requires the three holes to be located to each other and oriented to the face within 0.12. The next callout requires the holes to be to the face and located to the pilot within 0.25. The final callout requires the holes to be to the face, pilot and rotated to the slot within 0.85.

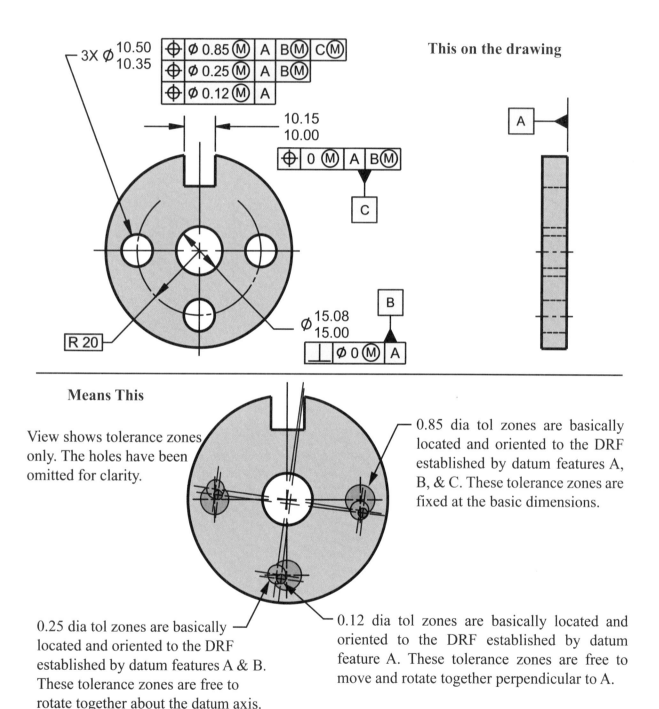

This on the drawing

Means This

View shows tolerance zones only. The holes have been omitted for clarity.

0.85 dia tol zones are basically located and oriented to the DRF established by datum features A, B, & C. These tolerance zones are fixed at the basic dimensions.

0.25 dia tol zones are basically located and oriented to the DRF established by datum features A & B. These tolerance zones are free to rotate together about the datum axis.

0.12 dia tol zones are basically located and oriented to the DRF established by datum feature A. These tolerance zones are free to move and rotate together perpendicular to A.

The axes off the holes must lie in all zones simultaneously. The features also have additional tolerances as they depart from MMC.

1. On the part below, apply postion controls to allow the holes to move 0.4 side to side but only 0.1 in the up and down direction. Include references to datum features A, B, and C.

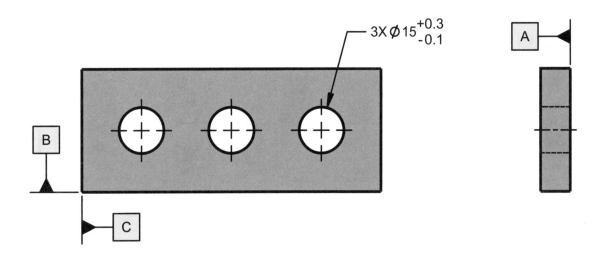

2. On the part below, show the size and shape of the tolerance zones that were defined on the previous drawing.

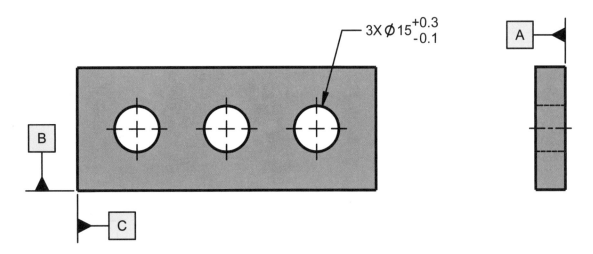

3. What is the perpendicularity tolerance for the 3 holes in relation to datum A? Explain.

_____

_____

4. On the rectangular hole in the part below, apply a profile of 0.1 all around in relation to datum A to control the size and shape. In addition, apply a position to the hole within a 0.5 boundary at MMC in relation to the datums A, B, C to control the location.

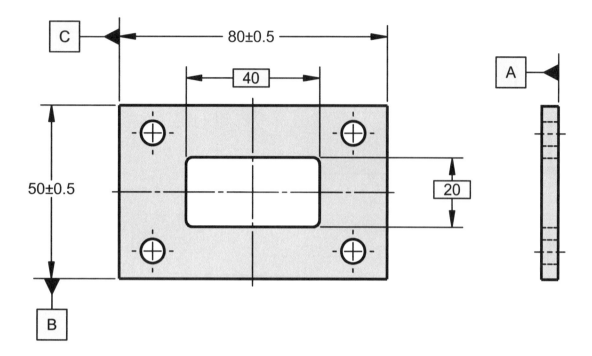

5. On the part below, calculate the allowable min and max hole sizes as well as the size of the MMB position boundary created.

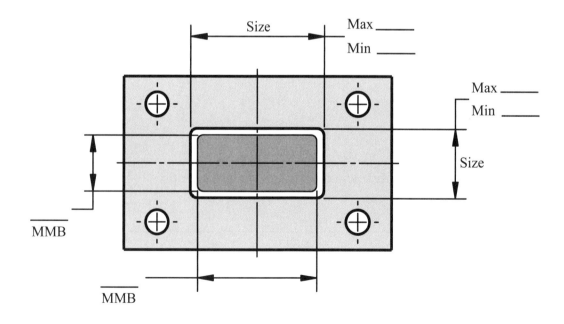

6. Below are two sets of position feature control frames, one of them is composite position the other is simply two position callouts. The upper entries of the two specifications are identical. However, there is a difference in the lower segments. In the space provided, explain the meanings of these two lower segments.

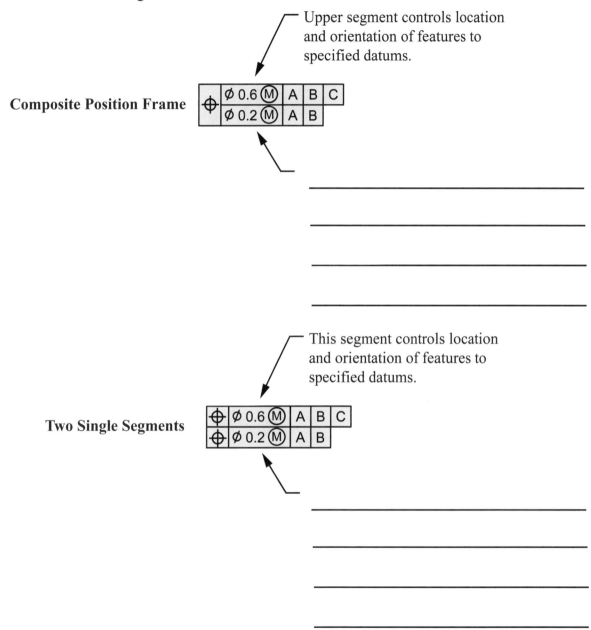

Upper segment controls location and orientation of features to specified datums.

Composite Position Frame

This segment controls location and orientation of features to specified datums.

Two Single Segments

_____

_____

_____

_____

_____

_____

_____

_____

7. The three functional gages below can be used to verify the three segment position specification that is shown earlier in this unit. Calculate the pin sizes for these gages to help explain the callout on the drawing. See product drawing earlier in this unit.

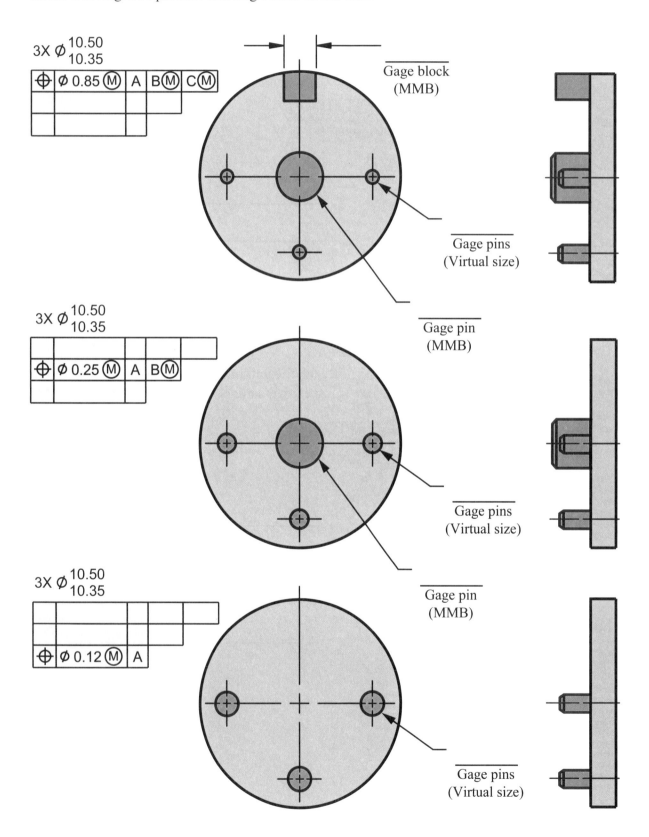

3X ⌀ 10.50 / 10.35

⊕ | ⌀ 0.85 Ⓜ | A | BⓂ | CⓂ

Gage block (MMB)

Gage pins (Virtual size)

Gage pin (MMB)

3X ⌀ 10.50 / 10.35

⊕ | ⌀ 0.25 Ⓜ | A | BⓂ

Gage pins (Virtual size)

Gage pin (MMB)

3X ⌀ 10.50 / 10.35

⊕ | ⌀ 0.12 Ⓜ | A

Gage pins (Virtual size)

Unit 14

Coaxial Controls
Runout, Concentricity, Position,
Profile, and Symmetry

The shaft to the right mounts in the assembly on two journals. Both journals share the same mounting importance. The following geometric callout is very common for establishing a single datum axis from two datum features.

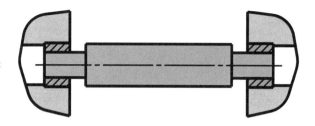

This on the drawing

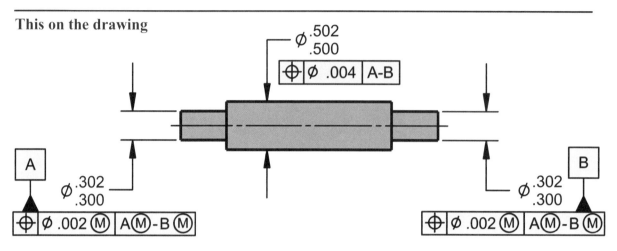

Means this   First establish the simulated datum axes A and B.

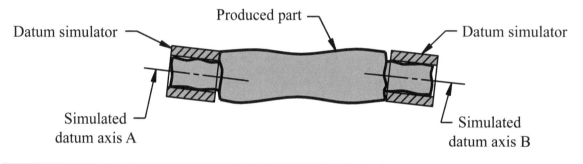

Then, collapse the smallest cylinder around both datum axes A and B. The axis of this cylinder is the simulated A-B axis. The A-B axis is the average of the two datum axes.

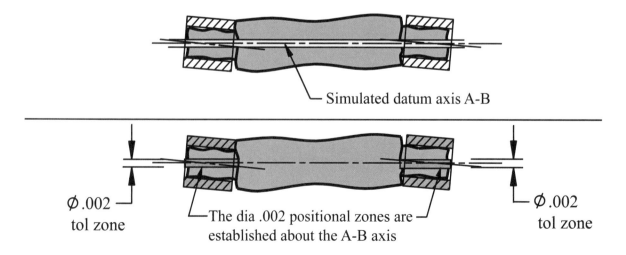

Virtual size boundaries for coaxial diameters

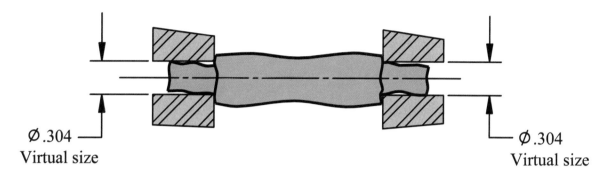

Ø.304
Virtual size

Ø.304
Virtual size

When designing the size openings for the mating part, make sure consideration is given to the virtual sizes on the shaft. The above datum features are applied at MMB. This allows the datum features to tilt within the virtual size boundary as they depart from MMB. If bearings are pressed on the journals, the datum features are usually referenced at RMB.

If the size of the journals, and the tilt/coaxiality of the journals are all combined in one value, consider the use of zero position tolerancing at MMC for the relationship between the features.

Coaxial control using a single feature control frame

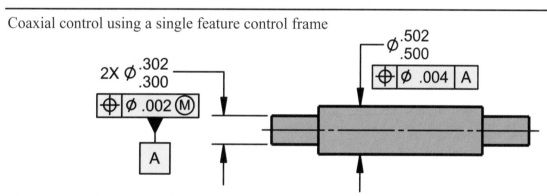

The drawing above is another way to control the relationship between two coaxial diameters. The end result of the above geometric control is identical to the result of the A-B call out on the previous page. Either specification may be used. If the diameters are a different size, the A-B specification may be more appropriate. See the seat latch part in Unit 9 for another similar example with two coaxial holes creating a single datum axis.

Sample inspection

The part may be set up in two vee blocks to simulate the two mating journals. An indicator is used to check the position requirements. A CMM could also be used by setting a cylinder alignment on both journals simultaneously to establish the A-B axis. The individual journals are then verified for position conformance to the A-B axis.

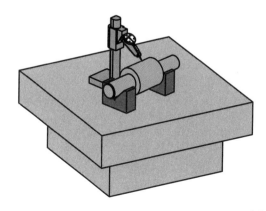

There are four main geometric characteristics that can be used to control a coaxial relationship between features: position, runout, concentricity, and profile. Each characteristic has a different meaning and will yield a different result when applied for coaxial relationships. The goal for applying any tolerancing control is obtaining maximum manufacturing tolerance while maintaining functional requirements. The geometric specification applied depends on the application.

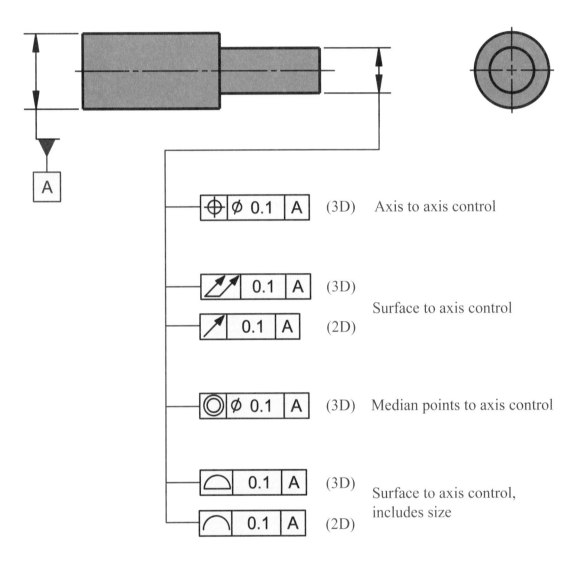

The part above illustrates the application and comparison between the four possible coaxial specifications. In order to explain the difference between these characteristics, the same part is shown four times over the next pages with each specification used. In addition to the detailed interpretation of each characteristic, a sample practical application for each is shown.

Runout Tolerances

Runout tolerances control the form, orientation, and location of surfaces to a datum axis. A reference to a datum axis is always required.

Symbol	Runout Tolerance	Common Shape of Tolerance Zone	2D / 3D	Application of Feature Modifier
 ↗	Circular Runout	2 Concentric circles about a datum axis  2 Concentric circular line elements about a datum axis 	2D	No It is a Surface Control
 ↗↗	Total Runout	2 Concentric cylinders about a datum axis 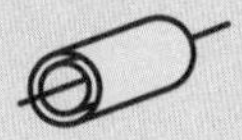 2 Parallel planes perpendicular to a datum axis 	3D	No It is a Surface Control

Overview:

Circular runout and total runout controls the form, orientation, and coaxiality of features to a datum axis. The common shapes of the tolerance zones are listed above.

Runout tolerances controls surfaces constructed around a datum axis and those constructed at a right angle to a datum axis. Runout may or may not have a plane surface referenced as part of the datum reference frame but must always be referenced to a datum axis.

Circular and total runout are identical controls except that total runout is a 3D control and circular runout is a 2D control. On a surface of revolution, total runout will control taper and straightness of the feature, but circular runout does not.

The two runout tolerances are both surface controls. Therefore, the feature modifiers MMC, LMC, and RFS are not applicable. Datum modifiers are applied RMB because of the unique nature of the runout specifications, where a feature is verified by rotation about a datum axis.

In the past, runout tolerances were often used on all rotating components. This is not always needed and can unnecessarily tighten manufacturing tolerances, and increase cost. The other coaxial controls of position and profile of a surface should be considered as well.

Both runout specifications are surface to axis controls and are often used when a component rides on the surface, such as a wheel or pulley with adjustments for larger size tolerances. Examples of applications are shown later in this section.

Circular Runout - Coaxial Features

Circular runout is a two dimensional surface to axis control. The tolerance is applied independently at each circular cross section. When applied to a surface constructed around a datum axis, circular runout controls the variations of circularity, orientation, and coaxiality. Unlike total runout, it does not control straightness or taper. The runout tolerances does not control the size and therefore the feature requires a separate size tolerance.

This on the drawing

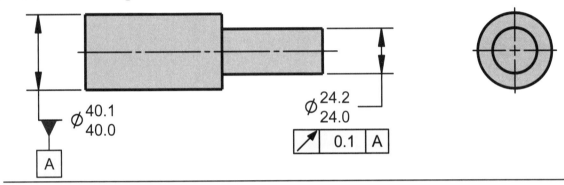

$\phi \begin{matrix} 40.1 \\ 40.0 \end{matrix}$

A

$\phi \begin{matrix} 24.2 \\ 24.0 \end{matrix}$

| ⌀ | 0.1 | A |

Means this

The theoretical datum feature simulator is the smallest circumscribed cylinder that contacts the high points of the feature. Depending on the accuracy required, this may be practically simulated by a collet, chuck, vee block, etc.

Datum axis A is the axis of the datum feature simulator

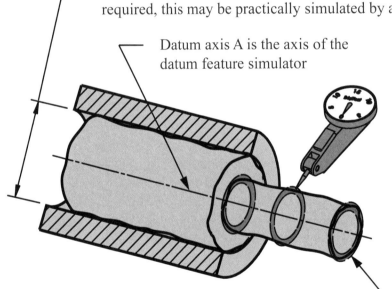

Circular runout is a 2D control and requires a series of circular cross-sectional checks of the surface.

This controls circularity, orientation, and location of the feature for a maximum of 0.1 full indicator movement. It does not control taper, straightness, or size. On this part, the outer boundary is 24.3. The inner boundary is 23.9.

Each circular element of the feature must lie between two circles, one having a radius of 0.1 larger than the other, perfectly concentric to the datum axis A. Additionally, the feature must be within the limits of size (24.0/24.2).

The circular runout specification may be verified with a dial indicator, CMM, or other methods. If a dial indicator is used, the indicator is placed on the surface. When the part is rotated 360 degrees about the datum axis, at each cross section the full indicator movement may be no more than 0.1. Because this is a 2D callout, the indicator is reset at each cross-section along the length of the feature.

Total Runout - Coaxial Features

Total runout is a three dimensional surface to axis control. Total runout provides a composite control of all surface elements. When applied to a surface constructed around a datum axis, total runout controls the variations of circularity, straightness, taper, orientation, and coaxiality. The runout tolerances do not control the size and therefore the feature requires a separate size tolerance.

This on the drawing

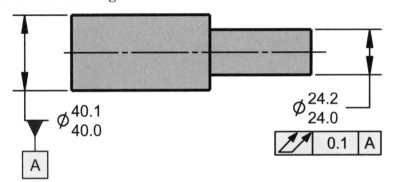 

Means this

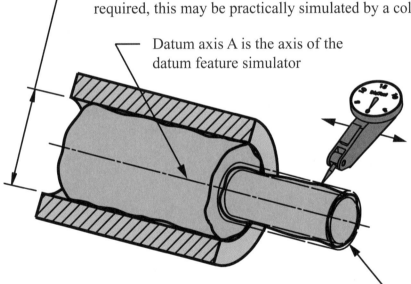

The theoretical datum feature simulator is the smallest circumscribed cylinder that contacts the high points of the feature. Depending on the accuracy required, this may be practically simulated by a collet, chuck, vee block, etc.

Datum axis A is the axis of the datum feature simulator

Total runout is a 3D control and requires a total sweep of the surface. This will control form, orientation, location, and taper (but not size) of the feature for a maximum of 0.1 full indicator movement (FIM). On this part, the outer boundary is 24.3. The inner boundary is 23.9.

The feature must lie between two cylinders, one having a radius of 0.1 larger than the other, perfectly concentric to the datum axis A. Additionally, the feature must be within the limits of size (24.0/24.2).

The total runout specification may be verified with a dial indicator, CMM, or other methods. If a dial indicator is used, the indicator is placed on the surface. When the part is rotated 360 degrees about the datum axis, the full indicator movement may be no more than 0.1. Because this is a 3D callout, the indicator is also swept along the entire length of the feature parallel to the axis.

Runout - Applied to Flat Surfaces

Both total runout and circular runout may be applied to surfaces at different angles to a datum axis. These callouts are refinements of the location tolerances. On the drawing below, perpendicularity and total runout will provide identical results and can be used interchangeably. Circular runout controls 2D circular elements about the axis. The surface may be concave or convex within the specified size or location tolerance.

This on the drawing

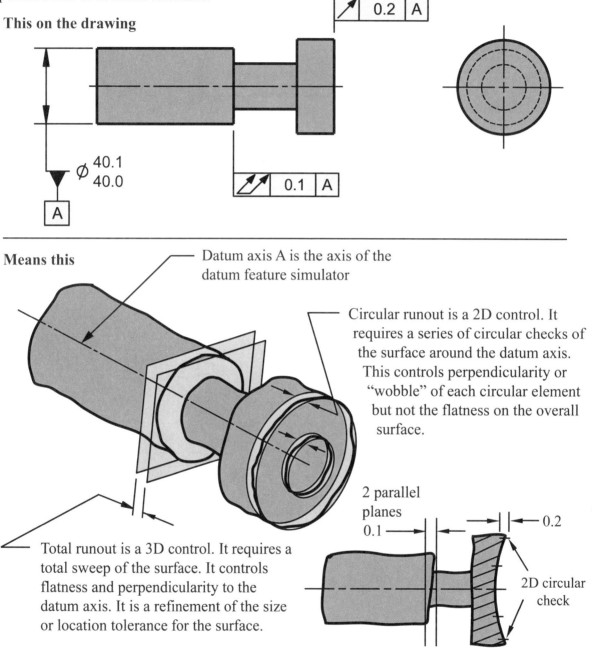

Total and circular runout may be verified with a dial indicator, CMM, or other methods. If a dial indicator is used, the surface or circular elements must be within the specified runout tolerance when the part is rotated 360 degrees. Total runout requires all elements to be within the full indicator movement without a reset on the indicator. With circular runout, the indicator is reset to zero after each circular check. In addition, the surfaces must also be within the limits of size or location.

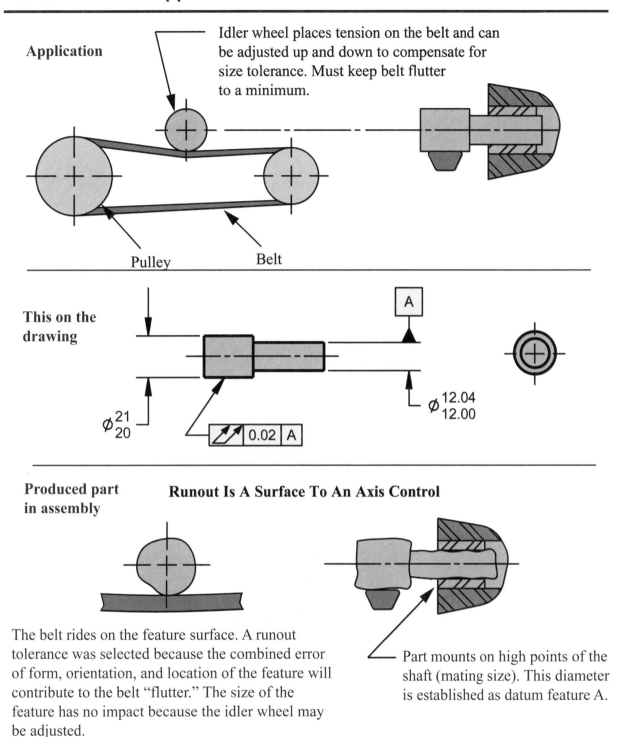

Application

Idler wheel places tension on the belt and can be adjusted up and down to compensate for size tolerance. Must keep belt flutter to a minimum.

Pulley   Belt

This on the drawing

A

$\phi \begin{smallmatrix} 21 \\ 20 \end{smallmatrix}$

0.02 | A

$\phi \begin{smallmatrix} 12.04 \\ 12.00 \end{smallmatrix}$

Produced part in assembly

Runout Is A Surface To An Axis Control

The belt rides on the feature surface. A runout tolerance was selected because the combined error of form, orientation, and location of the feature will contribute to the belt "flutter." The size of the feature has no impact because the idler wheel may be adjusted.

Part mounts on high points of the shaft (mating size). This diameter is established as datum feature A.

Position was not selected because the belt rides on the surface of the feature, not on the mating size (high points). The ovality of the OD will contribute to the belt "flutter." Position does not control the ovality or form.

Profile was not selected because the wheel may be adjusted for the size tolerance. Profile would require a basic dimension to the axis and would tighten the size of the feature more than necessary. If the idler wheel were not adjustable in location, profile would have been used.

Position - Coaxial Features

Position for coaxial features is a three dimensional, axis to axis control. It defines a cylindrical zone within which the axis of the actual mating size of the feature must lie. When applied to a feature constructed around a datum axis, position controls orientation and location. It has no effect on size, form, or other variations in the surface. If required, the MMC or LMC modifier may be applied to the feature. MMB, LMB or RMB may be applied to the datum feature.

This on the drawing

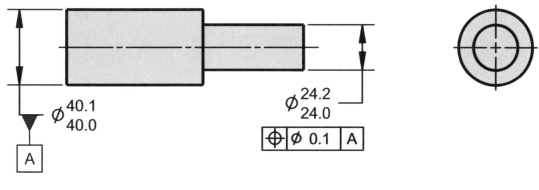

Means this

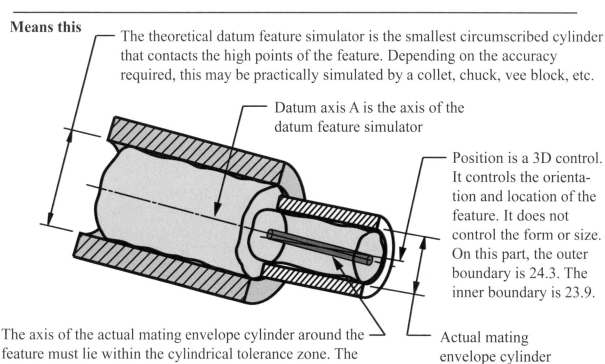

The theoretical datum feature simulator is the smallest circumscribed cylinder that contacts the high points of the feature. Depending on the accuracy required, this may be practically simulated by a collet, chuck, vee block, etc.

Datum axis A is the axis of the datum feature simulator

Position is a 3D control. It controls the orientation and location of the feature. It does not control the form or size. On this part, the outer boundary is 24.3. The inner boundary is 23.9.

The axis of the actual mating envelope cylinder around the feature must lie within the cylindrical tolerance zone. The cylindrical tolerance zone is concentric with the datum axis. In addition, the feature must be within the limits of size.

Actual mating envelope cylinder

The coaxial position specification may be verified with a dial indicator, CMM, or other methods. If a dial indicator is used, and the full indicator movement (FIM) does not exceed 0.1, the feature is good. If the FIM exceeds 0.1, the feature may still be good. The reason for this is, as an indicator rides on the surface, it inadvertently picks up form errors (ovality). Since position is only an axis to axis control, surface variations are not included in the requirement. A mapping of the surface to exclude these form errors may be necessary to accept features beyond the 0.1 FIM.

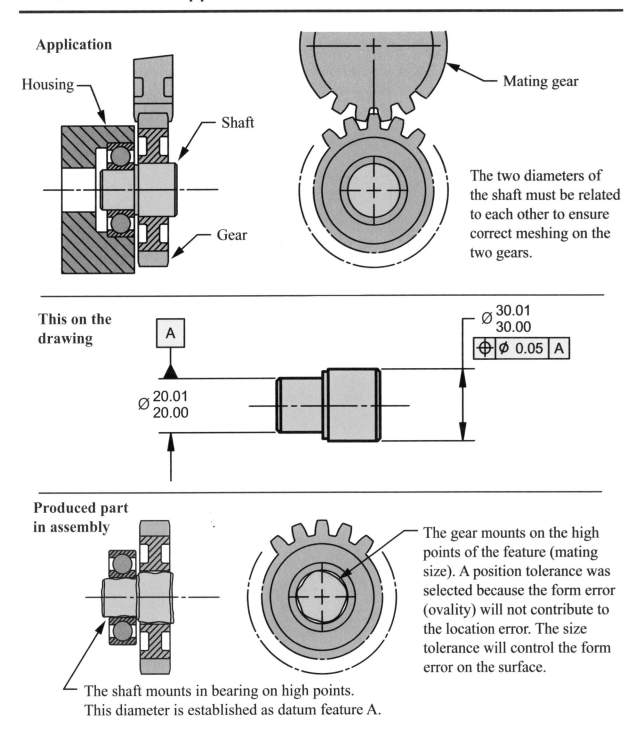

Application

Housing

Shaft

Gear

Mating gear

The two diameters of the shaft must be related to each other to ensure correct meshing on the two gears.

This on the drawing

A

$\varnothing$ 20.01
20.00

$\varnothing$ 30.01
30.00

$\oplus$ | $\varnothing$ 0.05 | A

Produced part in assembly

The gear mounts on the high points of the feature (mating size). A position tolerance was selected because the form error (ovality) will not contribute to the location error. The size tolerance will control the form error on the surface.

The shaft mounts in bearing on high points. This diameter is established as datum feature A.

A runout tolerance was not selected because the form error on the OD of the shaft will not effect the location of the gear center. Runout would require the form error to be included in the location check. If the position callout is checked with an indicator, the form error is calculated out of the reading. Position will provide maximum manufacturing tolerance while still preserving the functional requirements.

A profile tolerance was not selected because the size of the feature is important for the fit in the gear, and a separate larger location tolerance is needed. Profile would combine both of these requirements into a tighter profile tolerance and the diameter defined with a basic dimension.

14.11

Profile - Coaxial Features

Profile of a surface for coaxial features is a three dimensional surface to an axis control. It defines a zone of tolerance (0.2 wide on each side) that is equally disposed about a basic diameter. When applied to a surface constructed around a datum axis, profile of a surface will control the variations of size, circularity, straightness, coaxiality, orientation, taper and variations in the surface.

This on the drawing

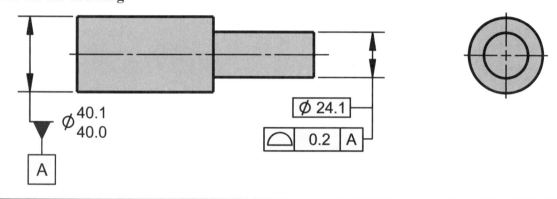

Means this

The theoretical datum feature simulator is the smallest circumscribed cylinder that contacts the high points of the feature. Depending on the accuracy required, this may be practically simulated by a collet, chuck, vee block, etc.

Datum axis A is the axis of the datum feature simulator

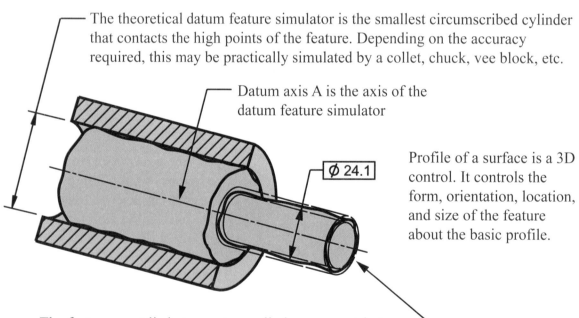

Profile of a surface is a 3D control. It controls the form, orientation, location, and size of the feature about the basic profile.

The feature must lie between two cylinders concentric to the datum axis A. The two cylinders are equally disposed about a basic cylinder of 24.1.
   The outer boundary is a diameter of 24.3
   The inner boundary is a diameter of 23.9

Profile of a surface specification may be verified with a dial indicator, CMM, or other methods. If a dial indicator is used, the indicator is "mastered" or set at the basic dimension of 12.05 (half of 24.1) from the datum axis. The entire surface, including the size, must lie within the profile zone (0.2 full indicator movement) when the part is rotated 360 degrees about the datum axis. Because this is a 3D callout, the indicator is also swept along the entire length of the feature parallel to the axis.

Application

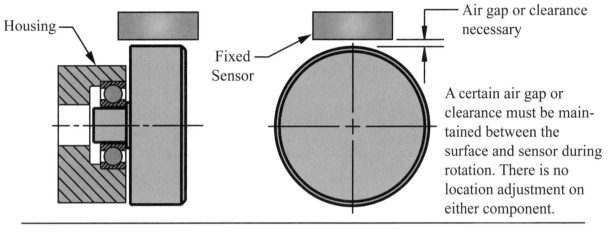

Housing

Fixed Sensor

Air gap or clearance necessary

A certain air gap or clearance must be maintained between the surface and sensor during rotation. There is no location adjustment on either component.

This on the drawing

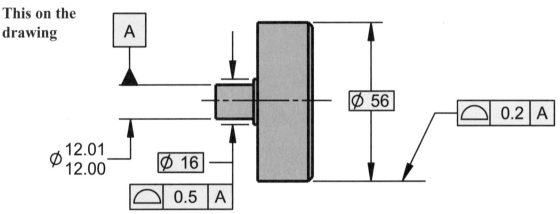

Produced part in assembly

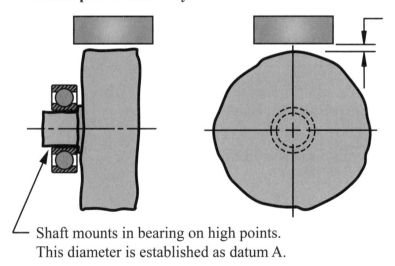

Shaft mounts in bearing on high points. This diameter is established as datum A.

The air gap required is totally dependent on the location of the surface from the mounting axis in the bearing. A profile was selected because it controls size, form, orientation, and location of the OD. It is simple to calculate and verify. It also provides maximum manufacturing tolerance while still preserving the functional requirement.

OB = 56.2
IB = 55.8

A runout or position tolerance was not selected because neither would control the location of the surface of the feature. A separate size and location would have to be specified. Since the two components are nonadjustable, the size, form, orientation, and location all contribute to the gap.

Coaxial Tolerances Review

This is a review of the three recommended general coaxial tolerances and their controls.

Position   ⊕   **Axis to axis**

      Controls: Orientation, Location

      Feature diameter must have a separate size tolerance*

Runout   ↗ ↗↗   **Surface to axis**

      Controls: Form, Orientation, Location

      Feature diameter must have a separate size tolerance*

      Circular Runout   ↗

          Controls: 2-D Form (Circularity)
                              Orientation, Location

      Total Runout

          Controls: 3-D Form (Cylindricity, Straightness, Taper)
                              Orientation, Location

Profile   ⌓   **Surface to axis, includes size**

      Controls: Form, Orientation, Location, Size

      Feature diameter dimension is basic

---

*The size tolerance controls the size as well as the form of the feature.
  Rule #1 requires perfect form at MMC.

Concentricity and symmetry tolerances control the opposing median points of a feature. A datum axis or median plane is required.

Symbol	Tolerance	Common Shape of Tolerance Zone	2D / 3D	Application of Feature Modifier
◎	Concentricity	Cylindrical about a datum axis 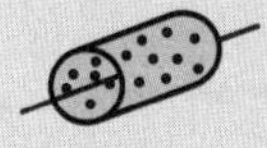	3D	No
⟌	Symmetry	2 Parallel planes about a datum median plane 	3D	No

Overview:

Concentricity and symmetry are 3D tolerances. The common shapes of the tolerance zones are listed in the chart above. The tolerance defines a zone in which the opposing median points must lie. The concept of concentricity and symmetry are identical except that concentricity is used for controlling opposing median points on cylindrical features, and symmetry is used for controlling opposing median points on non-cylindrical features.

Datums are always required with both concentricity and symmetry. The datum must always be an axis or median plane and are always applied on a RMB basis. The feature tolerance is always applied at RFS.

Concentricity and Symmetry are often misapplied and confusing for the entry level user of geometric tolerancing. The words "concentricity" and "symmetry" are very generic sounding terms, but in geometric tolerancing per the ASME Y14.5-2009 standard they are very complex requirements. The dilemma is that often untrained personnel will apply these controls but really want to apply position, runout, or profile tolerances. If you study the interpretations on the following pages, you will find that concentricity and symmetry require a very detailed analysis of the opposing median points of a feature to ensure conformance.

Concentricity and symmetry in the ISO standard do not have the unique interpretation as in the ASME standard. ISO interprets concentricity and symmetry the same as a position tolerance. In ISO, if the feature is coaxial, as on a shaft, then concentricity is used. If it is offset, then position is used. The ASME standard does not discriminate between coaxial or offset, position is used in either case.

14.15

Concentricity

Concentricity is a three dimensional control, controlling opposing median points to an axis. It will control location and have some effect on the form and orientation of a feature. It will not control the form of perfectly oval parts but will have an impact on irregular and "D" shaped features.

The application for concentricity is rare and is commonly misused. Irregularities in the form can make it difficult to find the opposing median points. Therefore, unless there is a definite need to control the opposing median points, consider position, runout, or profile instead. Concentricity may only be applied on a RFS basis, and the datum feature reference only at RMB.

This on the drawing

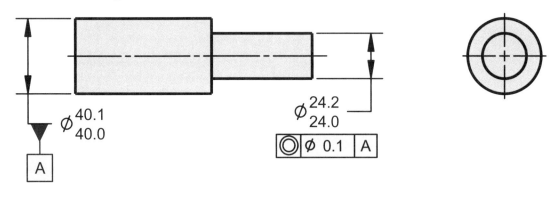

$\phi \begin{smallmatrix} 40.1 \\ 40.0 \end{smallmatrix}$

A

$\phi \begin{smallmatrix} 24.2 \\ 24.0 \end{smallmatrix}$

| ◎ | ϕ 0.1 | A |

Means this — The theoretical datum feature simulator is the smallest circumscribed cylinder that contacts the high points of the feature. Depending on the accuracy required, this may be practically simulated by a collet, chuck, vee block, etc.

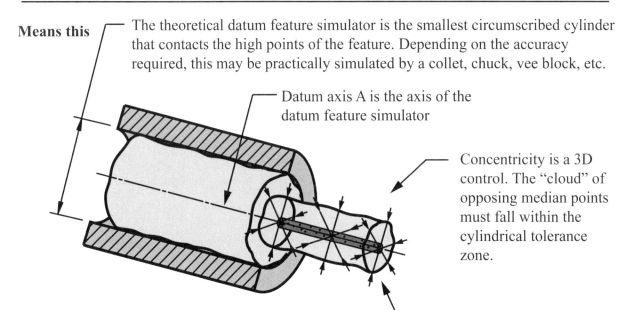

Datum axis A is the axis of the datum feature simulator

Concentricity is a 3D control. The "cloud" of opposing median points must fall within the cylindrical tolerance zone.

The median points of all diametrically opposed elements of the feature must lie within a 0.1 cylindrical tolerance zone. This tolerance zone is concentric with the datum axis. In addition, the feature must be within the limits of size.

The concentricity specification may be verified with a dial indicator, CMM, or other methods. If a dial indicator is used, the part is mounted on the datum and two diametrically opposed, mastered indicators are placed on either side of the toleranced feature. The part is then rotated about the datum axis while watching the readings on the indicators. See next page for more information on these indicator readings.

Concentricity Verification

If both indicators simultaneously read plus an equal amount or minus an equal amount, the feature's opposed median point at that specific location is perfectly concentric. If one indicator reads plus 0.1 and the other indicator reads minus 0.1, the median point of the feature at that specific location is off perfect concentricity by 0.2 dia. To conform, all opposing median points must fall within a 0.1 dia zone. The indicators are placed at every location along the surface in a position normal to the true geometric shape without a reset of the indicators. The feature must also be within the limits of size.

Concentricity vs. Runout verification at one cross section on a feature produced within size limits and perfectly oval.

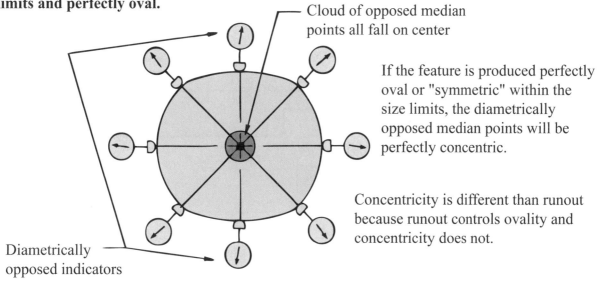

Cloud of opposed median points all fall on center

Diametrically opposed indicators

If the feature is produced perfectly oval or "symmetric" within the size limits, the diametrically opposed median points will be perfectly concentric.

Concentricity is different than runout because runout controls ovality and concentricity does not.

Concentricity vs. Position verification at one cross section on a feature produced within size limits but "D" shaped.

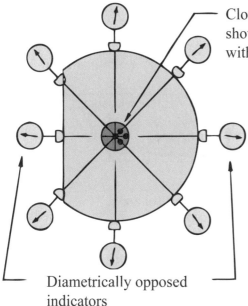

Cloud of opposed median points are shown displaced. They must all fall within a cylindrical zone of 0.1.

If the feature is produced within size limits but happens to be "D" shaped, the diametrically opposed median points will not be concentric.

Concentricity differs from position. If this part were verified for position, it would be reported as perfect. This occurs because position is specified as the axis of the mating envelope (collapsing cylinder) related to the datum axis. The axis of the mating cylinder for this feature is perfectly on center.

Diametrically opposed indicators

Concentricity is a complex control, consider the use of position, runout, or profile instead. If a complex fitting routine is required, a note may be added under a position feature control frame, such as least squares, maximum inscribed, three point centers, etc.

Symmetry

Symmetry is the condition where the median points of all opposed elements of a feature are congruent with the datum axis or center plane. Symmetry is the same concept as concentricity but for non-cylindrical features. Symmetry differs from position in that it controls the opposing points (derived median plane) of the surfaces, while position controls the center plane of the actual mating envelope.

The application for symmetry is rare and it is commonly misused. Irregularities in the form can make it difficult to find the derived median plane. Therefore, unless there is a definite need to control the opposing median points, consider position or profile instead.

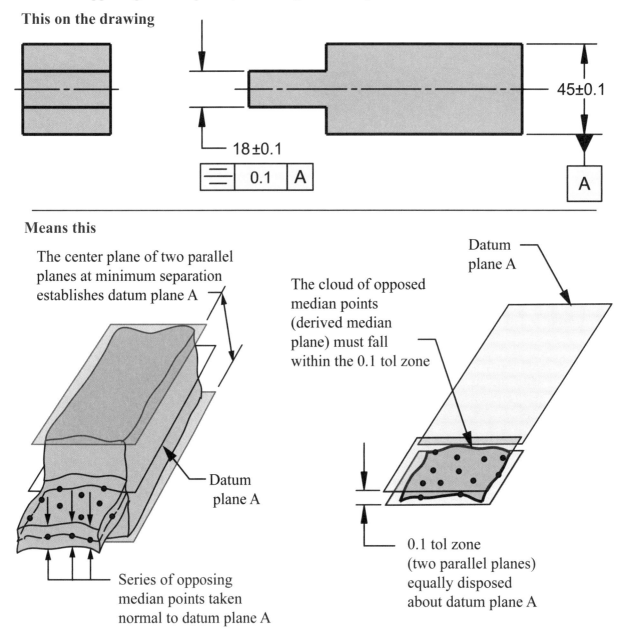

This on the drawing

45±0.1

18±0.1

0.1 | A

A

Means this

The center plane of two parallel planes at minimum separation establishes datum plane A

Datum plane A

Series of opposing median points taken normal to datum plane A

The cloud of opposed median points (derived median plane) must fall within the 0.1 tol zone

Datum plane A

0.1 tol zone (two parallel planes) equally disposed about datum plane A

The feature must be within the limits of size. In addition, all median points of opposed elements of the feature must lie within two parallel planes 0.1 apart. The two parallel planes are equally displaced about datum plane A. The specified tolerance applies only at RFS and the datum reference can only apply on an RMB basis.

Position Instead of Symmetry

Position tolerancing can be applied to non cylindrical features such as slots and tabs. A position tolerance is shown below locating the center plane of the tab. The tolerance value represents a distance between two parallel planes. This tolerance zone controls the location of the feature and also the orientation (parallelism in this example).

The tolerance on the feature below applies at RFS and datum feature is referenced at RMB. Position tolerancing also allows the feature tolerance to be applied on an MMC or LMC basis and datum feature reference at MMB or LMB.

This on the drawing

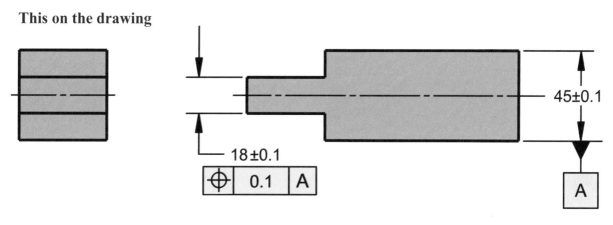

45±0.1

18±0.1

| ⊕ | 0.1 | A |

A

Means this

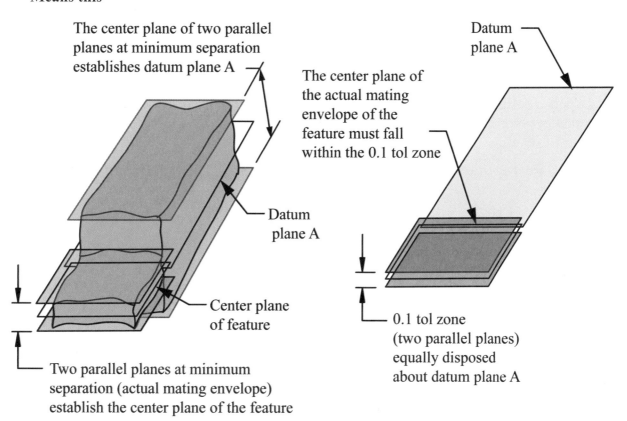

The center plane of two parallel planes at minimum separation establishes datum plane A

The center plane of the actual mating envelope of the feature must fall within the 0.1 tol zone

Datum plane A

Datum plane A

Center plane of feature

0.1 tol zone (two parallel planes) equally disposed about datum plane A

Two parallel planes at minimum separation (actual mating envelope) establish the center plane of the feature

The center plane of the actual mating envelope of the feature must lie between two parallel planes 0.1 apart which are equally disposed about the datum center plane A. In most cases, position is recommended over symmetry.

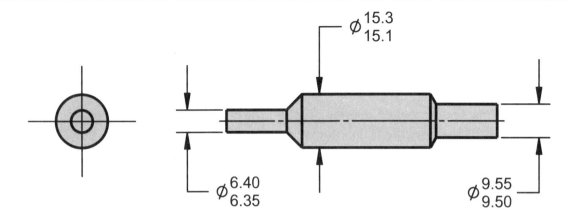

1. On the drawing above, before any geometric controls are applied, what is the coaxiality tolerance is implied between the diameters?

_____

2. On the part above, apply datum feature symbols to create a datum axis A-B by using the right and left diameters.

3. Apply a position coaxiality between the right and left diameter of 0.1 at MMC.

4. Apply a position tolerance to the large center diameter within a diameter of 0.3 in relation to the single datum axis established by both datum features A and B.

5. Calculate the virtual sizes for the right and left diameters and label them on the virtual gage below.

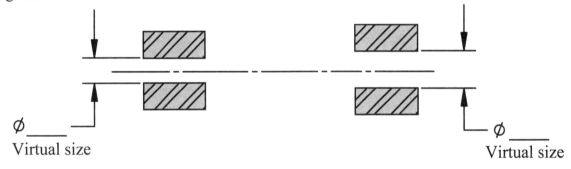

ø_____
Virtual size

ø_____
Virtual size

6. To define the A-B datum axis, do the datum feature simulators collapse to make contact with the datum features or are the simulators fixed at the MMB (virtual) size?

_____

_____

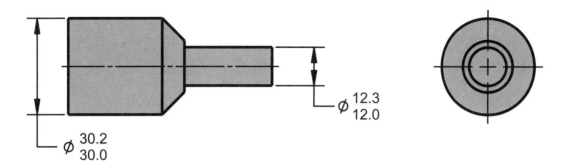

7. On the drawing above, what coaxiality tolerance is implied between the two diameters before any geometric callouts are applied?

_____

8. On the drawing above, identify the large diameter as datum feature A. Specify a 0.1 circular runout specification to locate the small diameter in relation to datum A.

9. With the circular runout specification applied, how much taper (on a diameter) is allowed on the small diameter?

_____

10. If the 0.1 circular runout specification were replaced with a 0.1 total runout specification, how much taper would be allowed on the small diameter?

_____

11. With the circular runout specification applied, how much circularity, on a radius, is allowed on the small diameter?

_____

12. With the circular runout specification applied, how much location, on a diameter, is allowed on the small diameter relative to datum A?

_____

13. On the drawing above, apply a runout requirement to make the left face perpendicular to the datum axis within 0.2 total.

14. On the drawing above, apply a runout requirement to make the circular elements of the right face perpendicular within 0.2 total.

15. What is the technical difference between the runout specifications on the two faces?

_____

Workshop Exercise 14.1

There are three recommended coaxial controls: position, runout, and profile. Answer questions 16-18 to show the fundamental differences in each of these requirements.

16. **Position** ⊕   Axis Control or Surface Control?

Controls what types of variation on the feature? Circle all that apply.

Form   Orientation   Location   Size

Is a basic dimension used for the diameter or is a size tolerance required?

---

17. **Runout** ↗ ↗↗   Axis Control or Surface Control?

Controls what types of variation on the feature? Circle all that apply.

Form   Orientation   Location   Size

Is a basic dimension used for the diameter or is a size tolerance required?

---

18. **Profile** ⌓   Axis Control or Surface Control?

Controls what types of variation on the feature? Circle all that apply.

Form   Orientation   Location   Size

Is a basic dimension used for the diameter or is a size tolerance required?

---

19. Explain the difference between total runout and position on a cylindrical part. Assuming the numbers are the same, which is a more restrictive requirement?

_____

_____

20. Only one of these controls can have a feature modifier, MMC or LMC applied. Which one?

_____

21. Which two geometric characteristics control opposing median points of a feature? Are these common controls used in geometric tolerancing?

_____

This Application

The wheel mounts on the axle which may be adjusted up and down to compensate for the variation in size. The "ride" of the shaft axis must be controlled to move no more than .005 total.

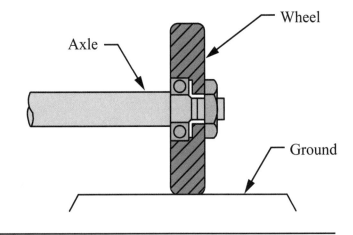

Axle

Wheel

Ground

This on the Drawing

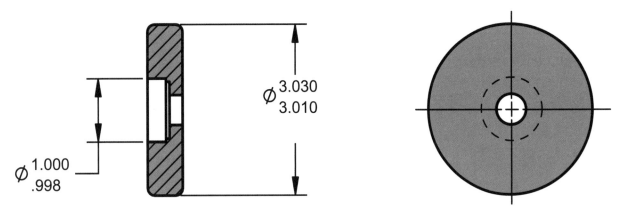

ϕ 3.030 / 3.010

ϕ 1.000 / .998

22. Read and study the functional requirements of the application above. On the product drawing, select the datum feature and apply one of the coaxial controls discussed in this unit. Make sure your feature control frame meets the functional requirements stated above. Explain your decision below.

This Application

The wheel mounts on the axle and can not be adjusted up and down. The air gap is critical and must remain within .010 to .020.

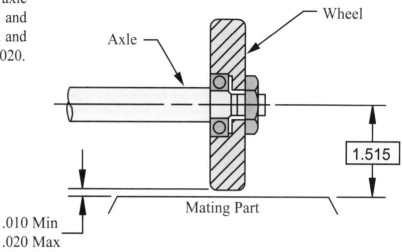

Axle

Wheel

1.515

Mating Part

.010 Min
.020 Max

This on the Drawing

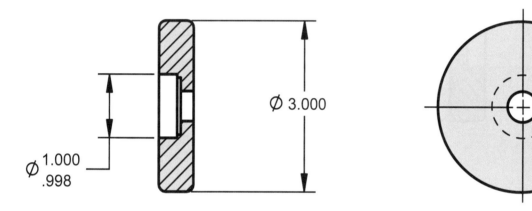

Ø 3.000

Ø 1.000
  .998

23. Read and study the functional requirements of the application above. On the product drawing, select the datum feature and apply one of the coaxial controls discussed in this unit. Make sure your feature control frame and tolerance meets the requirements stated above.

24. Consider another example. Suppose there was a mating fit requirement with a tight size tolerance on the 3.000 OD above and then a separate location requirement to be met. Which coaxial control would you consider using? Explain?

Unit 15

Fastener Formulas and Screw Threads

Floating Fastener Formula

The floating fastener formula is an important design tool. It is used by the designer for calculating position tolerance between mating parts. A floating fastener assembly is one in which both parts have clearance holes. This assembly is called a floating fastener because the fastener can "float" around in the holes and is not fixed in either part. Common floating fastener examples are assemblies that use bolts with nuts or rivets.

The two flat plates on the following page fasten together using four screws and nuts. This type of assembly is categorized as a floating fastener application. The designer must decide the size of the bolts and how many are required. The dimensions defining the location of the holes are designated as basic dimensions.

In order to determine the position tolerances, the designer will calculate the difference between the largest (MMC size) bolt and the smallest (MMC size) hole. The difference between these two sizes is the allowable tolerance that can be applied in the feature control frame of each part. The MMC modifier is applied to the feature tolerance which allows additional positional tolerance as the features depart from MMC.

The floating and fixed fastener formulas are similar to calculating virtual size but in reverse. After using these formulas to calculate the position, figuring virtual sizes for the hole and bolt can be used as a check for the calculations.

Use standard hole sizes: In the following floating fastener assembly, the example uses four 1/4-20 UNC screws. To accommodate these screws, the designer has selected clearance holes for the fasteners at .279/.287 diameter. The size of clearance holes were selected based on standard manufacturing drill sizes. This will help reduce cost in the manufacturing process as special tools will not be required.

If you will notice, standard drilled holes usually have a nominal size, with the plus tolerance larger than the minus tolerance. This is because as the drilling tool wears after it is sharpened, it will tend to wobble and wander inside the hole making the hole larger. If the designer uses a standard hole size and tolerance, it will give more life to the tools and reduce product manufacturing cost.

Floating Fastener Formula

Where two or more parts are assembled using bolts with nuts or rivets, and all parts have clearance holes, it is considered a floating fastener case. The fastener "floats" inside the holes.

This formula is used for determining the position tolerances of mating features to ensure assembly. It is valid for all types of features including patterns and will give a "no interference, no clearance" fit when all features are at MMC and at the extreme position. Additional consideration may have to be given for variables not accounted for in this formula, such as non matching datums to mating parts.

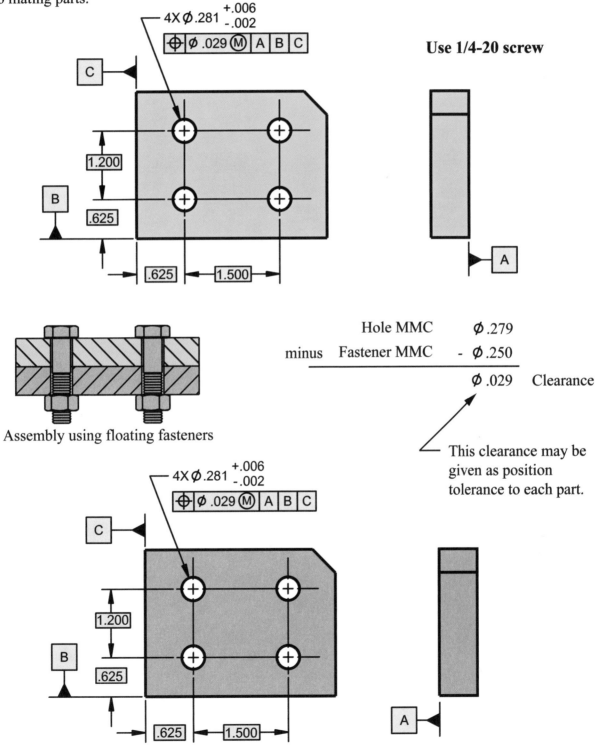

Use 1/4-20 screw

Assembly using floating fasteners

	Hole MMC	ϕ .279	
minus	Fastener MMC	- ϕ .250	
		ϕ .029	Clearance

This clearance may be given as position tolerance to each part.

Fixed Fastener Formula

A fixed fastener assembly is where one of the holes is a clearance hole, and the other hole is not a clearance hole but a press fit or a tapped hole. It is called a fixed fastener assembly because the fastener is "fixed" in the assembly. Common fixed fastener examples are assemblies that use dowels or tapped holes.

The two flat plates on the following page are fastened together using four screws that assemble with one part having tapped holes. The dimensions between the holes and the dimensions from the datums are designated as basic dimensions.

To determine the positional tolerances between the parts, the designer will calculate the difference between the largest (MMC size) bolt and the smallest (MMC size) hole. The difference between these two sizes is the allowable tolerance. This tolerance must be divided between the two parts in any combination that does not exceed the allowable tolerance.

The tapped hole is usually given a larger share of the tolerance for two possible reasons. First, the tapped hole is usually two operations - the tap drill and then the tap. This is a more difficult manufacturing operation and therefore usually needs more tolerance.

Second, and probably the most important reason for allotting the tapped hole more position tolerance is that the clearance hole allows the benefit of MMC. Whereas the position for the tapped hole is left at RFS. When the MMC modifier is applied to the clearance hole, it will allow additional positional tolerance as the feature departs from MMC. According to our rules, unless otherwise specified, positional tolerance for a tapped hole applies to the pitch diameter of the thread. When the bolt is tightened, because of the incline on the threads, it creates a centering effect, negating any extra position tolerance from an MMC departure.

The clearance hole usually has a large size tolerance and because of the MMC effect, it results in more positional tolerance at nominal and LMC sizes. To make the positional tolerance more equal, a larger positional tolerance is given to the tapped hole. As a general rule, split the tolerance 60/40. The tapped hole receives 60%, and the clearance hole receives 40%.

To account for perpendicularity problems with the mating bolt, a projected tolerance zone is usually applied in a fixed fastener assembly. The projected tolerance zone length is equal to the thickness of the mating part. More explanation of this projected zone is shown later in this unit.

More on Screw Threads, MMC or RFS?

The pitch diameter of a screw thread has size and therefore a feature modifier may be applied. As the screw is torqued down in the assembly, there is a centering effect. It is for this reason that functionally, the position tolerance on a screw thread applies at RFS.

It is often stated that MMC was applied to threads to make it easier to gage. But remember, the drawing usually does not define the method for checking the part. It describes the functional requirements. If MMC is applied to the thread, a reduction in the position tolerance would have to occur to account for the centering effect of the bolt. Instead it should be left at RFS with the max calculated position available. Quality then analyzes the measuring equipment available and determines the verification method with the corresponding measurement uncertainty to determine conformance.

Fixed Fastener Formula

Where two or more parts are assembled using restrained fasteners such as bolts into tapped holes, studs or pressed fit pins, it is considered a fixed fastener case. The fastener is "fixed" in one part.

This formula is used for determining the position tolerances of mating features to ensure assembly. It is valid for all types of features including patterns and will give a "no interference, no clearance" fit when all features are at MMC and at the extreme position. Additional consideration may have to be given for variables not accounted for in this formula, such as non matching datums to mating parts.

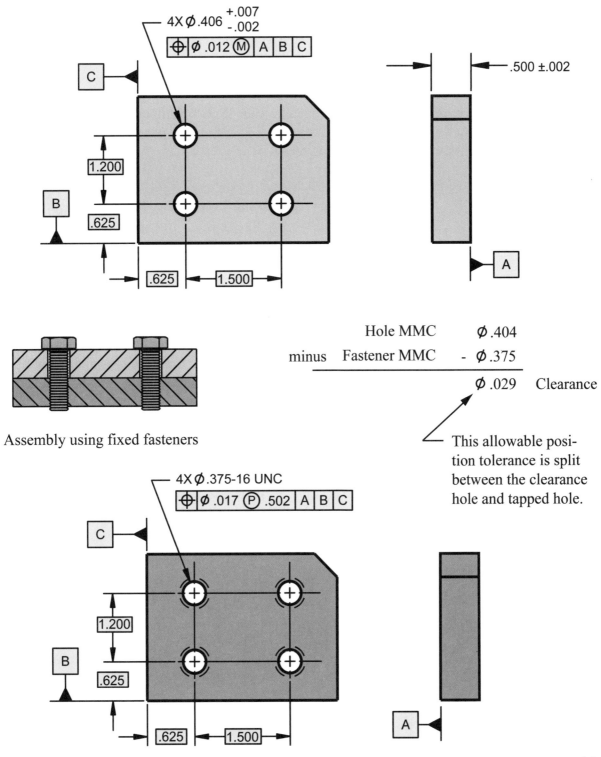

Assembly using fixed fasteners

	Hole MMC	ϕ .404
minus	Fastener MMC	- ϕ .375
		ϕ .029 Clearance

This allowable position tolerance is split between the clearance hole and tapped hole.

Projected Tolerance Zone

All geometric tolerances applied to features extend for the full length and depth of the feature. In some cases it may be necessary to "project" a tolerance zone. The projected tolerance zone concept will "project" the tolerance zone out of the feature by a specified amount.

Unlike the floating fastener, mating with only clearance holes, the attitude of a fixed fastener (for example, screw thread or press fit dowel) is governed by the orientation of the fixed hole in which it assembles inside. The fixed fastener formula does not ensure the complete interchangeability between parts by itself. The formula does not take into consideration a possible interference condition if the threaded or press fit holes were to tilt or be out of perpendicular within the allotted position tolerance zone. The projected tolerance zone concept is recommended where it is necessary to limit the variation of perpendicularity on threaded or press fit holes to ensure interchangeability between parts.

The location and perpendicularity of the threaded holes are only of importance as they affect the extended portion of the engaging fastener. The projected tolerance zone will project the tolerance out of the hole to the extent of the engaging fastener.

The height of the projected tolerance zone is usually the thickness of the mating part including any gaskets or shims. The thickness of washers are usually not included in the calculations as the washers may float in any direction and will not cause interference. If a stud or pin is toleranced as an assembly with the pin or stud installed, a projected tolerance zone is not necessary, as the tolerance extends for the full length of the feature. If the hole in the detail is toleranced without the studs or pins installed, then the hole should have a projected zone applied equal to the height of the studs or pins.

Where the direction of the projected tolerance zone is clear, only a symbolic call-out and height dimension is required. The symbolic callout is placed in the feature tolerance compartment of the feature control frame following the feature tolerance and any modifier. The height of the tolerance zone follows the projected tolerance zone symbol.

Where the direction of the projected tolerance zone is not clear, the symbolic call-out for the projected tolerance zone is placed in the feature control frame following the feature tolerance and any modifier. The dimensioned height value of the projected zone is included on the drawing view with a heavy chain line that is drawn closely adjacent to the center line of the hole.

Functional gaging of threads positioned at RFS.
A functional gage can be used to check positioned threads modified at RFS. The threaded gage has "go" threads built at MMC of the pitch diameter. The shoulder on the threaded gage pin is increased to reduce the allowable position tolerance by the amount of the pitch diameter tolerance.

4X ⌀.375-16 UNC

⊕	⌀ .017	A	B	C

— No projected tol zone
modifier is applied

The tilting on the tapped
hole causes interference
when the bolt is inserted.

By default, the cylindrical
tolerance zone extends the
length of the feature.

.625   1.500

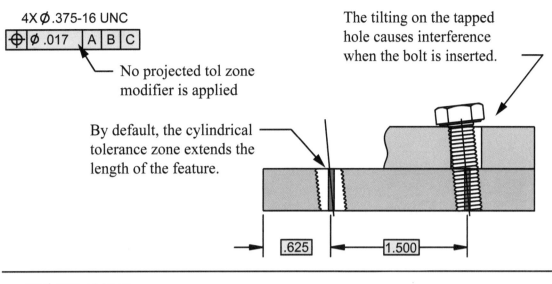

4X ⌀.375-16 UNC

⊕	⌀ .017 Ⓟ .502	A	B	C

— Projected tol zone
modifier is applied

The tilting on the tapped hole is accounted for with
the projected tolerance zone which allows the bolt
to clear the virtual condition of the clearance hole.

.625   1.500

.502

With the projected tolerance
zone modifier applied, the
tolerance zone is projected from
the datum a minimum of .502

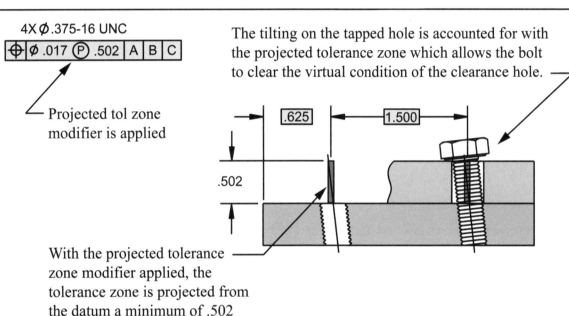

Functional Gage

Inspection uses a thread gage to simulate
a "perfect" bolt in the threaded hole.

Clearance between gage shoulder
pin and hole = ⌀ .017
(tapped hole position tol)

.502 min

Thread gage with
slits to expand in the
tapped hole. (RFS)

Fixed MMC size thread
"go" gage, may "wiggle"
in actual tapped hole.

Gage has edges to
simulate datums
A, B, & C

4X ⌀.375-16 UNC
thread

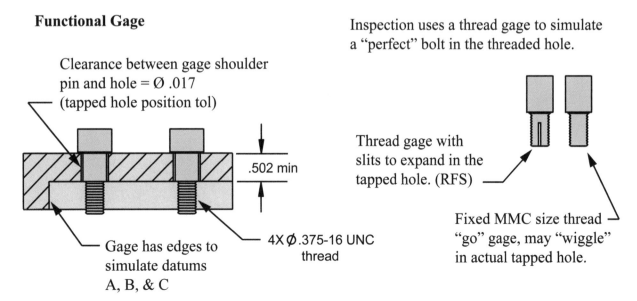

15.7

Workshop Exercise 15.1 - Shims

Below are two drawings of shims. The upper shim fits in the floating fastener assembly shown on page 15.3. The lower shim fits in the fixed fastener assembly shown on page 15.5. Use the formulas shown earlier in this unit to calculate the position for the holes to ensure the screws fitting through all the holes. Also, make sure to apply the correct feature modifer.

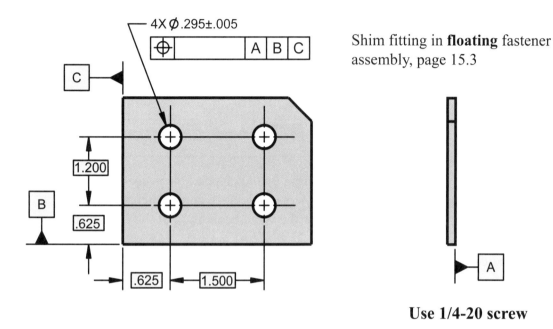

Shim fitting in **floating** fastener assembly, page 15.3

Use 1/4-20 screw

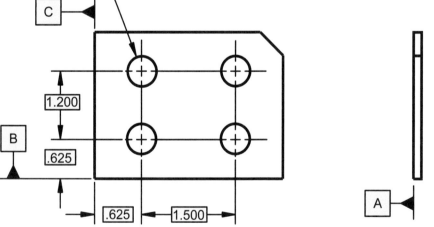

Shim fitting in **fixed** fastener assembly, page 15.5

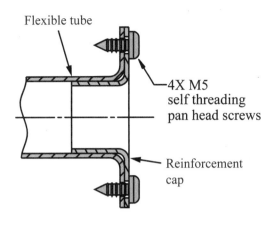

Flexible tube

4X M5
self threading
pan head screws

Reinforcement
cap

Study the sheet metal assembly to the left. It consists of two main parts, a flexible tube and reinforcement cap. They are attached with 4 self threading fasteners.

Select datum features based on the mating condition. Make sure to note the fit between the parts.

Calculate and apply position tolerances using the fastener formulas on the previous pages.

Apply other necessary geometric tolerances to ensure assembly and minimize mismatch on outside edges.

Flexible tube

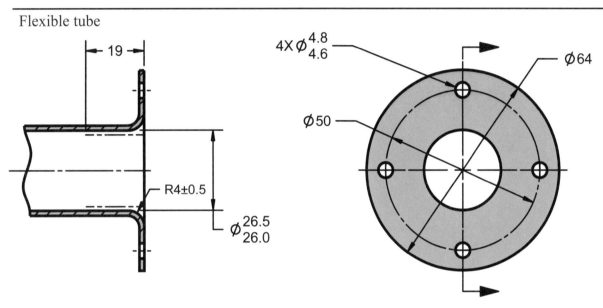

Reinforcement cap

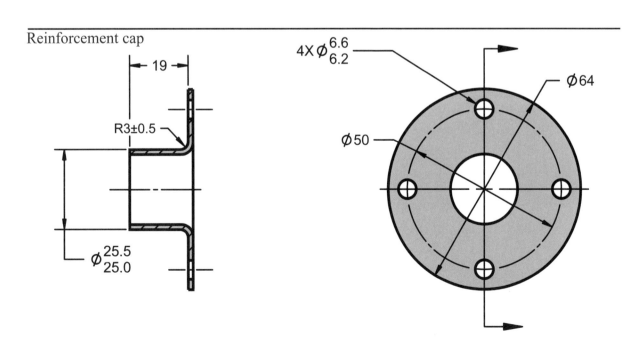

Unit 16

Boundary Formulas

Boundaries Overview

When geometric tolerances are applied to a feature, a variety of boundaries are created. These boundaries are located or oriented to themselves or relative to a datum reference frame based on the datum features referenced in a feature control frame. These boundaries are used by engineers to calculate fits and stack-ups between mating parts. The boundaries are essential and useful for purposes of discussions, recording, collection and interchange of data between engineering, manufacturing and quality control.

The Maximum Material Boundary (MMB), Least Material Boundary (LMB) and Regardless of Material Boundary (RMB) are new terms for the ASME Y14.5-2009 standard. They are powerful tools in understanding the effects of geometric tolerancing. These boundaries provide us with a clear mathematical description of what happens when geometric tolerancing is applied to features. They define how much a feature may grow, shrink, bow, tilt, shift or move around when it is contained by this boundary.

The MMB and LMB are most generally used and can be used in place of the Inner and Outer Boundary (IB and OB). The terms virtual and resultant condition are used in the specific cases when geometric tolerances are applied in conjunction with the MMC and LMC modifiers. The use of the virtual and resultant condition terms are not appropriate when using RFS, runout, profile etc.

The terms MMB, LMB and RMB are more general, and can be used with all geometric tolerances such as profile, runout, position, orientation, form and size. In certain cases, they can substitute for the terms virtual and resultant conditions.

The terms used for describing these boundaries are shown below. Graphical descriptions and mathematically calculated boundaries and their associated terms are shown on the following pages. Consult the ASME Y14.5-2009 standard for more information.

Maximum Material Boundary (MMB) is the limit defined by a tolerance or combination of tolerances that exists on or outside the material of a feature(s).

Least Material Boundary (LMB) is the limit defined by a tolerance or combination of tolerances that exists on or inside the material of a feature(s).

Regardless of Material Boundary (RMB) indicates that a datum feature simulator progresses from MMB toward LMB until it makes maximum contact with the extremities of a feature(s).

Inner Boundary (IB) is a worst-case boundary generated by the smallest feature (MMC for an internal feature and LMC for an external feature) minus the stated geometric tolerance and any additional geometric tolerance (if applicable) resulting from the feature's departure from its specified material condition.

Outer Boundary (OB) is a worst-case boundary generated by the largest feature (LMC for a internal feature and MMC for an external feature) plus the stated geometric tolerance and any additional geometric tolerance resulting from the feature's departure from its specified material condition.

Virtual Condition is a constant boundary generated by the collective effects of a considered feature of the size's specified MMC or LMC and the geometric tolerance for that material condition.

Resultant Condition is a single worst-case boundary generated by the collective effects of a feature of the size's specified MMC or LMC, the geometric tolerance for that material condition, the size tolerance, and the additional geometric tolerance derived from the feature's departure from its specified material condition.

The terms used for these boundaries can be a little confusing at first. Below is a guide to help explain the differences. Note the terms "virtual condition" and "resultant condition" are used only when an MMC or LMC modifier is used. The terms MMB, LMB, IB and OB may be used for all features that have a boundary. The terms MMB and LMB are new terms for the Y14.5-2009 standard and are used to explain the effect of datum feature modifiers.

For an Internal Feature (**Hole**) Controlled with a Geo Tol at **RFS**
      IB = MMB   smallest – geo tol
      OB = LMB   largest + geo tol

For an External Feature (**Pin**) Controlled with a Geo Tol at **RFS**
      IB = LMB   smallest – geo tol
      OB = MMB   largest + geo tol

For an Internal Feature (**Hole**) Controlled with a Geo Tol at **MMC**
      IB = MMB = Virtual Condition   smallest – geo tol
      OB = LMB = Resultant Condition   largest + geo tol + bonus

For an External Feature (**Pin**) Controlled with a Geo Tol at **MMC**
      IB = LMB = Resultant Condition   smallest – geo tol - bonus
      OB = MMB = Virtual Condition   largest + geo tol

For an Internal Feature (**Hole**) Controlled with a Geo Tol at **LMC**
      IB = MMB = Resultant Condition   smallest – geo tol - bonus
      OB = LMB = Virtual Condition   largest + geo tol

For an External Feature (**Pin**) Controlled with a Geo Tol at **LMC**
      IB = LMB = Virtual Condition   smallest – geo tol
      OB = MMB = Resultant Condition   largest + geo tol + bonus

*bonus = max departure from stated material condition
*geo tol = stated value in feature control frame

Boundaries are established when features are modified at RFS or RMB. The graphical representation and name of these boundaries are shown below. These boundaries are important tools to understand and evaluate the worst case conditions created by the combination of size and geometric tolerances. The terms and description for these boundaries are shown on the preceeding pages.

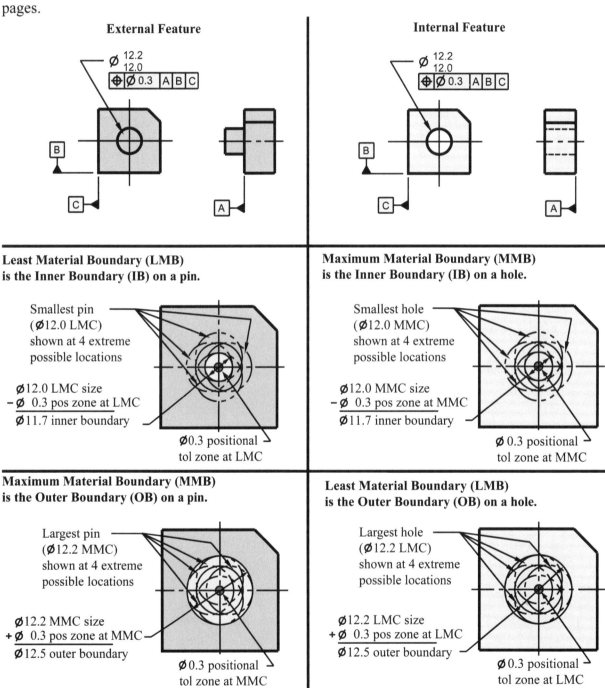

Least Material Boundary (LMB)
is the Inner Boundary (IB) on a pin.

Smallest pin
(⌀12.0 LMC)
shown at 4 extreme
possible locations

⌀12.0 LMC size
− ⌀ 0.3 pos zone at LMC
⌀11.7 inner boundary

⌀0.3 positional
tol zone at LMC

Maximum Material Boundary (MMB)
is the Inner Boundary (IB) on a hole.

Smallest hole
(⌀12.0 MMC)
shown at 4 extreme
possible locations

⌀12.0 MMC size
− ⌀ 0.3 pos zone at MMC
⌀11.7 inner boundary

⌀ 0.3 positional
tol zone at MMC

Maximum Material Boundary (MMB)
is the Outer Boundary (OB) on a pin.

Largest pin
(⌀12.2 MMC)
shown at 4 extreme
possible locations

⌀12.2 MMC size
+ ⌀ 0.3 pos zone at MMC
⌀12.5 outer boundary

⌀0.3 positional
tol zone at MMC

Least Material Boundary (LMB)
is the Outer Boundary (OB) on a hole.

Largest hole
(⌀12.2 LMC)
shown at 4 extreme
possible locations

⌀12.2 LMC size
+ ⌀ 0.3 pos zone at LMC
⌀12.5 outer boundary

⌀0.3 positional
tol zone at LMC

Formula to calculate the IB and OB for features applied at RFS.

If RFS is applied to a feature or RMB is applied to a datum feature:

OB/IB = Mean Size ± (½ Tst +Gtol)

Tst = Total size tolerance
GTol = Geometric Tolerance

Boundaries - MMC Internal/External

Boundaries are established when features are modified at MMC or MMB. The graphical representation and name of these boundaries are shown below. These boundaries are important tools to understand and evaluate the worst case conditions created by the combination of size and geometric tolerances. The terms and description for these boundaries are shown on the preceeding pages.

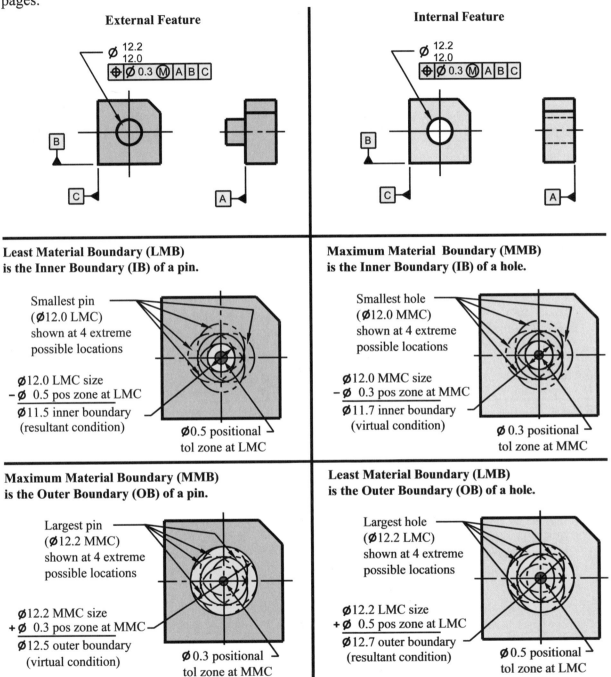

External Feature

Ø 12.2 / 12.0
⊕ Ø 0.3 Ⓜ A B C

B

C

A

Internal Feature

Ø 12.2 / 12.0
⊕ Ø 0.3 Ⓜ A B C

B

C

A

**Least Material Boundary (LMB)
is the Inner Boundary (IB) of a pin.**

Smallest pin
(Ø12.0 LMC)
shown at 4 extreme
possible locations

Ø12.0 LMC size
− Ø 0.5 pos zone at LMC
Ø11.5 inner boundary
(resultant condition)

Ø0.5 positional
tol zone at LMC

**Maximum Material Boundary (MMB)
is the Inner Boundary (IB) of a hole.**

Smallest hole
(Ø12.0 MMC)
shown at 4 extreme
possible locations

Ø12.0 MMC size
− Ø 0.3 pos zone at MMC
Ø11.7 inner boundary
(virtual condition)

Ø 0.3 positional
tol zone at MMC

**Maximum Material Boundary (MMB)
is the Outer Boundary (OB) of a pin.**

Largest pin
(Ø12.2 MMC)
shown at 4 extreme
possible locations

Ø12.2 MMC size
+ Ø 0.3 pos zone at MMC
Ø12.5 outer boundary
(virtual condition)

Ø 0.3 positional
tol zone at MMC

**Least Material Boundary (LMB)
is the Outer Boundary (OB) of a hole.**

Largest hole
(Ø12.2 LMC)
shown at 4 extreme
possible locations

Ø12.2 LMC size
+ Ø 0.5 pos zone at LMC
Ø12.7 outer boundary
(resultant condition)

Ø0.5 positional
tol zone at LMC

Formula to calculate the IB and OB for features applied at MMC.
If MMC is applied to feature or MMB is applied to a datum feature:

OB/IB = LMC ± (Tst + Gtol)

Tst = Total size tolerance
GTol = Geometric Tolerance

Boundaries are established when features are modified at LMC or LMB. The graphical representation and name of these boundaries are shown below. These boundaries are important tools to understand and evaluate the worst case conditions created by the combination of size and geometric tolerances. The terms and description for these boundaries are shown on the preceeding pages.

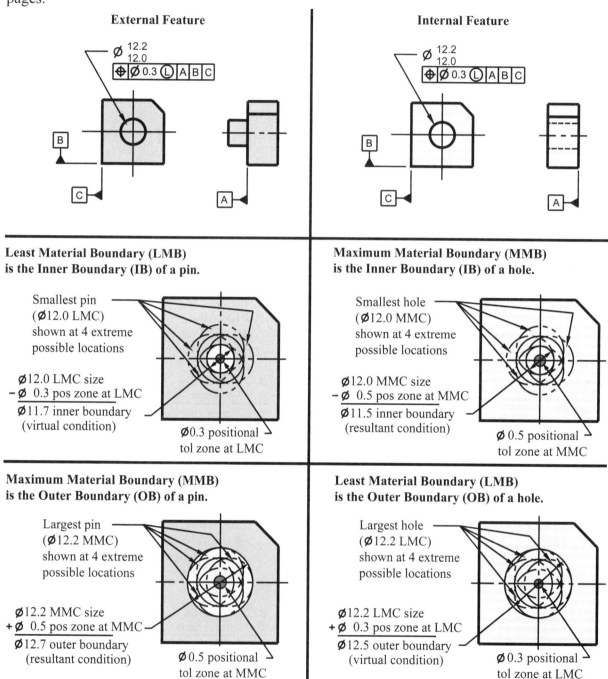

External Feature

Internal Feature

Least Material Boundary (LMB)
is the Inner Boundary (IB) of a pin.

Smallest pin
(∅12.0 LMC)
shown at 4 extreme
possible locations

∅12.0 LMC size
− ∅ 0.3 pos zone at LMC
∅11.7 inner boundary
(virtual condition)

∅0.3 positional
tol zone at LMC

Maximum Material Boundary (MMB)
is the Inner Boundary (IB) of a hole.

Smallest hole
(∅12.0 MMC)
shown at 4 extreme
possible locations

∅12.0 MMC size
− ∅ 0.5 pos zone at MMC
∅11.5 inner boundary
(resultant condition)

∅ 0.5 positional
tol zone at MMC

Maximum Material Boundary (MMB)
is the Outer Boundary (OB) of a pin.

Largest pin
(∅12.2 MMC)
shown at 4 extreme
possible locations

∅12.2 MMC size
+ ∅ 0.5 pos zone at MMC
∅12.7 outer boundary
(resultant condition)

∅0.5 positional
tol zone at MMC

Least Material Boundary (LMB)
is the Outer Boundary (OB) of a hole.

Largest hole
(∅12.2 LMC)
shown at 4 extreme
possible locations

∅12.2 LMC size
+ ∅ 0.3 pos zone at LMC
∅12.5 outer boundary
(virtual condition)

∅0.3 positional
tol zone at LMC

Formula to calculate the IB and OB for features applied at LMC.

If LMC is applied to feature or LMB is applied to a datum feature:

Tst = Total size tolerance
GTol = Geometric Tolerance

$$OB/IB = MMC \pm (Tst + Gtol)$$

Calculate the inner and outer boundaries for the following problems. Consult the IB and OB formulas in the text if you have any problems.

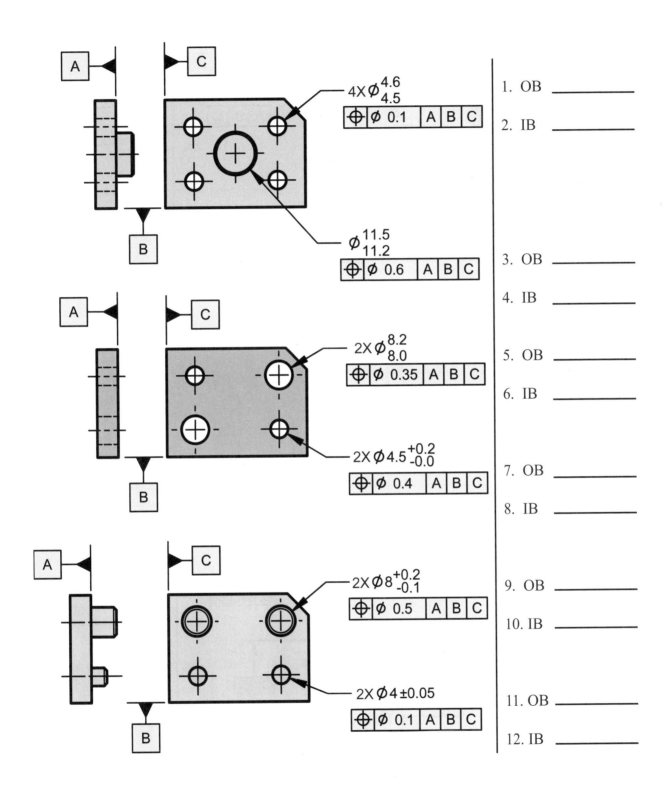

1. OB _____

2. IB _____

3. OB _____

4. IB _____

5. OB _____

6. IB _____

7. OB _____

8. IB _____

9. OB _____

10. IB _____

11. OB _____

12. IB _____

Calculate the inner and outer boundaries for the following problems. Consult the IB and OB formulas in the text if you have any problems.

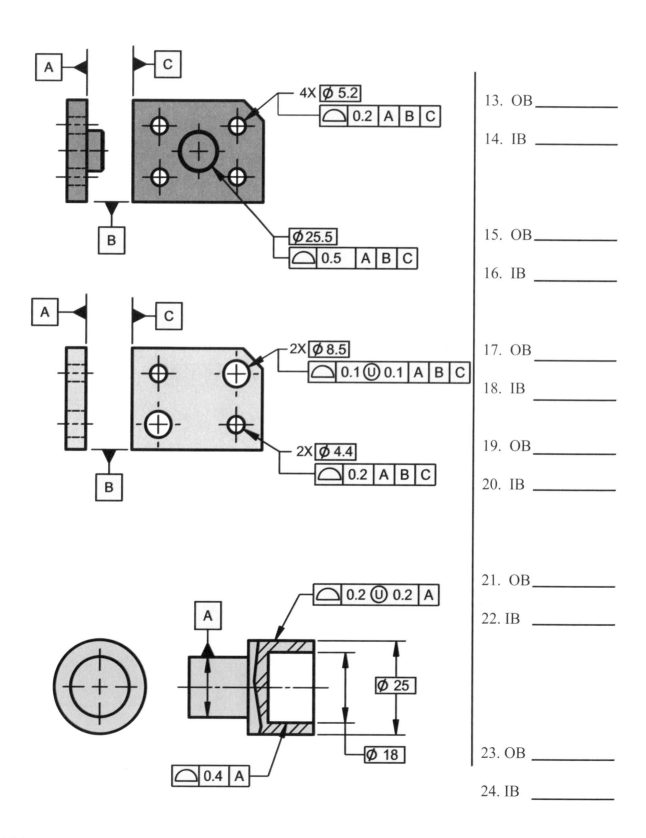

13. OB _____

14. IB _____

15. OB _____

16. IB _____

17. OB _____

18. IB _____

19. OB _____

20. IB _____

21. OB _____

22. IB _____

23. OB _____

24. IB _____

16.8

Calculate the inner and outer boundaries for the following problems. Consult the IB and OB formulas in the text if you have any problems.

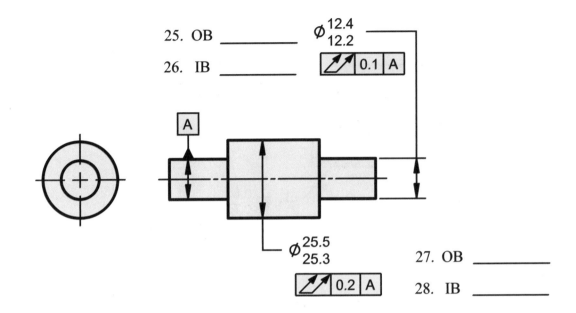

25. OB _____

26. IB _____

27. OB _____

28. IB _____

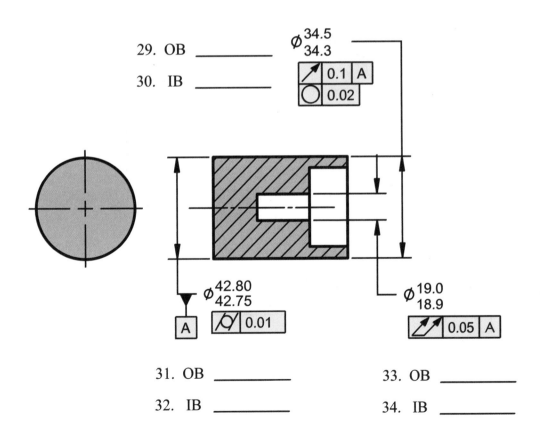

29. OB _____

30. IB _____

31. OB _____

32. IB _____

33. OB _____

34. IB _____

Calculate the inner and outer boundaries for the following problems. Consult the IB and OB formulas in the text if you have any problems.

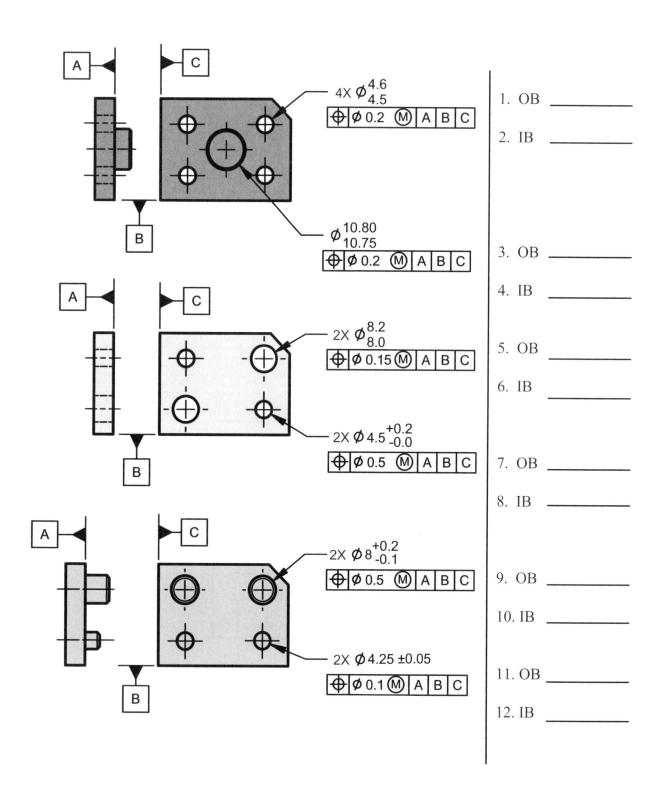

1. OB _____

2. IB _____

3. OB _____

4. IB _____

5. OB _____

6. IB _____

7. OB _____

8. IB _____

9. OB _____

10. IB _____

11. OB _____

12. IB _____

Workshop Exercise 16.3 - LMC Internal/External

Calculate the inner and outer boundaries for the following problems. Consult the IB and OB formulas in the text if you have any problems.

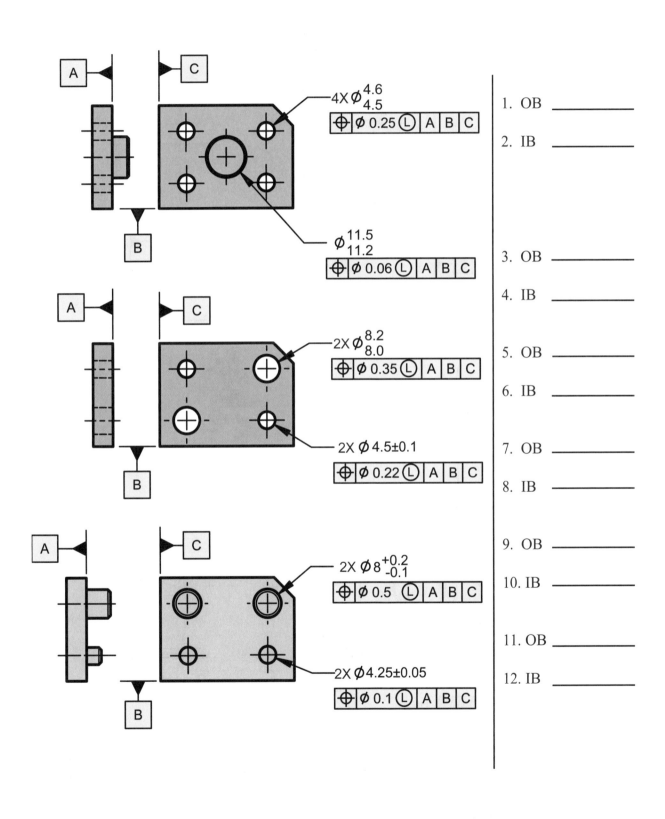

1. OB _____

2. IB _____

3. OB _____

4. IB _____

5. OB _____

6. IB _____

7. OB _____

8. IB _____

9. OB _____

10. IB _____

11. OB _____

12. IB _____

Minimum Wall Calculations

The illustration at the bottom of the page shows a simple min/max wall thickness calculation allowed by the product drawing. The calculations were done using IB/OB on the hole and the consideration of the form and orientation on the datum features.

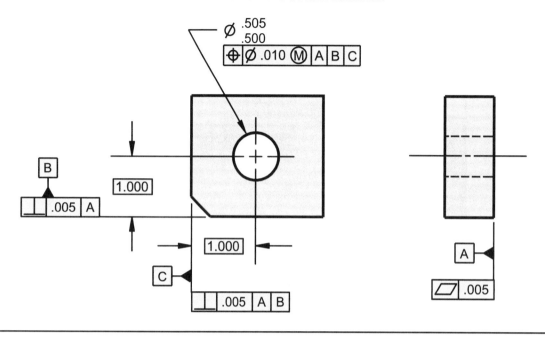

Min/max wall calculations to a DRF vs. to a datum feature

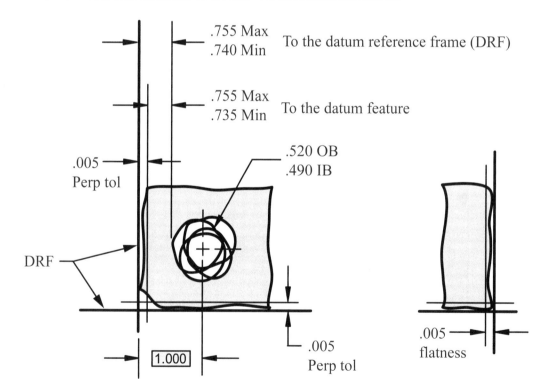

Remember that the high points of the datum features make contact with the DRF. All basic dimensions originate from the DRF not the datum features. The datum features are qualified with the flatness and perpendicularity tolerances. When making min/max wall calculations, consideration should be given to the applicable flatness or orientation tolerences.

16.12

IB and OB Stack-Up Calculations

The illustration at the bottom of the page shows some simple min/max wall thickness calculations allowed by the product drawing at the top. The calculations were done using IB/OB on the holes and the consideration of the profile on the outside edges.

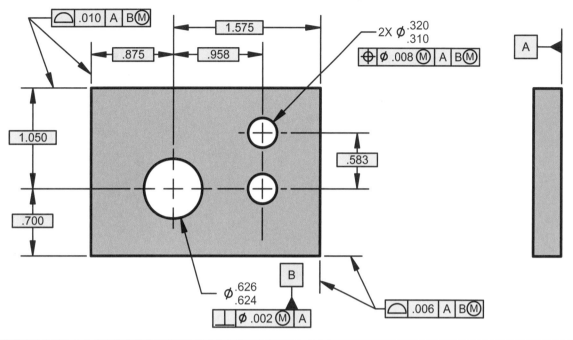

Stack up between features

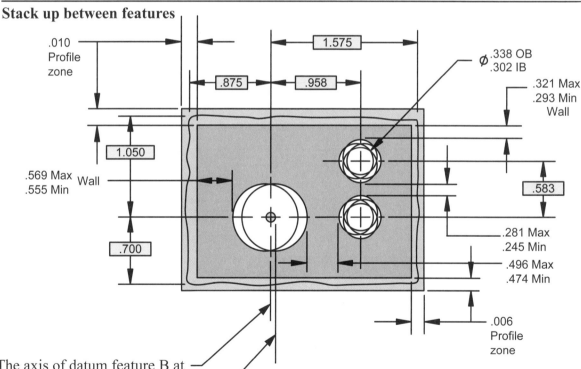

The axis of datum feature B at MMB (virtual) establishes the axis of the DRF. All basic dimensions originate from this point.

The axis of the actual datum feature B is allowed to be displaced relative to the datum axis (axis of datum feature B at MMB). It may displace in a zone equal to the difference between the MMB (virtual) and the actual MMB to a maximum of .004.
The OB of the hole is .630, the IB is .622.

16.13

1. The top drawing is the product drawing. The bottom illustration requires the calculation of the IB/OB for the holes and the min/max wall thicknesses.

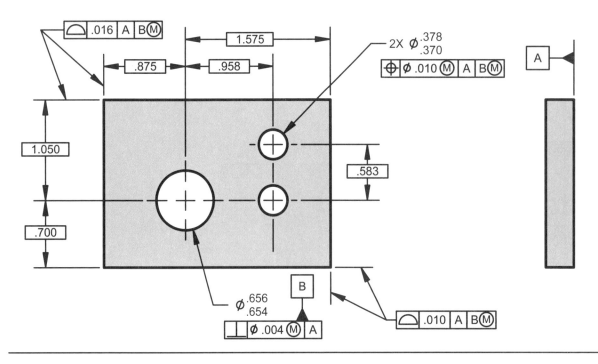

Stack up between features

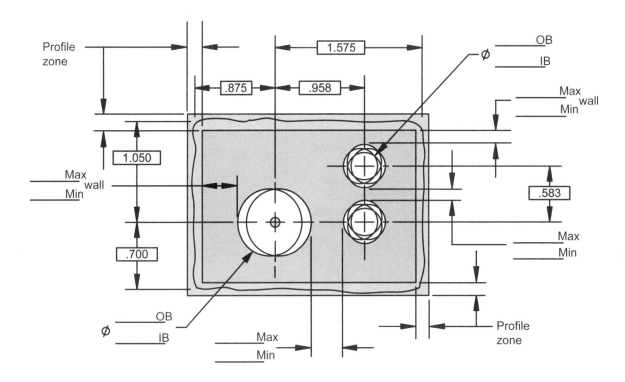

Find the IB and OB or min and max as required on this drawing and record your answers on the worksheet. For this example, make all calculations from the DRF or the datum features as specified. Generally, the higher the number the harder the problem. After number 9, watch for multiple DRF's.

Note: All untoleranced dimensions are basic.

16.15

The drawing is so bad, it looks like a ransom note. See if you can find any mistakes. Look for missing symbols, misused symbols, improper datum feature symbol placement, incorrect feature control frames, etc. If you think the drawing looks good, take the course again.

Unit 17
Application Exercises

Workshop Exercise 17.1 - Corner Bracket
Workshop Exercise 17.2 - Flange Bearing
Workshop Exercise 17.3 - Powder Case
Workshop Exercise 17.4 - Link

This corner bracket reinforces the table assembly as shown below. The 12 mm surface radius was designed to clear the head of a pin that is pushed through the assembly. The four 8 mm holes mount with 7.2 mm max rivets and the three 9 mm holes mount to the mating cover with M7 screws into tapped holes that have a 1 mm position tolerance. Specify necessary datums, relate the datum features, and apply necessary geometric tolerances to this part to insure a functional fit. Make sure your work is clear and neat.

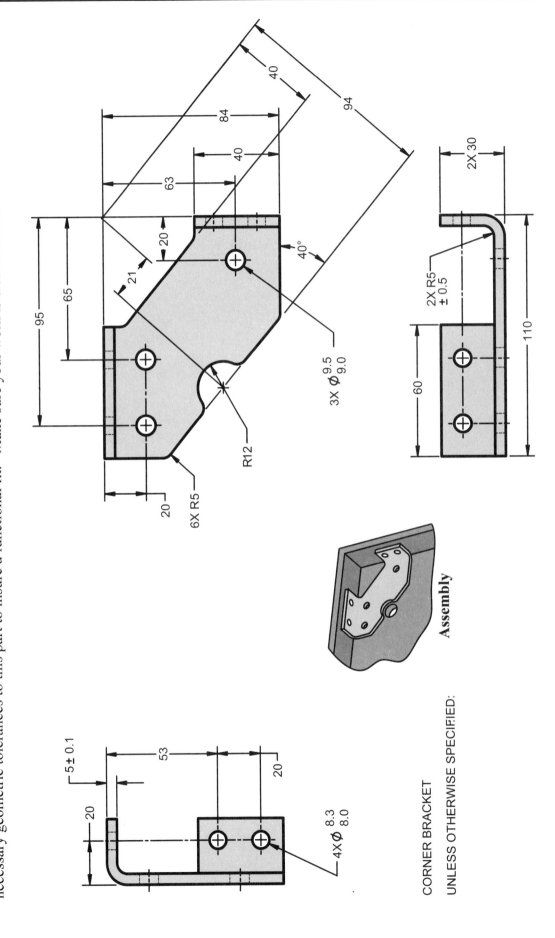

Assembly

CORNER BRACKET

UNLESS OTHERWISE SPECIFIED:

The flange below has a bearing pressed into the large 2.2 dia hole. Four .4375 bolts attach the flange to a frame. Select datum features to establish a DRF and qualify the datum features with any necessary form, orientation and/or location tolerances. Make necessary dimensions basic. Calculate the allowable position tolerance for the 4 holes. The mating frame has tapped holes with .017 position tolerance. In this problem, just estimate the remaining tolerances based on realistic expectations.

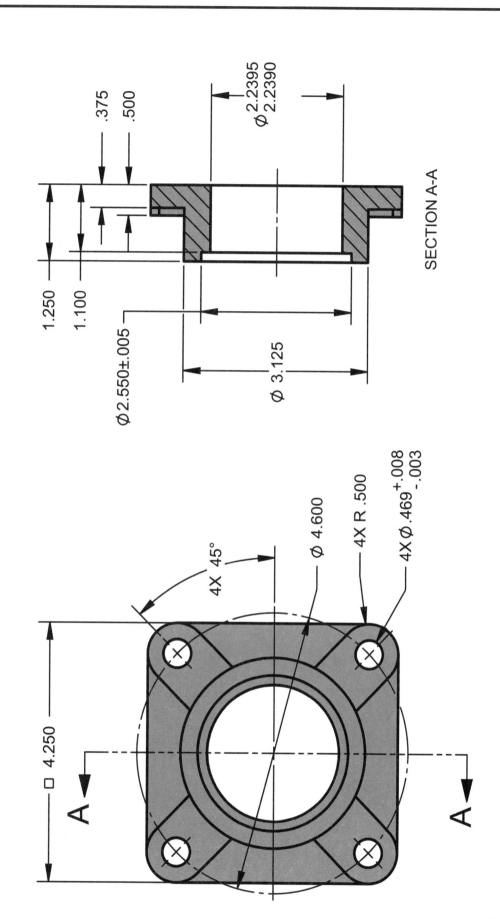

SECTION A-A

17.3

Workshop Exercise 17.3 - Powder Case

The two parts below are mating parts. Select datum features and apply any necessary geometric tolerancing on each part to ensure parts will fit together.

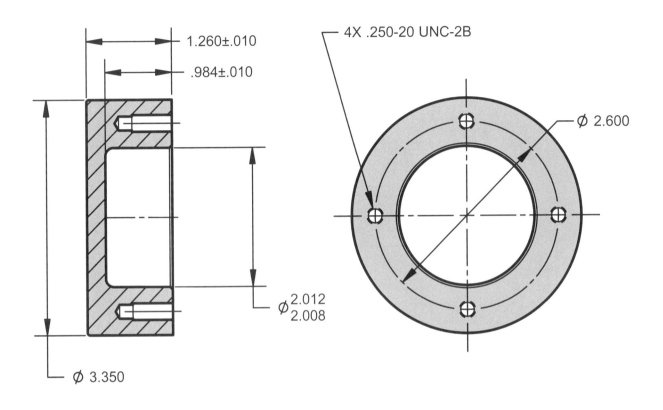

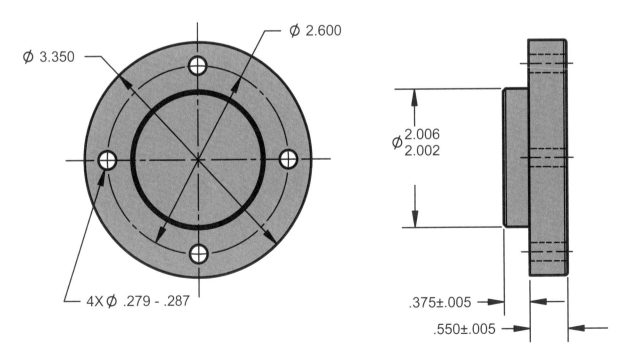

Study the assembly at the bottom of the page. Identify the datum features on the drawing to create a datum reference frame. Then fill in the feature control frames to locate the features to the established reference frame. The tolerance value for the 6 mm holes can be calculated, but other values in this exercise may be engineering estimations. Be sure to make necessary dimensions basic. When using position tolerances, choose the most functional feature modifiers (MMC, LMC, and RFS) and datum feature modifiers (MMB or RMB) for the application.

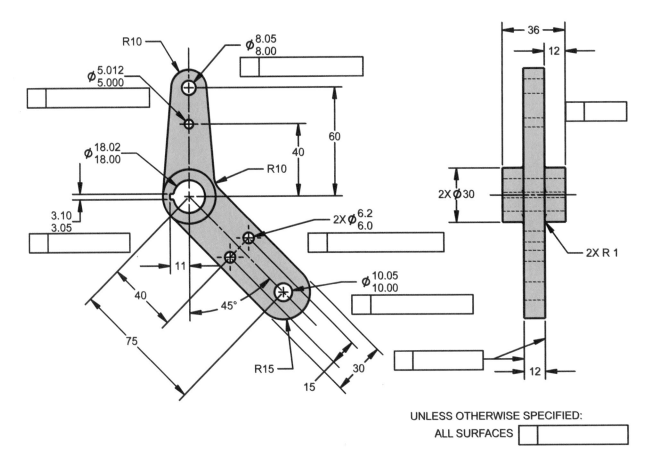

UNLESS OTHERWISE SPECIFIED:
ALL SURFACES

Use the switch drawing below and virtual size calculations to find the position tolerance for the 6 mm holes.

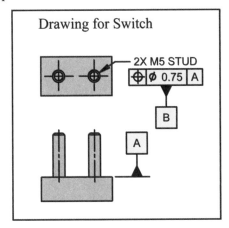

Drawing for Switch

2X M5 STUD

⊕ ⌀ 0.75 A

B

A

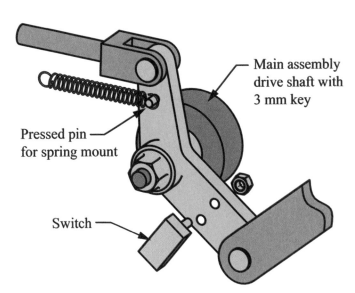

Main assembly drive shaft with 3 mm key

Pressed pin for spring mount

Switch

Appendix A

Principle changes, improvements, clarifications and enhancements from the ASME Y14.5M-1994 standard to the ASME Y14.5-2009 standard

Appendix A

Principle changes, improvements, clarifications and enhancements in the ASME Y14.5-2009 from the earlier ASME Y14.5M-1994 standard

The new ASME Y14.5-2009 standard on dimensioning and tolerancing reflects a culmination of effort extending over 15 years. It is a revision of the ASME Y14.5M-1994 standard. The M in the title which reflected the standard was metric compatible was deemed no longer necessary and was dropped. There are many changes, improvements and enhancements from the earlier 1994 standard.

The chapters in the standard have been reorganized to present the foundational principles first and then build on the foundations. This reorganization has made the concepts much easier to read and understand. Many of the changes are simple rearrangements, or definition changes. The reader should understand that only the major concepts are outlined. Many of the changes, expansions, revisions, clarifications and term definitions are woven through the fabric of the document.

The standard has been revised to emphasize/encourage the use of basic dimensions and geometric tolerancing as the preferred method of controlling the form, orientation, and location of features. The use of direct tolerancing methods, limit dimensioning and plus minus tolerancing should be used to control the size of features only. Size has also been expanded to include irregular features of size.

Many major enhancements and clarifications have occurred in the datum section. The degrees of freedom concept has been introduced and a table has been added defining the arrested degrees of freedom for primary datum features. This enhancement also allows any feature to become a datum feature. As a result of allowing any feature to become a datum feature, datum feature modifiers have been expanded using new datum feature modifier terms such as Maximum Material Boundary (MMB), Least Material Boundary (LMB) and Regardless of Material Boundary (RMB). These new terms replace the term virtual condition for datum features and provide a much clearer definition of the boundary established by referenced, related datum features.

A new set of datum feature simulator requirements along with a new datum feature translation modifier was added. These requirements along with the translation modifier clarifies the process of establishing datums and datum reference frames from datum features. It also clarifies the establishment of a secondary or tertiary datum relative to a higher precedence datum axis.

The profile section also has significant enhancements. A non-uniform profile zone was added where the user can specify the applicable tolerance zone with dimensions. The symbol, circle U, was added and can be used in as feature control frame to identify a unilateral or unequal unilateral tolerance zone.

Since the release of the former ASME Y14.5M standard there have been major advances in technology in CAD, CMM and CNC type equipment. The standard has also advanced to take advantage of these changes. There is a trend towards applying geometric tolerancing to the solid model. The ASME Y14.41 illustrates how to apply the geometric tolerancing symbology but the ASME Y14.5-2009 still defines the basic principles of tolerancing.

The ASME Y14.5.1-1994 Mathematical Definitions of Dimensioning and Tolerancing standard is still valid and provides a solid mathematical foundation. Work is also being planned on updating the ASME Y14.5.2 Certification of Geometric Tolerancing Professionals to the new ASME Y14.5-2009 standard. The Y14.43 Dimensioning and Tolerancing Principles for Gages and Fixtures also has a planned update to the new standard. Currently a new standard Y14.45 Measurement Data Reporting is in the works to standardize the reporting of geometric toleranced parts.

In this short outline, it is impossible to address the implication of all the changes and only the major changes are shown. In order to understand the impact of these changes, it is assumed the reader has knowledge of the earlier 1994 standard. The reader can find detailed information on the changes in this text. It is suggested the reader study the material in the workbook as it will provide a better understanding for the extension and expansion of the principles. In addition, appendix A of the ASME Y14.5-2009 standard there is also a listing of the principle changes.

Significant New and Revised Terms and Definitions

Least Material Boundary, (LMB)
Maximum Material Boundary, (MMB)
Regardless of Material Boundary (RMB)
Actual Mating Envelope
Unrelated Actual Mating Envelope
Related Actual Mating Envelope
Feature of Size
      (a) Regular Feature of Size
      (b) Irregular Feature of Size

Fundamental rules revised:
A zero basic dimension applies where axes, center planes or surfaces that are shown congruent on a drawing and geometric controls establish the relationship between the features.

General Tolerancing Principles

The standard has been revised to emphasize/encourage the use of basic dimensions and geometric tolerancing as the preferred method of controlling the form, orientation, and location of features. The use of direct tolerancing methods, limit dimensioning and plus minus tolerancing should be used to control the size of features only.

Rule #1 for individual features of size has been expanded to include irregular features of size.

Continuous feature concept and new symbol added.

The terms virtual condition and resultant condition to define datum feature boundaries has been replaced and clarified with the new terms Maximum Material Boundary (MMB, Least Material Boundary (LMB) and Regardless of Material Boundary (RMB).

New symbology added

All Over
Spotface
Continuous Feature
Independency
Unequally Disposed Profile
Datum Reference Frame
Moveable Datum Target
Datum Translation
Degrees of Freedom
MMB, LMB, RMB

Datum Reference Frame (DRF)

The concept of degrees of freedom constrained by the primary, secondary, and tertiary datum features has been added.

Degrees of Freedom Table for primary datum features at RMB.

New terms for the datum feature boundaries have been added.
    Maximum Material Boundary (MMB)
    Least Material Boundary (LMB)
    Regardless of Material Boundary (RMB)

The term datum feature simulator has been revised and expanded into two types.
    Datum Feature Simulator (Theoretical) is defined as theoretically perfect for use in the standard and has replaced and eliminated the term true geometric counterpart.
    Datum Feature (Physical) is defined as the practical embodiment of the theoretical datum feature simulator used in manufacturing, inspection or tooling.

Requirements for datum feature simulators are defined to clarify the establishment of a DRF.

Translation datum feature modifier and new symbol added.

Customized datum reference frames have been added to allow the user to override the requirements of a standard DRF and selectively and specifically state the degrees of freedom constrained by each datum feature.

Calculations of various datum feature simulator boundaries are explained and clarified.

Values of the datum feature simulator may be added in the feature control frame with dimensions or the term BASIC or BSC for clarification.

Datum Feature Shift is clarified.

A.4

The establishment of a secondary or tertiary datum relative to a higher precedence datum axis has been clarified and expanded.

Datum feature modifiers MMB, LMB and RMB may be applied to surfaces in certain cases.

Datum reference frame symbol added to identify the datum reference frame labeled with the x,y, z translational degrees of freedom.

Contoured surface established as a datum feature is clarified.

Moveable datum target symbol added and adopted from ASME Y14.8 and ASME Y14.41.

Form Tolerances

Flatness of a center plane has been added and replaces straightness of a center plane.

Orientation Tolerances

The use of the angularity symbol to control all orientation is introduced as an alternative practice to using the perpendicularity and parallelism symbols.

Position Tolerance

Clarification that the surface interpretation takes precedence over the axis interpretation for positional tolerance.

Multiple segments to a composite tolerance feature control frame are explained and clarified.

Position tolerance applied without a datum feature reference to control coaxial features is clarified.

The term BOUNDARY placed under a feature control frame is now optional when applying position tolerance to an irregular feature of size.

Profile Tolerance

The use of profile tolerancing is encouraged for locating surfaces in lieu of direct tolerancing methods. Profile, with a datum feature reference, has replaced plus/minus (+/-) for location of features to each other or other features on a part. The use of direct tolerancing methods which includes limit dimensioning and plus minus tolerancing should be used to control the size of features only.

A new symbol, circle U, is used in a feature control frame as a preferred method of indicating that a profile tolerance is either unilateral or unequally disposed bilateral.

The application of a profile all over can be applied with a new double circle symbol placed at the intersection of the horizontal and radial portions of the leader line from the feature control frame, or the term "ALL OVER" may be used beneath the feature control frame.

The application of a profile non-uniform tolerance zone has been added. The term "NON-UNIFORM" replaces the tolerance value in the feature control frame and the extent of the tolerance zone is dimensioned. This can also be used to control undesired abrupt changes in the tolerance zone.

The X symbol (example 2X) has replaced the note designation (example 2 SURFACES) for a number of surfaces when using profile tolerance.

Co-planar profile for controlling two or more offset surfaces has been enhanced and clarified.

Addition of multiple segments to a composite tolerance feature control frame are explained and clarified. Composite profile tolerancing has been expanded to include multiple features in a pattern.

Profile tolerancing is clarified as a method to control coaxiality.

Index